D0387461

VARIABILITY OF THE OCEANS

VARIABILITY OF THE OCEANS

ANDREY S. MONIN
VLADIMIR M. KAMENKOVICH
VLADIMIR G. KORT

Shirshov Institute of Oceanology
USSR Academy of Sciences

JOHN J. LUMLEY

Editor of the English Translation
Department of Aerospace Engineering
Pennsylvania State University
for Scripta Technica, Inc.

A WILEY-INTERSCIENCE PUBLICATION
JOHN WILEY & SONS, New York · London · Sydney · Toronto

877d25

Library of Congress Cataloging in Publication Data:

Monin, Andrei Sergeevich.
Variability of the oceans.

"A Wiley-Interscience publication."
Translation of Izmenchivost' Mirovogo okeana.
Includes bibliographies.

1. Oceanography. 2. Hydrodynamics. 3. Ocean circulation. 4. Ocean-atmosphere interaction.
I. Kamenkovich, Vladimir Moiseevich, joint author.
II. Kort, V. G., joint author. III. Title.
GC201.2.M6613 551.4'7 77-286
ISBN 0-471-61328-2

Printed in the United States of America

10 9 8 7 6 5 4 3 2 1

This book sheds light on a number of urgent problems in oceanology that fall under the general heading of variability of the ocean. Observational data and current theories in this field are subjected to systematic review. The variability of the ocean is discussed in its relation to the various areas of the economy and to the development of long-term weather forecasting; the mechanism of the air-sea interaction is set forth, and nonstationary processes are classified, with analysis of methods for observing them. The prospects for study of the variability of the ocean, based on world-wide observations and numerical modelling, are examined.

The book is intended for a broad readership (oceanologists, meteorologists, climatologists, fishing-industry experts, and others in related fields).

PREFACE TO THE ENGLISH EDITION OF "VARIABILITY OF THE OCEANS"

We are very pleased that the English translation of our book is being published in the United States and that it will thus become more accessible to oceanographers of other countries. We are deeply convinced that rapid progress in ocean research is no longer possible without close cooperation between oceanographers of all countries.

Entirely new phenomena have been discovered in oceanography in recent years, and a profusion of new theoretical work has appeared. All this new work requires some ordering. For that reason we have attempted to present briefly information on the new data and hypotheses, the knowledge of which is, in our opinion, necessary to develop a comprehensive, correct theory of the ocean as an unsteady physical system. Unfortunately, limitations of time have not permitted us to expand the English edition beyond the material presented in the Russian original. We have only taken the opportunity to correct typographical errors. However, we believe that the general concepts on which this book is based have not been changed significantly by the newest achievements in the field.

We are grateful to Professor J. Lumley for editing the translation of our book.

A. Monin
V. Kamenkovich
V. Kort

February 23, 1976

PREFACE

The oceanological observations of recent years have revealed such significant synoptic and year-to-year variability in the dynamic and thermodynamic state of the ocean (currents and temperature and salinity fields) that there is a clear need for a synoptic service capable of observing and forecasting this variability in support of various activities that man may undertake in the oceans. The creation of this service will require extensive prior oceanological research, technical developments, and organizational measures. In planning this work, it is desirable to have at the start a summary of the available observational data and theory of the most important nonstationary processes in the ocean. The object of this book is to provide such a summary.

Work oriented toward broad and practical, even though perhaps seemingly remote, goals of scientific and technical progress, the attainment of which requires profound new knowledge of the laws of nature, is in our opinion the very definition of basic scientific research. We think that such a dedication to goals should be the basic principle of planning in science, a principle that would provide the preconditions for scientific discovery, even though one cannot plan what such discoveries may bring. The knowledge and means needed to attain a major goal represent the intermediate missions that, in turn, are preceded by third-order objectives, etc. In this treatment, the plan for solution of a basic scientific problem then becomes a kind of a critical-path chart that may perhaps be subject to change as a result of new discoveries, but must be prepared anyway. We have attempted to adhere to this methodology in the present book.

In considering the ocean as a whole as our basic object of study, we shall analyze in each chapter the open- (or deep-) sea conditions, leaving aside the specific conditions in the littoral shallows, which require a separate analysis.

The overall plan of the book and the initiative behind its preparation are due to A. S. Monin. The various sections were written by different authors (the authorship is indicated in the Table of Contents), and then discussed jointly and revised. Overall, the book is a collective effort, which, in our view, is of much greater usefulness to the reader than collections of articles on specific topics, where each paper is written independently of the others.

An extensive list of references from the Soviet and non-Soviet scientific literature, originally scattered through many journals, is appended to each section of the book. However, we do not claim that the list of references herein cited is either historically consistent or complete. It contains both studies that we felt were important and interesting and sources that are only of illustrative value. The non-Soviet sources cited most extensively are O. M. Phillips' excellent book *The Dynamics of the Upper Ocean*, and a series of original papers by W. Munk. These sources contributed greatly to our knowledge in many of the areas represented in this volume.

We thank V. A. Burkov for his help in writing Chap. 2, S. S. Voyt for coauthoring Sec. 4-3, B. N. Filyushkin for Sec. 4-4, K. N. Fedorov for a helpful discussion of Sec. 3-4, and E. G. Agafonova, E. P. Belova, L. M. Belova, N. S. Bogdanova, A. N. D'yakonova, L. I. Lavrishcheva, V. A. Savsonova, and T. A. Yakusheva for their help in arranging the manuscript.

We are especially grateful to our reviewer, A. M. Yaglom, for numerous helpful recommendations and suggestions.

NOMENCLATURE

The definitions of the symbols given below are adhered to throughout the book.

x, y, z	Cartesian space coordinates
x, y	horizontal plane coordinates
$\mathbf{x}(x, y, 0)$	radius vector in the horizontal plane
z	vertical coordinate; unless otherwise stated, reckoned downward from the undisturbed surface of the ocean
t	time
Δ	Laplace operator in space
Δ_h	Laplace operator on the x, y plane
Δ_s	Laplace operator on the surface of a sphere of unit radius
∇	gradient operator in space
∇_h	gradient operator on x, y plane
div	divergence operator in space
curl	curl operator in space (right-hand rotation)
curl_z	vertical component of curl vector
g	acceleration of gravity
$\mathbf{\Omega}$	angular velocity vector of earth's rotation
$\Omega = \lvert\mathbf{\Omega}\rvert$	absolute value of angular velocity of earth's rotation
f	Coriolis parameter
β	latitudinal variation of Coriolis parameter
N	Väisälä frequency
$\mathbf{u}$	velocity vector
u, v, w	zonal, meridional, and vertical velocity components, respectively
p	pressure
ρ	density
T	temperature in °C (unless otherwise stated)
$\mathbf{k}(k_x, k_y, 0)$	horizontal wave vector
k_x, k_y	wave numbers on x and y axes, respectively
$k = \lvert\mathbf{k}\rvert$	absolute value of horizontal wave vector (horizontal wave number)

$\varkappa(k_x, k_y, l)$	wave vector
l	vertical wave number
$\varkappa = \lvert\varkappa\rvert$	absolute value of wave vector (wave number)
ω	frequency

CONTENTS

1 INTRODUCTION

1-1 THE PRACTICAL IMPORTANCE OF THE VARIABILITY OF THE OCEAN

All types of unsteady processes in the ocean affect human activity in one way or another. This applies even to processes on the smallest of scales. Thus, the formation and evolution of inhomogeneities in the thin-layer vertical microstructure of the ocean are apparently essential to the natural biological productivity of the ocean from which man benefits chiefly through the catches of ocean fisheries (which now account for about three-quarters of the value extracted from the ocean [1]). According to Vinogradov, et al. [2], without such inhomogeneities there would be no zooplankton, which is a vital link in the food chains of larger marine animals. If their food, i.e., the phytoplankton, were uniformly distributed, the copepod zooplankton would have to expend more energy moving about in a search for food than the food could provide.

The turbulence within the layers of the vertical microstructure causes all kinds of impurities, both useful (dissolved gases and mineral salts which are needed for phytoplankton growth) and harmful (radioactive contaminants, oil-derived pollutants, DDT, lead), to disperse much more rapidly in the ocean water than by molecular diffusion alone. Surface waves interfere seriously with ocean transport. Consequently, the USSR now has an operational system for optimum selection of ship routes on the basis of meteorological forecasts of wave conditions on shipping lanes [3]. Storm waves determine the dynamics of the shores and bottom in the littoral zone of the sea and may destroy structures along the shore. At the same time, there are some prospects for engineering utilization of wave energy: for example, wave-driven generators have been developed in Japan to supply electric power for ocean buoys.

The importance of the tides is generally known. They must be taken into account in the operation of seaports, in navigation and in fishing in littoral areas. In spite of large and slowly recoverable capital outlays, tidal power-generating stations may offer promise in certain areas. For example, the tidal cycle could be harnessed to generate peak power for pooled power systems. Experience in the design and operation of such plants has been acquired in France and in the USSR. A 240,000-kW commercial tidal power plant has been built at the mouth of the Rance River in France, and a 400-kW tidal power pilot plant at Kislogubsk on the Kola Peninsula in the USSR, built in sections that were floated into position by the Bernshteyn method, is now operational. Evaluation of the practical potential of internal waves remains a task for the future. Tidal and inertial oscillations make a strong contribution to the variability of currents, and it is sometimes helpful to take this into account in practical navigation.

Attention should now be concentrated on the large-scale (synoptic, annual, and year-to-year) variability of the oceans, forecasts of which are already needed for navigation of surface ships and submarines, fishing, and long-term weather forecasting practice [4].

The principal requirement of navigation is, of course, the knowledge of the actual currents. In the case of submarines, this need for precise data becomes more acute as their underwater residence time increases. But we definitely know from even the few and fragmentary observations now available that the range of the synoptic variability of the currents is comparable to the current vectors themselves, not only in the regions of the known meanders of strong currents, but perhaps also nearly everywhere in the ocean (see Fig. 1-1-1 for an example of the variability of the North Equatorial Current). Obviously, such strong variations make it impossible to rely on the average currents represented on climatic charts in the atlases of the ocean, and synoptic (multiweek) current forecasts are needed. Such forecasting will require, first, the development of a hydrodynamic theory of synoptic processes in the ocean (and, apparently, of Rossby waves to begin with). However, we still have very little observational data on these processes. Second, forecasting of this kind requires routine acquisition of initial data, i.e., synoptic global observation of the ocean. Solution of these two problems is one of the most important tasks of contemporary oceanology.

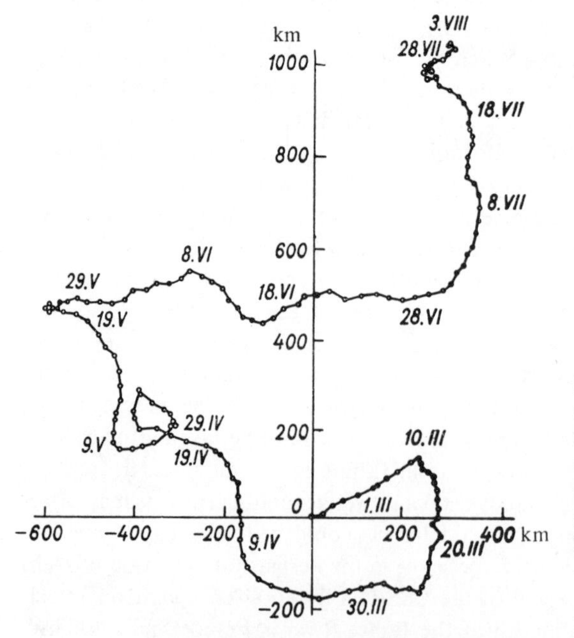

FIG. 1-1-1 Five-month sequence of diurnal average current velocity vectors (km/day) at depth of 50 m at the center of the 1970 hydrophysical test range in the tropical Atlantic (16°30′N, 33°30′W) (according to Brekhovskikh et al. [5]).

The fishing industry also needs synoptic forecasts of the currents, temperature, and salinity of the upper ocean, but their requirements are broader. They also require shorter-term forecasts of fish-shoal concentrations, as well as forecasts of the year-to-year variability of several oceanological fields that are ecological factors in the dynamics of commercial fish populations. These problems are in the province of fisheries-related oceanology, which Berenbeym [6] defines as the science dealing with the effects of oceanological factors on the productivity, numbers, and behavior of commercially valuable marine organisms, with the object of developing oceanological support for fisheries forecasts.

The first monographs on fisheries-serving oceanology were published by Knipovich (1938) and Izhevskiy (1961). The present state of this science is described by Hela and Laevastu [7]. *The temperature field is the most important of the oceanological fields studied in fisheries-related oceanology.* We know now that some species of fish are sensitive to temperature differences as small as 0.03°C. Temperature influences the maturation of the gametes in fish. In addition, temperature also effects the time and place of spawning, the incubation time of the eggs (which is roughly inversely proportional to the temperature), the larvae survival rate (which increases with temperature), the rate of development of the plankton food of the fish, their feeding activity, metabolism, and growth (which decrease at temperatures above the optimum). Large sudden temperature changes may result in massive fish kills.

It appears that each species of fish has a set of optimum existence temperatures that differ at different stages of development. This, together with the food-concentration fields, determines the areas and depths of the highest fish-shoal concentrations. (Here there may be a problem whether to base the forecast on the otpimum temperatures or the quantity and quality of fish fodder. This problem is probably solved with considerable statistical scatter.) As an example, Fig. 1-1-2 shows optimum temperatures for various commercial fish off the shores of Japan, according to Uda [8]. The narrow optimum-temperature ranges are easiest for the forecaster to find in the zones of thermal fronts, where there are large horizontal temperature gradients (and in "pockets" or partings of cold water that are formed, among others, where meanders separate out at thermal fronts). Indeed, it is at such points that shoals of pelagic fish are often observed (see, for example, Fig. 1-1-3, taken from Vyalov [9]). A practical method of forecasting fish concentrations in the Sea of Azov has now been developed [11] on the basis of Fel'zenbaum's theory [10] of wind-driven currents in shallow seas. Its introduction into routine fisheries practice made it possible to increase, as early as 1970, the efficiency of Clupeonella fishing by a factor of 1.3 and to catch an additional million kilograms of this fish [12].

Other oceanological factors are also important for fisheries-serving oceanology: the light field in the sea, currents, waves, and dissolved-oxygen concentration. The research institutions of the fishing industry now have much data on the effect of oceanological factors on the fecundity, development, migration, and food range of various commercial fish in several areas of the ocean. These data make it possible to utilize to some degree the ocean-variability forecasts to plan fishing operations and to rationally manage the movements of fishing fleets. However, full utilization of ocean-variability forecasts in fisheries, that is, their reduction to a procedure routinely

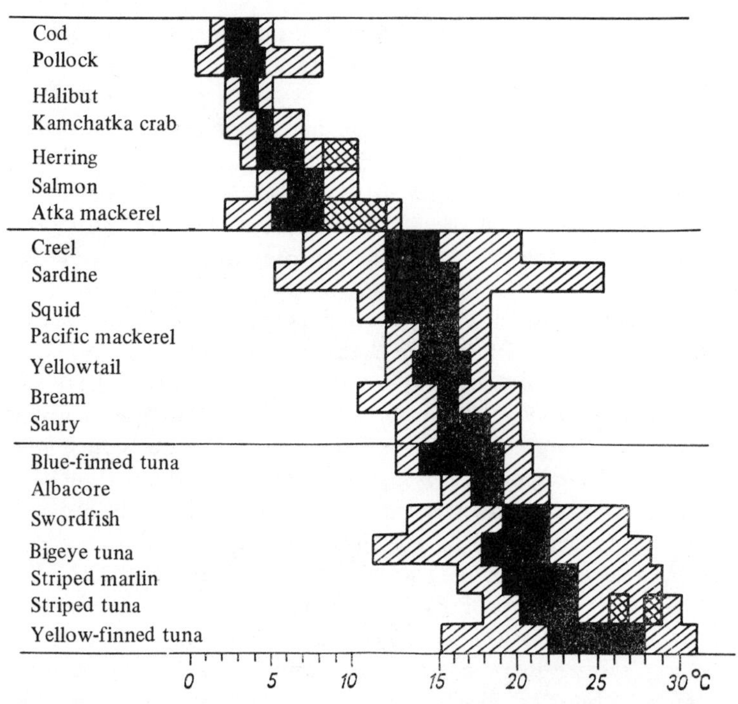

FIG. 1-1-2 Optimum temperatures for the most valuable commercial fish off the coast of Japan (according to Uda [8]). It is assumed that the number n of fish of a given species that can be caught depends on the ambient temperature as $n = n_0 \exp [-(T - T_0)^2/2\sigma^2]$, where T_0 is the optimum temperature for the given species. The temperature range at which fishing is profitable is determined on this basis (shaded areas); the interval $|T - T_0| \leqslant \sigma$ corresponds to the optimum temperature range (blackened areas). The temperature range indicated by oblique shading is optimum for subpopulations (size, age, etc.) [prepared from data of [8] and personal communications of Prof. Uda].

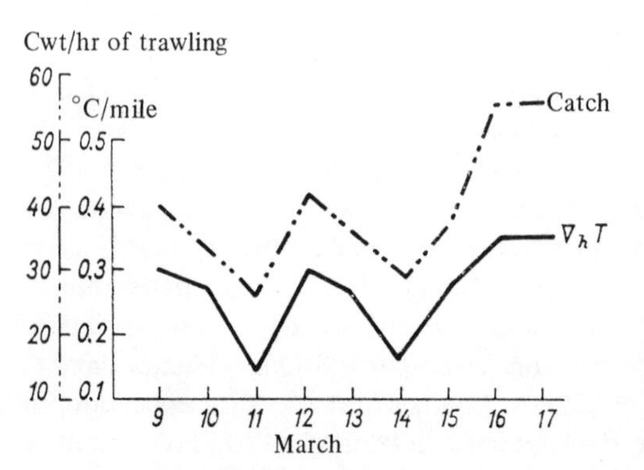

FIG. 1-1-3 Relation between horizontal temperature gradients $\nabla_h t$ in surface layers of water and variations of fish haul per hour of trawling in the southwestern part of the Georges Bank in March 1966, according to observations of V. D. Strogalev (after Vyalov [9]).

used in fisheries practice, will require mathematical description, or models, of fisheries. These models will need to incorporate, first, a quantitative description of the composition (primarily an age breakdown), evolution (including fecundity and reproduction, growth, diseases, losses due to natural causes and as a result of fishing) and behavior (with emphasis on migrations) of the fish populations of commercial interest in specific waters. Second, the models will contain a similar description of their food supply, as well as observational data on past and present oceanological factors that have influenced the population and prognostic data for the future; and finally, the economic planning indicators and profitability criteria, without which no forecast could be economically useful [13]. With updating to include new observational data and oceanological forecasts, such mathematical models will provide a scientific base for management of ocean fisheries, raising them to the level of a planned oceanic-fishing industry. The same models could be helpful in finding ways to control the biological productivity of the ocean. Several general theoretical principles for such modelling are set forth in Watt [14], who also includes special sections on fish populations. The Menshutkin monograph [15] is devoted to the mathematical fundamentals of modelling aquatic animal populations.

Finally, let us discuss the problem of long-term weather forecasting. In this case the large-scale variability of the sea is probably a controlling factor since the basic source of long-term weather anomalies must be the action of the sea on the atmosphere [16, 17]. The atmosphere, considered as a system isolated from external perturbations, does not appear to contain any active mechanisms with the time constants on the order of months that are characteristic of the long-term weather anomalies. The solar heat flux exhibits almost no irregular oscillations. Therefore, the source of long-period irregularities must be the interaction of the atmosphere with the surface underlying it, which is largely the sea.

The principal action that the sea exerts on the atmosphere is to supply it with moisture (water evaporates from the sea surface at a rate of over $3 \cdot 10^{20}$ g/yr) and with heat, including the latent heat of the water vapor (on the order of $2 \cdot 10^{23}$ cal/yr) and the sensible heat, imparted to the atmosphere by radiative and turbulent (contact) heat transfer. The rate of this moisture- and heat-transfer is governed primarily by the temperature of the sea surface, *which therefore is the main oceanological field that must be calculated for a long-term weather forecast* [16]. However, the properties of the atmosphere itself, for example, its temperature and humidity, and their vertical gradients in the air layer over the sea surface, as well as the wind velocity, also are factors controlling the transfer processes.

Certain data indicate that a significant, if not principal effect of the sea on the atmosphere is exerted via storms (see, for example, [18]). Together with other less sharply defined short-term synoptic atmospheric processes, storms constitute the mechanism via which the set of random large-scale atmospheric motions tends to adapt itself to the thermal state of the sea (the latter has a large thermal inertia, and its thermal state changes much more slowly than the state of the atmosphere). The distribution of the heat input from the sea over the atmosphere is regulated primarily by the clouds, in which the latent heat is released on condensation of water vapor and the radiant and turbulent heat fluxes are retained. Since they also regulate the input of solar heat to the surface of the sea, clouds are the feedback mechanism in the processes in which the sea acts on the atmosphere, and may thus endow these processes with oscillatory properties [19].

Therefore, for long-term weather forecasting, one needs to know, *a priori*, the pattern of evolution of the temperature field T_w of the sea surface. However, to predict the variability of this field one needs a long-term weather forecast, primarily of the wind and clouds (together with initial observational data on the T_w field), that is, the weather and the variability of the sea must be forecast *jointly* over the long term on the basis of an overall numerical model of the atmosphere and the ocean. We think that the development of such a model is the only scientifically sound approach to long-term forecasts for both the weather and the variability of the sea.

REFERENCES

1. Mikhaylov, S. V. Ekonomika Mirovogo okeana (*The Economics of the Ocean*), Ekonomika Press (1964).
2. Vinogradov, M. Ye., I. I. Gutel'zon, and Yu. I. Sorokin. Spatial structure of communities in the euphotoc zone of tropical ocean waters. In: Funktsyenirovaniye pelagicheskikh soobshchestv tropichestikh rayonov okeana (*Functioning of Pellagic Communities in the Tropical Regions of the Ocean*), Nauka Press, Moscow, 255–262 (1971).
3. Livshits, V. M. and Yu. A. Khovanskiy. Spravochnik dlya sudovoditeley po gidrometeorologii (*Handbook of Hydrometeorology for Navigators*), Transport Press, Moscow (1967).
4. Monin, A. S. General views on problems of physics and geophysics of the world ocean. In: *The Ocean World*, Proc. of Joint Oceanogr. Assembly IAPSO, CMG, SCOR, Tokyo, Kap. Soc. Promot. Sci., 87–88 (1971).
5. Brekhovskikh, L. M., G. N. Ivanov-Frantskevich, M. N. Koshlyakov, K. N. Fedorov, L. M. Fomin, and A. D. Yampol'skiy. Certain results of a hydrophysical experiment on a test range in the tropical Atlantic, Izv. Akad. Nauk SSSR, Fizika Atm. i Okeana, 7, No. 5, 511–528 (1971).
6. Berenbeym, D. Ya. Fisheries-serving oceanology as a science. In: *Commercial Ichthyology and Oceanology*, Series I, No. 2, Central Scientific Research Institute of Information and Technical and Economic Research for the Fishing Industry, Moscow, 4–5 (1970).
7. Hela, I., and T. Laevastu. *Commercial Oceanography* (Translated from English), Food Industry Press, Moscow, 1970.
8. Uda, M. A. A consideration of the long years trend of the fisheries fluctuation in relation to sea conditions. Bull. Japan Soc. Sci. Fish., **23**, Nos. 7–8, 368–372 (1957).
9. Vyalov, Yu. A. Use of atmospheric pressure data in predicting fish hauls. In: Atlanticheskiy Okean (*The Atlantic Ocean*), Fish Location Studies, No. 2, Kaliningrad, 65–73 (1969).
10. Fel'zenbaum, A. I. Teoreticheskiye osnovy i metody Rascheta ustanovivshikhsya morskikh techeniy (*Theoretical Bases and Methods for Calculation of Steady-State Ocean Currents*), USSR Academy of Sciences Press, Moscow (1960).
11. Vasil'yev, A. S., G. N. Deynega, and R. T. Fedotov. A method of short-term forecasting of the distribution of commercial concentrations of fish at sea. In: *Commercial Ichthyology and Oceanology*, Series I, No. 2, Central Scientific Research Institute of Information and Technical and Economic Research for the Fishing Industry, Moscow, 68–84 (1970).
12. Fedotov, R. T. A fisheries forecast service at the Computer center aids in fleet deployment, Rybnoye Khozyaystvo, No. 5, 82–83 (1971).
13. Monin, A. S. Use of unreliable forecasts, Izv. Akad. Nauk SSSR, Ser. Geofiz., No. 2, 218–228 (1962).
14. Watt, K. *Ecology and Resource Management: A Quantitative Approach*, Academic Press (1966).

15. Menshutkin, V. V. Matemachiskoye modelirovaniye populatsiy i soobshchestv vodnykh zhivotnykh (*Mathematical Modelling of Populations and Communities of Aquatic Animals*), Nauka Press, Leningrad (1971).
16. Monin, A. S. Physical mechanism of long-term weather changes, Meteorologiya i Gidrologiya, No. 8, 43–46 (1963).
17. Monin, A. S. Prognoz pogody kak tadacha fiziki (*Weather Forecasting as a Problem of Physics*), Nauka Press, Moscow (1969).
18. Gavrilin, B. L., and A. S. Monin. Calculation of climatic correlations from numerical models of the atmosphere, Izv. Akad. Nauk SSSR, Fizika Atm. i Okeana, **6**, No. 7, 659–666 (1970).
19. Gavrilin, B. L., and A. S. Monin. A model of long-term air-sea interactions, Dokl. Akad. Nauk SSSR, **176**, No. 4, 822–825 (1967).

1-2 AIR-SEA INTERACTION

The atmosphere could exist on the earth even in the absence of the oceans. Thus, if the surface of the earth were sufficiently moist (for example, covered by marshes and shallow seas), enabling the atmosphere to acquire sufficient moisture by evaporation, the *instantaneous* states of the weather in this atmosphere would be similar to those observed on the real earth. This is because the *basic specific feature of the weather of the earth*, that is, *the variable cloud cover*, would be preserved even under these conditions [1]. The existence of a deep sea changes these conditions, chiefly because the thermal state of the sea, with its high heat capacity and consequent thermal inertia, can vary slowly in time. The effects of these variations on the atmosphere are apparently the cause of the long-term weather anomalies, the mechanism of which was briefly described at the end of the preceding section.

On the other hand, the action of the atmosphere accounts for the very existence of the sea, and for most nonstationary processes in it. In fact, *nearly all motions of the water in the sea*, except for motions of tidal origin, local currents near the mouths of rivers, and sudden motions of the tsunami type (which are caused by tectonic and volcanic processes at the sea bottom), *are produced by atmospheric factors* [1] that involve momentum transfer from the atmosphere to the sea and, to some degree, by mass and heat fluxes on the sea surface that increment the available potential energy, convertible to kinetic energy of sea motion.

The momentum fluxes from the atmosphere to the sea, both turbulent and those caused by atmospheric-pressure variations, control various types of wave motions in the sea, including wind and internal waves, inertial oscillations, and Rossby waves. In some cases, for example, in that of wind-driven and high-frequency internal waves, the most important factors may be direct *resonant excitation*, that is, the effect of variations of the atmospheric variables with frequencies and wave numbers equal to those of the possible natural oscillations of the sea. In other cases there may be nonresonant excitation, especially in the case of long-period oscillations such as Rossby waves, in part as a result of nonlinear interactions of motions generated directly by atmospheric factors.

The most important factor in separation of quasistationary and nonperiodically varying currents is direct wind friction on the sea surface. This produces the *pure drift currents* in the upper layer of the ocean that reflect both the quasistationary wind field which creates the basic pattern of circulation of the surface waters of the ocean, including the gigantic anticyclonic gyres around subtropical centers of action in the atmosphere and the equatorial and monsoon currents, and also the transient wind systems of atmospheric lows.

Currents at all depths in the sea can be generated by horizontal pressure gradients that arise as a result of set-up and set-down of the pure drift currents. These produce horizontal variations of ocean level that lead to a corresponding restructuring of the density field at all ocean depths (*wind gradient* currents). Currents of all depths can also be generated by synoptic variations of atmospheric pressure at the sea surface (*pressure-gradient* currents). The pure drift and wind gradient currents, taken together, account for the wind-driven or *wind* currents in the sea, but the pressure-gradient currents are generally not too significant in the sea. Finally, rises or drops in the density of the surface waters caused by cooling, heating, and increase or decrease in salinity due to exchange of heat and moisture with the atmosphere, may give rise to *thermohaline* currents.

Atmospheric factors not only generate motion in the sea, but also control its stratification and the variability of its thermodynamic (primarily thermal) state. In contrast to the atmosphere heated mainly from below, that is, from the underlying surface, a process which often causes unstable stratifications in the lower atmosphere, the sea is warmed from above by the direct and atmosphere-scattered solar radiation, and exchanges heat with the atmosphere over its surface. The geothermal heat flux through the sea floor is very small and has no significant effect on its thermal state. On the other hand, waters cooled from above in the polar regions descend to the bottom and spread across it. As a result, the density stratification of the ocean water is hydrostatically very stable almost everywhere. The net effect of this stability on the nature of the flow in the sea is very pronounced and shows up in the weak development of turbulence, strong development of internal waves, and apparently the jetlike nature of the principal currents [1].

The nonuniformity in the heating of the upper ocean is due primarily to the variability of the cloud cover that screens the direct solar radiation. This suggests that long-term anomalies in the cloud cover over a given area of the sea contribute to shaping of the enthalpy anomalies of the upper ocean in that area. In addition, turbulent sea-air heat exchange and transport of latent heat from the sea to the atmosphere by evaporation of moisture, the two most important factors in the cooling of the upper sea, depend strongly on the wind velocity. Therefore the greatest cooling of the upper ocean should occur in the areas of storms. Thus the large-scale variability of the thermodynamic state of the ocean is governed directly by atmospheric factors.

Given the high inertia of the sea, its response to the action of the atmosphere is generally *lagging* (see the discussion of this problem in Lighthill's paper [2]). Even practically instantaneous response carries slowly decaying traces. Nonresonant response, even including response to stationary perturbations (with zero frequencies), is often associated with chains of internal interactions which require time. Resolution of internal instabilities also requires a certain amount of time. An example of this kind of process is the formation of the Rossby waves that appear to be a very important factor in the long-term synoptic variability of ocean currents. Because of the lag of its

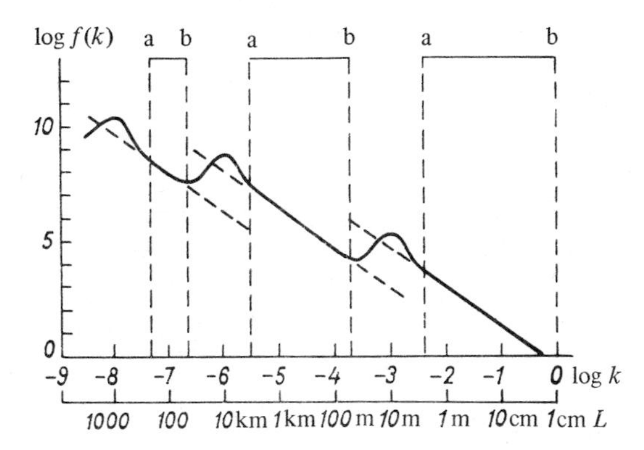

FIG. 1-2-1 Schematic kinetic-energy spectrum of ocean motions (after Ozmidov [3]). *a-b*–Zones in which the 5/3 law is satisfied.

response, the sea acts as a kind of "storage device" that preserves the imprint of atmospheric perturbations that have occurred some time in the past.

The atmosphere feeds energy to sea motions over a variety of spatial scales. Ozmidov [3] distinguishes three main energy-input ranges. First, there are the scales L_1 of about 10^6 m, that is, of the order of the dimensions of the oceans and of quasistationary and moving atmospheric lows and highs. These are the scales of both the currents of the quasistationary oceanic circulation and of nonstationary currents, such as drift currents, that are the responses of the sea to the action of synoptic processes in the atmosphere. In this scale range, the rate ϵ_1 of kinetic energy input into the ocean (per unit of mass) appears to be on the order of 10^{-5} cm^2/sec^3, and the effective coefficient k_1 of horizontal mixing, produced by nonstationary currents, is of the order of 10^4 m^2/sec. Second, inertial fluctuations are generated at scales L_2 of about 10^4 m; the tidal oscillations have comparable periods. The total kinetic energy input rate ϵ_2 in this case is on the order of 10^{-3} cm^2/sec^3, and the effective horizontal mixing coefficient k_2 is on the order of 10^{-1} m^2/sec. Third, the sea receives kinetic energy from the atmosphere at wind-wave scales L_3 of about 10^1 m, the rate ϵ_3 of this apparently being on the order of 10^{-1} cm^2/sec^3. Here the mixing coefficient k_3 is of the order of 10^{-3} m^2/sec. Figure 1-2-1 taken from Ozmidov's paper [3], gives a schematic presentation of the complete kinetic-energy spectrum of sea motions.

Forecasts of the synoptic, annual, and year-to-year variability of the thermodynamic state of the sea and of the wind-driven and thermohaline currents in it, are needed even now for a number of practical purposes. However, these forecasts can only be generated on the basis of global mathematical models of sea-air interaction that account in sufficient detail for all the basic interactions, the various responses to each action, and the possible feedbacks.

Some of the qualitative statements made here will be formulated quantitatively in later sections of the book.

REFERENCES

1. Monin, A. S. Fundamental consequences of air-sea interaction, Izv. Akad. Nauk SSSR, Fizika Atm. i Okeana, 5, No. 11, 1102–1114 (1969).

2. Lighthill, M. J. Time-varying currents, Phil. Trans. Roy. Soc. London, A270, No. 1206, 371–390 (1971).
3. Ozmidov, R. V. Certain features of the oceanic turbulent energy spectrum, Dokl. Akad. Nauk SSSR, **161**, No. 4, 828–832 (1965).

1-3 CLASSIFICATION OF NONSTATIONARY PROCESSES IN THE OCEAN

The time-dependent variations of oceanological quantities, that is, the variations of distribution of different characteristics of the sea, are produced by a number of physical processes in the atmosphere and the sea. Among such distributions are, first, those of current velocities, temperature and salinity, then those of the thermodynamic properties reflecting the state of the water (density, entropy, electrical conductivity, sound velocity, refractive index), which are controlled by temperature, salinity, and pressure (depth), and also those of the concentration of organic and mineral suspended matter and dissolved gases and the characteristics of the sea surface, such as sea level, ice, and sea state. In classifying these fluctuations, we think it useful to proceed, by analogy to the classification of types of variability of meteorological quantities [1], from the concept of periods in the spectrum. This yields a classification with the following seven ranges.

1. *Small-scale phenomena.* These have periods ranging from fractions of a second to tens of minutes. Among the phenomena that fall into this range are surface and internal waves, turbulence, and the processes of evolution of the vertical microstructure of the sea (some elements of the latter may also have longer lifetimes). These phenomena are controlled by phenomena of larger scale (i.e., they "adapt" to them and must be parameterized by the large-scale characteristics) and at the same time serve as an internal mechanism shaping the large-scale variability of the ocean.

Waves on the surface of the sea are a very important factor in the dynamics of its upper layer, much more important than in the air surface layer over the sea. Wind waves are especially widespread and are dynamically the most important. (We shall not consider long waves of the tsunami type.) The wind waves are highly variable, being controlled by the force and length of action, or fetch of the wind. The spectra of periods of developed waves are quite narrow; the periods τ of the waves with the largest amplitudes vary from 1 to 3 sec in weak winds to 8 to 10 sec in moderate winds, and 20 to 30 sec in very strong winds. The lengths L (roughly speaking, $L = g\tau^2/2\pi$) range from a few meters to several hundred meters, the heights (doubled amplitudes) from centimeters and decimeters in weak winds to several meters in moderate winds and 20 to 30 m in very strong winds in the open sea. (The wave amplitudes decrease with depth z as $e^{-2\pi z/L}$.) A review of the available information on this topic will be found, among others, in Kinsman's book [2].

Internal gravity waves within the sea are very widespread and are dynamically important, much more so than in the atmosphere where gravity waves of finite amplitude are fairly uncommon. They may be generated by several factors such as tidal forces, variations of atmospheric pressure and fluctuations of wind action on the sea surface, surface waves, effects of flow over roughness on the sea floor, and inhomogeneities of large-scale currents. Their *natural* frequencies (in the case of not very long waves) lie between the inertial frequency $2\Omega \sin \varphi$ (φ is the local geographic latitude) and the peak Brunt-Väisälä frequency N, so that their longest possible period is equal to one day divided by $2 \sin \varphi$, and the shortest to $2\pi/N_{max}$ (in the thermocline, this figure is about 10 min, but it may be even smaller on "steps" of the vertical microstructure of the sea).[1] Since the vertical water-density differences in the sea are small, vertical displacements do not require large energy, so that the waves often grow to very large amplitudes. Thus, for example, Bockel [3] recorded an internal wave with an amplitude of about 100 m in the straits of Gibraltar. The theory of internal waves is not very complicated, but it is difficult to observe them. Consequently, our observational data are still deficient in many respects. A review of this topic can be found, for example, in Krauss [4].

Turbulence is the chief mechanism for vertical mixing in the sea and, in particular, for momentum and heat exchange with the atmosphere. This exchange gives rise to the large-scale variability of the sea. Turbulence is most highly developed in the upper mixing layer, but even there it is quite weak, that is, velocity and temperature fluctuations are of the order of 1 cm/sec and 10^{-1} °C, respectively. The spectrum of the turbulent-fluctuation periods is broad, apparently covering the range from 10^2 to 10^{-3} sec. Not enough is known about turbulence in the sea. The same also holds for its vertical microstructure, that is, for the relatively short-lived quasi-homogeneous layers of water, with thicknesses of the order of meters and tens of meters, separated by thin "sheets" (thicknesses in centimeters) with sharp velocity, temperature, and salinity gradients.

2. *Mesoscale phenomena.* These have periods from hours to a day, and encompass tidal and inertial fluctuations that are produced (because of the rotation of the earth on its axis) by the gravitational attraction of the moon and sun, and by inertial forces attendant to that rotation. Long-period tides are also possible, but these have very small amplitudes. Diurnal thermally induced fluctuations, that is, those caused by the diurnal variation of insolation, are also mesoscale phenomena.

The tidal oscillations of sea level and also the tidal currents have a dominant period of half a lunar day (averaging 12 hr 25 min). In addition, a diurnal period may also be quite pronounced. Because of friction-induced lag, the tides always reach their maximum height somewhat later than the culmination of the moon. In the open sea, this lag averages only a few minutes, but it may amount to several hours in shallow waters. In the open sea, the height of the tides varies from very small values to 1 to 2 m. The maximum tidal-current velocities are only a few centimeters per second. However, these currents are generated over the entire depth of the ocean, and on large areas may be the basic component of bottom currents. In shallow waters, near the coasts, and especially in certain narrow straits, gulfs, and bays, the height of the tides may range to several meters: for example, 14.5 m in the mouth of the River Severn in England, 13.5 m in the mouth of the Rance River in France, 12.9 m at the head of Penzhina Bay in the USSR. Moreover, the velocities of the tidal currents may range to several knots (one top figure recorded is 16 knots in Skjerstad Fjord, Norway). Although tides have been studied fairly thoroughly, the theory of tidal currents which requires integration of the equations of motion for the real ocean, is complex and not yet useful in practical computation. Also, observations in the open sea are difficult and therefore few data are available. For a review of tides, see Dietrich [5, Chap. 9].

Inertial fluctuations of current velocity with a period of 24 hr/2 sin φ are often generated in the ocean; the velocity variations may range to tens of centimeters per second. Here, however, few observational data exist. For a review of the available data, see Webster [6].

Because of the large thermal inertia of the sea, the diurnal variations of water temperature in it are much smaller than in the atmosphere. At the sea surface they have an amplitude on the order of 10^{-1} °C and decrease rapidly with depth, that is, by at least an order of magnitude in the first 50 meters.

---

[1] The parameter N is defined as the frequency of the natural oscillation (in a field of buoyant restoring forces) of fluid parcels displaced adiabatically along the vertical from their equilibrium level [see relation (2-2) in Chap. 2].

3. *Synoptic variability* has periods that range from a few days to several months. Its principal manifestation is aperiodic formation of sea eddies with scales on the order of 100 km, analogous in many respects to atmospheric lows and highs but with much longer lifetimes and much slower movement. The major factors inducing synoptic variability are the cumulative effect of atmospheric disturbances, primarily the action of the variable wind on the sea surface, as well as the thermal action of the atmosphere, and processes induced by hydrodynamic instability of large-scale ocean currents.

The synoptic variation of ocean-current velocities may be comparable to, and sometimes even larger than, the multiyear-average velocities of these currents. Therefore, they are the strongest of all variations of oceanological fields. Thus, observations on the 1970 Soviet test range in the North Equatorial Current in the Atlantic [7] showed that the current at a given point may for periods of weeks be opposite to the usual west-southwestward direction. This implies, among other things, that almost all the ocean-current charts now in use that were constructed by the dynamic method (see Chap. 2) or from isolated one- or two-day instrumental measurements, may be nonrepresentative and should be replaced by synoptic current charts. In local regions, for example, in the regions of the Gulf Stream or Kuroshio meanders, the synoptic temperature variations of the upper ocean may be of the order of several degrees.

Given the almost complete lack of long-term, continuous series of observations on the sea, we know practically nothing about its synoptic variability. Study of the synoptic variability will be the main objective of continuous worldwide oceanic observations to be organized during the next few years. In the future, prediction of the synoptic variability will be the basic function of sea forecasts.

4. *Seasonal variations* (yearly period and its harmonics). These are most conspicuous in high latitudes and in the monsoon zone of the Indian Ocean.

Thus, the range of the annual sea-surface temperature variations exceeds 14 to 18°C in the northwestern Atlantic and Pacific. It exceeds 20 to 25°C in the waters off North Korea and northeastern China, and ranges upward from 2°C in the equatorial zones. In regions with strong annual variation of water-surface temperature, we also observe significant seasonal variations in the thickness of the upper mixed layer, that is, fluctuations in the thermocline depth. In the summer, the thermocline forms at minimum depths ranging from 15 to a few tens of meters, but in winter the thickness of the mixed layer ranges upwards of 200 to 300 m because of the convection set up by cooling from the surface. Sometimes the convection current penetrates to the sea floor. Over the course of a year, the ice cover of the seas clearly varies significantly in high latitudes. The largest annual variation of surface-water salinity, on the order of 1 to 3 percent, is observed in the Bay of Bengal and the Australasian seas, as a result of the sharp annual variation of monsoon rainfall. In the open part of the North Atlantic they amount to only 0.2 percent, but near Newfoundland they range to 0.7 percent due to the melting of the ice in summer. Several currents also exhibit significant seasonal variations due primarily to the seasonal variations of the atmospheric circulation. In particular, there are conspicuous seasonal variations of the Somali current and an intensification of the Equatorial countercurrents in the Atlantic and Pacific oceans in the northern summer. On the other hand, the Equatorial countercurrents in the Indian Ocean appear to disintegrate at that time.

5. *The year-to-year variability*, that is, the synchronous, year-to-year variations of the state of large areas of the ocean and the entire atmosphere. Examples are the self-oscillation of the northern branch of the Gulf Stream with a period of about 3.5 years (Shuleykin [8]), and the quasi-two-year El Niño phenomenon in the eastern equatorial zone of the Pacific (Bjerknes [9]). A probable significant factor in the

year-to-year variability is the movement of thermal anomalies along gigantic oceanic gyres, around which water eddies over periods of several years. Prediction of the year-to-year variability will be an important element in forecasting the reserves (population dynamics) and permissible hauls of the basic commercial fish of various ages in various waters of the ocean.

6. *The secular variability*, with periods in the tens of years, is interrelated with the secular variations of climate. It is exemplified by the variations of the state of the ocean (warming by several degrees in the Arctic and a slight chilling in low latitudes) during the climatic warming of the first half of this century, reported by Mitchell [10]. The most noticeably warmer winters during this period were in northern latitudes of the Northern Hemisphere.

7. *The century-to-century variability*, with periods of a hundred years or more, is related to the century-to-century variations of climate. An example of this variability was reported by Bjerknes [11]. Thus, the state of the northern half of the Atlantic during the "Little Ice Age" of the 17th to 19th centuries differed from the present state in that the Sargasso Sea was 2 to 3°C warmer and the waters around Iceland 1°C cooler than the present norm. Bjerknes thinks that this long-period variation was produced by an anomalous decrease in heat exchange in the air-sea interaction.

One of the most readily studied manifestations of the variability of the state of the ocean during the postglacial epoch is the global variations of its level through tens of meters as recorded, for example, in coastal terraces and at the mouths of rivers. The variability of the state of the ocean during the glacial periods of the Pleistocene, and in the even more remote past, can be judged from the micropaleontology of the layers in ocean-sediment cores (preferably in combination with isotopic determinations of the absolute age of these layers) and by the Urey paleotemperature method (i.e., from the oxygen isotope ratio $^{18}O/^{16}O$ in biogenic carbonates). However, this problem will not be dealt with in our book.

REFERENCES

1. Monin, A. S. Prognoz pogody kak zadacha fiziki (*Weather Forecasting as a Problem of Physics*), Nauka Press, Moscow (1969).
2. Kinsman, B. *Wind Waves, Their Géneration and Propagation on the Ocean Surface*, New York (1965).
3. Bocket, M. Traveaux océanographiques de l' "Origny" à Gibraltar, Cahiers Océanographiques, **14**, No. 4, 325–329 (1962).
4. Krauss, W. *Internal Waves* (Translated from the German), Gidrometeoizdat Press, Leningrad (1938).
5. Dietrich, G. *General Oceanography*, Wiley (1963).
6. Webster, F. Observations of inertial-period motions in the deep sea, Rev. Geophys, **6**, No. 4, 473–490 (1968).
7. Brekhovskikh, L. M., G. N. Ivanov-Frantskevich, M. N. Koshlyakov, K. N. Fedorov, L. M. Fomin, and A. D. Yampol'skiy. Certain results of a hydrophysical experiment on a test range in the tropical Atlantic, Izv. Akad. Nauk SSSR, Fizika Atm. i Okeana, 7, No. 5, 511–528 (1971).
8. Shuleykin, V. V. Fizika morya (*Physics of the Sea*), Fourth edition, Nauka Press, Moscow (1968).
9. Bjerknes, J. Large-scale air-sea interaction. In: *Basic Problems of Oceanology*, Nauka Press, Moscow, 7–9 (1968).
10. Mitchell, J. M. On the world-wide pattern of secular temperature change. In: *Changes of Climate (Arid Zone Research, XX)*, UNESCO, Paris; 161–179 (1963).

11. Bjerknes, J. Atmosphere-ocean interaction during the "Little Ice Age" (seventeenth to nineteenth centuries AD), WMO-IV66 Symposium on Research and Development Aspects of Long-Range Forecasting, WMO Tech. Note, No. 66, 77–88 (1965).

1-4 METHODS OF OBSERVING THE VARIABILITY OF THE OCEAN

The variability of the oceanological quantities has time scales ranging from seconds at least into the millennia and in space from centimeters to tens of thousands of kilometers. Stommel's sketch [1], which appears in Fig. 1-4-1, gives a clear picture of the space-time spectrum of this variability for the case of the temperature field. Here the periods P of the fluctuations and the horizontal dimensions L of the spatial inhomogeneities are plotted on logarithmic scales on the two horizontal axes. The vertical axis carries the average values of the squared amplitude of the water-temperature variations per unit frequency and per unit wave number. Our knowledge of the ocean is not yet sufficient for conversion of this sketch into a true quantitative graph even in the case of specific areas of the sea about which we know the most. However, the concept of the space-time spectrum of a given oceanological quantity, treated as a random function of the spatial coordinates and time, is very useful for developing a strategy for oceanological measurements.

The different kinds of measurements of the variation of a given oceanological quantity are best made in different space-time scales of this variation. To study the small-scale phenomena, for example, turbulence, surface and internal waves, the fine

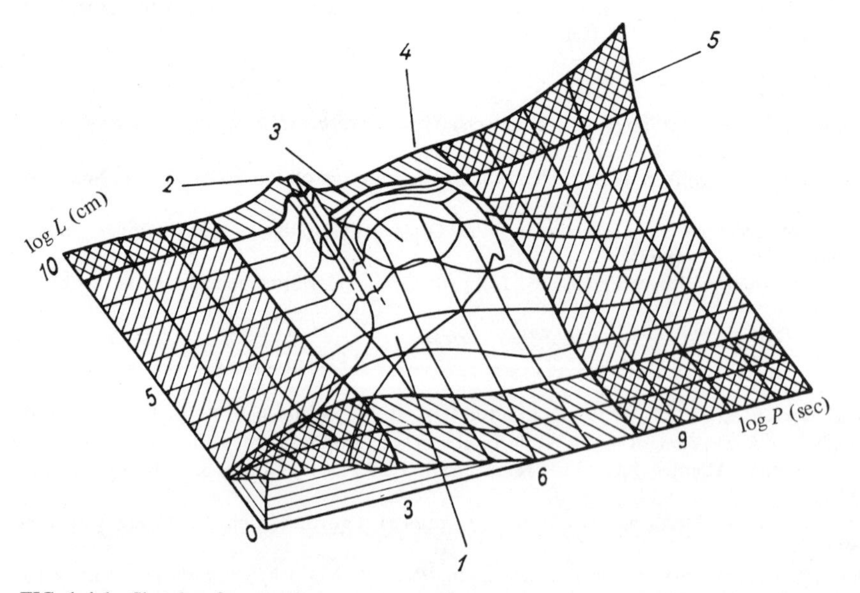

FIG. 1-4-1 Sketch of space-time spectrum of water-temperature variations in the ocean (after Stommel [1]). 1–Internal waves; 2–tides; 3–geostrophic eddies; 4–seasonal variations; 5–climatic variations.

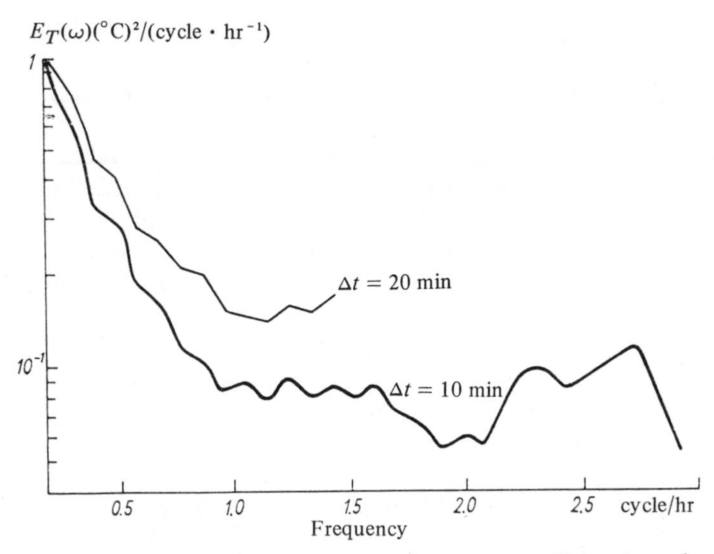

FIG. 1-4-2 Spectral densities (spectra) of temperature fluctuations, derived from the same series of observations but with different sampling intervals Δt (photothermographic observations in the thermocline of the tropical Atlantic). The diagram was made available to us by K. D. Sabinin.

structure of the thermohalocline, one employs low-lag instruments and small space-time observation scales. However, to investigate the synoptic variation, one can use standard oceanographic instruments and observation scales on the orders of months and hundreds of kilometers. The measurements must be planned so as to yield statistically significant results. This is ensured by proper selection of the sampling interval and duration of the observations. Usually it is assumed that $4\Delta t \leqslant P \leqslant n\Delta t/10$, where P is the period of the oscillations to be studied, Δt is the sampling interval, and n is the total number of observations.

Here it must be remembered that discretization of observations results in undesirable aliasing, whereby the estimate of the true amplitude of oscillations with period P is distorted by oscillations with shorter periods $P_m = P\Delta t/(mP \pm \Delta t)$, $m = 1, 2, \ldots$ (Fig. 1-4-2) that are present in a given process. In practice, this effect can be eliminated by choosing a certain relationship between the sampling interval and the time constant of the instrument [2].

The effort and money required, that is, the cost, is an important factor in planning and selecting a method for study of variability of the ocean. For example, to study the year-to-year variability of the ocean it is natural to follow the path taken by meteorologists on land, that is, to cover the entire ocean with an optimally dense network of permanent oceanographic stations that will operate for many years. In this case the first estimates of the year-to-year variability would be obtained not earlier than in 10 to 15 years. If this network were to cover every two-degree square of the ocean, then one would need about 7000 oceanographic stations, and the annual operating cost would be several billion dollars. This method would obviously be too expensive. Thus it seems realistic to plan a network of only a few hundred stations located on waters in which the variability is presumed to be highest and which are most important from a practical standpoint.

At the present time, the Intergovernmental Oceanographic Commission (IOC) of UNESCO and the World Meteorological Organization (WMO) are planning an International Global Oceanographic Station System (IGOSS), whose function will be to monitor conditions in the sea on a routine basis and assist the World Weather Watch. Measurements of the following characteristics will be most important from the standpoint of developing a picture of ocean variability for the purpose of meeting practical needs: currents, temperature and salinity of the water, wind and internal waves, turbulence, sonic velocity in and the electrical conductivity of the water, its optical characteristics, sea level, tides, and ice conditions.

The International Working Committee for IGOSS, which was organized in 1969 by the IOC, has already done much preparatory work. The following development principles have been set for IGOSS:

1. IGOSS is to be a global oceanic system. It is to consist of systems operated by the participating countries, with coordination and assistance by the IOC and other divisions of UNESCO, the WMO, and other international and regional organizations;
2. IGOSS is to service operational and research requirements to be agreed upon by the participating countries, and is to use the most modern observational, communications, and data processing equipment;
3. IGOSS is to be a dynamic system with sufficient flexibility to adapt to scientific and engineering advances;
4. IGOSS is to be planned and operated in close collaboration with the World Weather Watch and with work being done under the Global Atmospheric Research Program (GARP). The information collected by IGOSS is to be placed at the disposal of all countries in a form convenient for use.

Global observation, telecommunications, data processing, storage, and distribution systems are to be component parts of IGOSS. The observation system will include coastal and island stations and manned offshore stations, as well as floating beacons and towers, weather ships, automatic buoy stations, stations on shipping lanes, drift stations (ice fields, drifting buoys, etc.), satellite and submarine-cable systems that can be used for oceanographic observations, and enroute observations from ships and aircraft. The IGOSS remote control and communication systems must ensure systematic technical control of the automatic observation equipment, collection of oceanographic and synchronous meteorological information at the observing stations, and its transmission to data collection and processing centers. The allocation of the frequencies reserved for oceanography is to be based on the actual distribution of the global observing network, the available communications facilities, and radio propagation conditions. The World Administrative Radio Conference (WARC) of July 1969 allocated the following six frequency bands for transmission of oceanographic data: 4162.5 to 4166.0, 6244.5 to 6248.0, 8328.0 to 8331.5, 12479.5 to 12483.0, 16636.5 to 16640.0, and 22160.5 to 22164.0 kHz.

In 1969, the Sixth Session of the IOC approved the program for the first stage of IGOSS. The program includes organization of systematic enroute shipboard hydrometeorological observations from all research, commercial, and fishing vessels, the performance during 1972 of an experiment in the collection and exchange of bathythermograph data and development of instructions and codes for collection and exchange of oceanographic data. The organization of a permanent network of oceanographic buoy stations was deferred to the second-stage IGOSS program, implementation of which will be determined by the availability of the proper equipment. A no less important problem that must be solved prior to the second IGOSS stage is the development of principles and methods for planning the optimum placement of the oceanographic buoy stations. It is our hope that this book will serve as a starting theoretical base for the planning of IGOSS studies.

We shall not deal here with the instruments, technical facilities, and methods of oceanographic research. Instead we refer the reader to the *Handbook of Hydrological Research in the Oceans and Seas* [3]. We shall consider only some of the tactical problems of investigating the variability of the ocean in conformity to the classification set forth above.

To study the small-scale variability, it is best to measure the fluctuations of the temperature, sound velocity, electrical conductivity, optical-property, and wind-wave fields. This does not require measurements over large areas of the ocean, and statistically significant data series can be accumulated in periods ranging from a few minutes to a few hours. The small-scale phenomena should be studied at several points of the sea that are climatically and dynamically typical. Convenient installations for use in such studies could be artificial islands or oceanographic towers. In addition, drifting or anchored oceanographic stations such as FLIP or Cousteau's habitable buoy, which exhibit very small vertical fluctuation amplitudes (no more than a few centimeters) in storm waves, may also be used. The basic constraints on instruments for such measurements are quick response and small size of the sensors.

The mesoscale variability of the oceanological quantities is coupled with the diurnal variation of insolation of the sea surface, the tide-forming forces exerted by the moon and sun, and inertial oscillations which depend on geographic latitude. Thus the mesoscale phenomena must be studied in all climatic zones of the ocean. Multiday observations from artificial islands or anchored buoy stations, equipped with instruments to measure the principal hydrological characteristics, are the most rational approach to the analysis of mesoscale phenomena. It would appear that free-drifting subsurface floats with neutral buoyancy and telemetry equipment, of the type developed by Swallow, would be useful precisely for study of mesoscale phenomena. Long-term operation of wave recorders on the sea floor is highly promising for study of the tides.

The synoptic variability of the ocean is interrelated with the formation of oceanic Rossby waves and eddies with scales in the tens and hundreds of kilometers. The hydrodynamic instability of the currents and the effects of sea-floor relief cause the currents to meander. Study of these phenomena requires long-term (at least 3 to 6 months) measurements over large areas of the ocean. Such measurements are best performed on hydrophysical test ranges, that is, by emplacement for long periods of time of clusters of anchored buoy stations equipped with current and temperature meters. Each test range of this type could serve as a model for the future network of stationary oceanographic stations, and provide continuous observations of the synoptic variability of the sea. Past experiments of this kind (which will be described in Chap. 5), and especially the 1970 Soviet hydrophysical test range experiment in the tropical Atlantic, have indicated high promise for this method of studying the synoptic variability of the sea.

Because of the large cost of studies of this type, oceanographic research by the long-term hydrophysical test range method should be done selectively and in areas of the ocean that differ climatically and dynamically. Zones with strong boundary currents, oceanological fronts, and upwelling and sinking of water are among the most important of this kind.

Study of the seasonal and year-to-year variability of the physical quantities in the sea requires multiyear observations at fixed stations or in fixed regions of the sea. In practice, these observations are of a sampling nature, with sampling intervals on the order of a month or even 3 to 6 months, depending on the climatic zone. For example, in the equatorial zone, where the annual variations of the hydrophysical characteristics are small, observations in the summer and winter seasons are sufficient for purposes of analysis of the year-to-year variability. This raises the problem of eliminating the distortion caused by short-period oscillations. Thus, to eliminate tidal and inertial oscillations, it is sufficient to make observations once an hour for a few days. When measurements are being made across oceanographic sections, short-period perturbations can be eliminated by placement of stations at specific observation depths, and by smoothing of the data, that is, by the method of Koshlyakov [4].

Finally, the seasonal and year-to-year variabilities of the main hydrophysical quantities (temperature, salinity, current velocity), can be estimated from the integrated enthalpy and salinity characteristics of the waters, determined, for example, once a month or season on oceanographic sections. This method is described by Kort [5]. Its forte is that the effect of small-scale variability is automatically eliminated when the integrated characteristics are computed. In [6], Kort submitted a plan for study of the year-to-year variability of the thermal regime of the North Atlantic, employing "standard thermal sections." Figure 1-4-3 shows a layout of standard thermal sections for the entire ocean; it consists of standard sections that already exist and where observations are made more or less systematically, together with additional sections along which observations would be made in order to obtain a fairly complete picture of heat exchange in the oceans. It is essential that observations along the thermal sections be made during the same periods of time. The sections should be long enough to accommodate the maximum amplitudes of horizontal current meanders. Temperature observations, and salinity observations when temperature-salinity probes are used, should be made down to the lower boundary of the baroclinic layer of the ocean. Collection of the data from a thermal-section network of this kind will require a minimum of effort and money, and could be done by research and fishing and commercial vessels on a schedule of "hydrological days."

In recent years, the preparation and radio transmission of various operational charts of ocean surface temperature, waves, etc., has become routine. As an example,

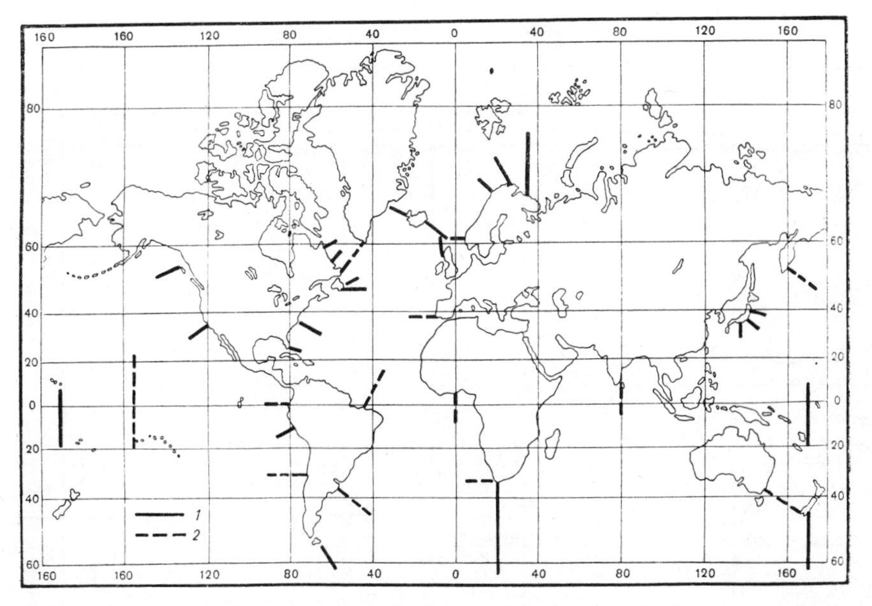

FIG. 1-4-3 Diagram of location of standard thermal sections in the ocean. 1–Existing standard sections; 2–planned standard sections.

Fig. 1-4-4 shows a chart of the ten-day-mean ocean surface water temperature from bathythermograph observations collected underway by fishing and commercial ships.

Table 1-4-1 gives a brief summary of the data transmitted by various countries in 1972–1973. It should be noted, however, that the table is not complete and is given only to provide a general picture.

Satellites offer great promise for study of the ocean and development of its resources and are extremely important for ship navigation (satellite navigation can be used to determine the position of a ship to within a few dozen meters). Satellites are used to determine the meteorological situation over the sea from televised cloud-cover photographs (the latter make possible tracing of storm paths by observing their cloud patterns), and to relay measured data from unmanned oceanological buoys and stations. In principle, satellites can be used for direct remote measurements of several important characteristics of the sea. The first of these is the temperature of the sea surface, which is accessible to remote measurement with airborne or satellite radiometers. These record the infrared radiation from the sea surface, for example, in the 10.5 to 12.5 μm range. Satellite surveys of the sea-surface temperature have already produced some interesting results. As an example, Fig. 1-4-5 shows a surface-temperature chart taken by the Nimbus-2 satellite on 27 July, 1966. Such charts can be used to direct fishing fleets to places with the largest horizontal temperature gradients, where shoals of fish often gather. Daily satellite surveys (even surveys made once a week or every ten days might suffice) of the temperature field would be an excellent means for recording the synoptic variability of temperatures and currents in the upper ocean. Similarly, satellite observations are highly promising for mapping the ice cover of the polar seas.

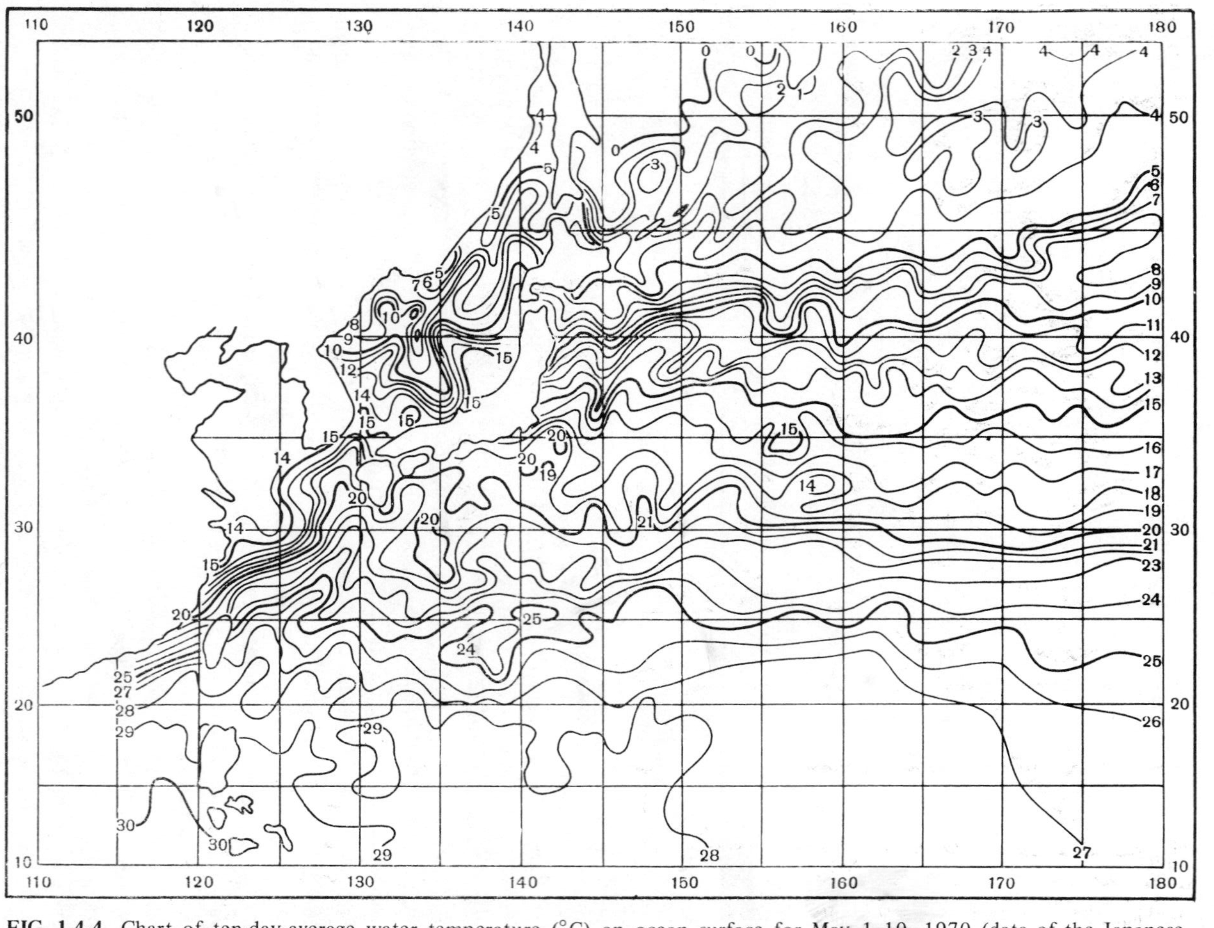

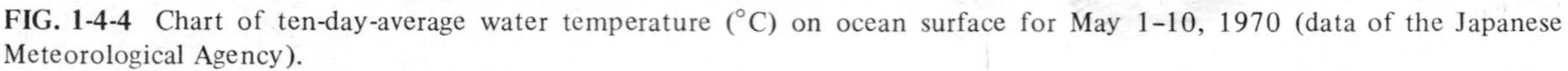
FIG. 1-4-4 Chart of ten-day-average water temperature (°C) on ocean surface for May 1–10, 1970 (data of the Japanese Meteorological Agency).

TABLE 1-4-1 Summary of Oceanographic Information and Forecasts Provided by Various Countries

Country	Radio-meteorological center	Broadcast area	Type of information	Forecast
USSR	Moscow	North Atlantic (down to 20°N)	Wind-wave charts (heights in meters); daily	24-hr waveheight forecast Surface water temperature forecast for the cold season (for five-day periods referred to the middle of the month)
		North Pacific Ocean (down to 15°N)	Same	Same
	Vladivostok	Same	Same	Forecast of 10-day-average surface water temperature on the 2nd, 12th, and 22nd of the month
				24-hr waveheight forecast
		Bering Sea	Same	Same, on 7th, 17th, and 27th days of the month
	Murmansk	Barents and Norway seas	Charts of temperature, wind waves, and position of the ice edge (daily)	24-hr wave forecast (daily)
Canada	Halifax	Davis Strait-Newfoundland	Ice situation charts, twice a day	
		North Atlantic	Wind-wave charts (heights, period, direction); daily	12-, 24-, and 36-hr forecasts of wave height, period, and direction
		Northwestern Atlantic Ocean (30–55°N, 45°W)	Surface water temperature charts and charts of thermocline depth (five-day-average values); daily	
Great Britain	Bracknell	Barents, Norway, and Greenland Seas, North Atlantic (to 45°N)	Ice-situation and surface-water-temperature charts (five-day average values); daily	24- and 48-hr wave-height forecasts
West Germany	Quickborn	North Atlantic	Iceberg-line charts; daily	
			Wind-wave and wind charts (with synoptic analysis); daily	24-hr wave-height forecast

TABLE 1-4-1 Summary of Oceanographic Information and Forecasts Provided by Various Countries—*Continued*

Country	Radio-meteorological center	Broadcast area	Type of information	Forecast
USA	Washington	Western North Atlantic (from 30°N to equator)	Water-temperature and thermocline-depth charts (five-day-average values); daily	
			Wind-wave charts; daily	12-, 24-, and 36-hr wave-height forecasts
	Kodiak	Bering Sea, Gulf of Alaska	Ice-situation and ice-edge charts (from satellite data)	
		North Pacific Ocean (to 40°N)	Wind-wave charts	24-hr wave-height forecast
	Pearl Harbor	Central Pacific Ocean (from 40°N to 40°S)	Hydrometeorological navigation charts (temperature, waves, rainfall zones, aircraft icing zones); daily	24-hr wave-height forecast
	Guam	Southwestern part of North Pacific Ocean	Water-temperature charts for the surface and at depths of 200, 400, 800, and 1200 feet; charts of temperature gradients in thermocline; wind waves; daily	Same
	Rota	Eastern part of North Atlantic	Wind-wave charts; daily	12-, 24-, and 36-hr wave-height forecasts
		Mediterranean Sea	Same	Same
Japan	Tokyo	Northwestern Pacific	Ten-day-average surface water temperature charts	Water temperature forecast for Japanese coastal waters
			Oceanographic charts (temperature, thermocline, monthly average currents) on 9th, 19th, and 29th of the month	
Union of South Africa	Pretoria	Southeastern part of South Atlantic (from Capetown to the Gulf of Guinea)	Surface water temperature charts; daily	

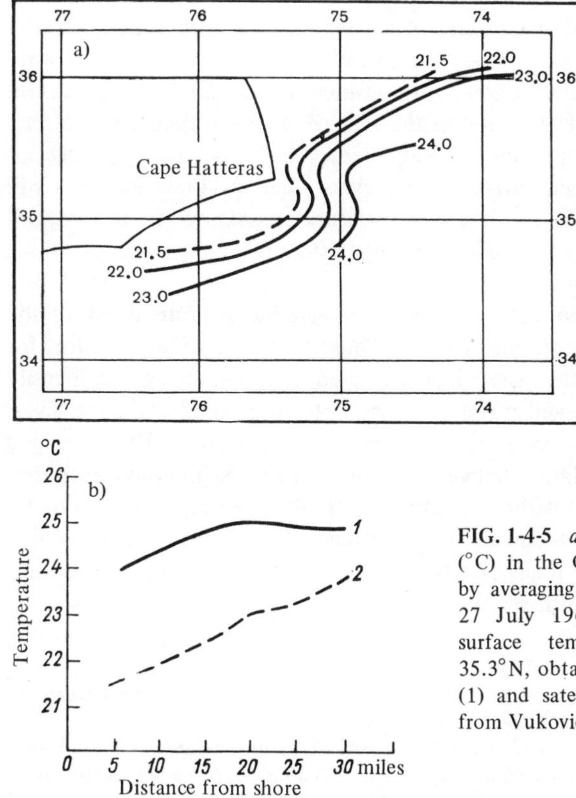

FIG. 1-4-5 *a*–Sea-surface temperature (°C) in the Gulf Stream zone, obtained by averaging Nimbus-2 satellite data for 27 July 1966. *b*–Comparison of sea surface temperatures along latitude 35.3°N, obtained from ship observations (1) and satellite data (2). Figure taken from Vukovich [7].

However, since infrared radiation cannot penetrate clouds, the above method is useful only for sea areas that are not obscured by clouds. This limitation can be circumvented by recording the microwave emission of the sea surface at centimeter wavelengths, to which clouds are practically transparent. The USSR Academy of Sciences Institute of Atmospheric Physics has run experiments of this kind, both from an aircraft over the Caspian Sea [8] and from the Kosmos-243 satellite in 1968 [9]. The results indicated that the microwave method is promising. At the same time it was observed that patches of foam formed on the sea surface in heavy waves emit stronger microwave radiation and appear as hot spots on microwave radiometer traces. Specifically, they significantly interfere with the registration of the temperature of the sea surface (on the other hand, this noise carries information on the state of the sea. The development of methods for utilization of this information requires estimation of the amount of foam on the sea surface from aboard ships). The development of the microwave techniques would seem to be the most important current problem in satellite-based oceanological methods.

The direct measurement of the sea state from satellites is of great interest. The development of satellite-borne microwave altimeters with differential error on the order of a few decimeters has already made it possible to measure the heights of ocean waves under satellite orbits. Installation of active radars on satellites will make it possible to measure two-dimensional spectra of the sea state. The establishment of a relationship between the sea state, mainly its anisotropy and the direction of the wave principal axes, with the local wind in areas with developing and developed waves will also make it possible to estimate the wind in the surface air layer, that is, will offer a unique opportunity for remote wind measurement, based on the presence of wind-generated waves on the sea surface. The development of these methods will require correlation of satellite and shipboard wave measurements, which, in turn, will require development of methods for measuring the sea state from moving ships, which do not now exist.

It will also be necessary to determine whether satellite measurements of the biological characteristics of the sea are feasible. Thus, the color of the sea (or, for greater detail, the spectrum of the visible light scattered by the water), which is easily recorded from satellites, is governed mainly by the matter suspended in the top layer, primarily phytoplankton, which imparts to the water its green color. Therefore it is necessary to establish the correlation between the color of the sea and phytoplankton concentration. Sophisticated satellite measurements of spectral absorption by chlorophyll and other photosynthetically active pigments of the phytoplankton may also prove successful. Moreover, it may also be possible to measure from satellites films of biological and other origin on the sea surface.

REFERENCES

1. Stommel, H. Some thoughts about planning the Kuroshio survey. In: *Proceedings of Symposium on the Kuroshio*, Tokyo, 22–23 (1963).
2. Sabinin, K. D. Selection of correspondence between periodicity of measurements and instrument time constant, Izv. Akad. Nauk SSSR, Fizika Atm. i Okeana, **3**, No. 5, 473–480 (1967).
3. Rukovodstvo go gidrologicheskim rabotam v okeanakh i moryakh (*Handbook of Hydrological Research in Oceans and Seas*), Gidrometeoizdat Press, Leningrad (1967).
4. Koshlyakov, M. N. Smoothing of oceanographic data, Okeanologiya, **4**, No. 3, 488–496 (1964).
5. Kort, V. G. The role of heat advection by ocean currents in the large-scale air-sea interaction, Dokl. Akad. Nauk SSSR, **182**, No. 5, 1059–1062 (1968).
6. Kort, G. An international project for studying North Atlantic dynamics and hydrology, Proc. Scor., **3**, No. 1, 48–54 (1967).
7. Vukovich, F. M. Detailed sea-surface temperature analysis utilizing Nimbus HRIR data, Monthly Weather Review, **99**, No. 11, 812–817 (1971).
8. Gurevich, A. S., and S. T. Yegorov. Determining the surface temperature of the ocean from its thermal radio emission, Izv. Akad. Nauk SSSR, Fizika Atm. i Oceana, **2**, No. 3, 305–307 (1966).
9. Basharinov, A. Ye., A. S. Gurvich, and S. T. Yegorov. Determination of geophysical parameters from measurements of the thermal radio emission on the Kosmos 243 satellite, Dokl. Akad. Nauk SSSR, **188**, No. 6, 1273–1276 (1969).

2 THE CIRCULATION OF THE OCEAN

The instantaneous state of the sea, characterized by the set of oceanological quantities enumerated in Sec. 1-3, can be represented as the sum of the multiyear-average state. That sum consists of a quasistationary part, whose time variations have periods in the tens of years and longer and can be called climatic, and of deviations from that state, produced by short-period nonstationary processes, such as those of types 1 to 5 in the classification of Sec. 1-3. The similarity between the instantaneous states of the sea and its multiyear-average state is much greater than is the case with the corresponding states of the atmosphere (in other words, the resultant relative intensity of the short-period nonstationary processes is lower in the sea than in the atmosphere, since the sea exhibits a greater inertia.

Indeed, the instantaneous states of the atmosphere, as reflected, for example, on the daily synoptic charts, bear little resemblance to the multiyear-average picture of its global circulation (which contains several not very pronounced circulating cells or "wheels" in meridional sections). The ocean, on the other hand, always exhibits the same *global* system of principal currents with relatively slowly varying geographic distribution and intensity. For this reason the quasistationary part of the state of the ocean, as represented on maps of the ocean, also gives a fairly good representation of the instantaneous states of its *global* circulation. In this chapter, we shall describe the quasistationary state of the overall oceanic circulation as the background against which short-period nonstationary processes develop.

Dimensions and shape of the ocean. From the hydrodynamic standpoint, the ocean is first of all a basin with complex boundaries filled with a large mass of salt water. This mass is $1.370323 \cdot 10^{24}$ g, or 258 times larger than the mass of the atmosphere. Note that glaciers on land contain $2.9 \cdot 10^{22}$ g of ice; if this ice were to melt, the sea level would be raised by 80 m. The ocean has a surface area of $3.61059 \cdot 10^{8}$ km^2, covering 70.8% of the surface of the earth. The average depth of the ocean is 3795 m; its greatest depth is the 11,022 m observed in the Mariana Trench from the R/V *Vityaz'* in 1957.

The ocean is divided somewhat arbitrarily into four parts: the Pacific Ocean (52.8% of the mass and 49.8% of the total area of the ocean, average depth 4028 m), the Atlantic Ocean (24.7% of the mass and 25.9% of the area, average depth 3627 m), the Indian Ocean (21.3% of the mass and 20.7% of the area, average depth 3897 m), and the Arctic Ocean (1.2% of the mass and 3.6%

of the area, average depth 1296 m). For details see the paper by Stepanov [1]. These oceans are defined to include the corresponding sectors of the Antarctic and also the marginal seas, which account for a total of 3% of the mass and 10% of the area of the ocean; the Mediterranean, Black and Caspian seas are considered to belong to the Atlantic Ocean. The ring of Antarctic waters bordering the Pacific, Atlantic, and Indian Oceans on the south is sometimes called the Antarctic or Southern Ocean; the Arctic Ocean is sometimes included in the Atlantic as an Arctic Inland Sea. The hydrodynamicists also subdivide the sea further into ***basins*** separated by sea-floor rises (in particular, by the mid-ocean ridges), which disrupt the continuity of the deep water circulation. A certain analogy to the water circulation at depth in sea basins can be found in the atmospheric circulation on Mars, on which (from data on the concentration of radiation-absorbing CO_2 in a vertical column of the atmosphere) there are differences of up to 15 km in the heights of the surface relief. The net result is that the bulk of the thin atmosphere of Mars appears to be concentrated in the individual depressions on its surface.

Sea level. Generally speaking, the surface of the sea is a *geoid*, namely a figure defined by the equilibrium between the attractive forces of all the masses of the earth and the centrifugal forces set up by its rotation. However, the geoid is superposed by systematic anomalies (the periodic anomalies, that is, the tides, obviously do not contribute to the multiyear average figure) generated by the set-up and set-down action of systematic winds on the sea surface, as well as by systematic differences in atmospheric pressure and between the amounts of moisture evaporated from and precipitation on the various areas of the sea (these are of the order of several decimeters of water per year). Figure 2-1 is a chart of the rainfall-evaporation differences, derived from the data of [2]. Note that the available precipitation and evaporation data are not too reliable. They are probably too low, since precipitation is systematically measured only on certain islands and the very few existing weather ships, while we have no data at all on the evaporation rate on the sea (all that we have are computations based on meteorological characteristics, produced by means of very suspect equations).

Figure 2-2 shows a so-called dynamic chart of the sea-surface topography and cited after Burkov, et al. [3]. This can be interpreted roughly as a chart of the multiyear average deviations of sea level from the geoid. The procedure used in constructing dynamic charts will be discussed later in this chapter.

Wind-driven circulation. The sea acquires momentum and kinetic energy directly from the atmosphere because of the friction of the wind against its surface. The frictional stress τ is of the order of 1 dyn/cm^2, and the friction velocity in water is $u_* = \sqrt{\tau/\rho} \sim 1$ cm/sec. This momentum is imparted both to *pure drift currents* in the upper ocean and to surface and internal waves (the latter also acquire some momentum from currents and surface waves). The waves, in turn, appear capable of transferring momentum to the currents. These momentum-redistributing mechanisms are still not well understood. The nonuniformity of the wind field, characterized by the curl of the frictional stress ($\mathrm{curl}_z\, \tau$) which is of the order of 10^{-7} to 10^{-8} dyn/cm^3 at horizontal scales of the windfield inhomogeneities ranging from 100 to 1000 km, as well as horizontal barriers, that is, coasts, produce set-up and set-down of the surface waters, that is, are partly responsible for the sea-level anomalies mentioned above. The corresponding horizontal pressure gradients give rise to wind-driven *gradient currents*. As a result, the pattern of the currents in the upper ocean reflects to

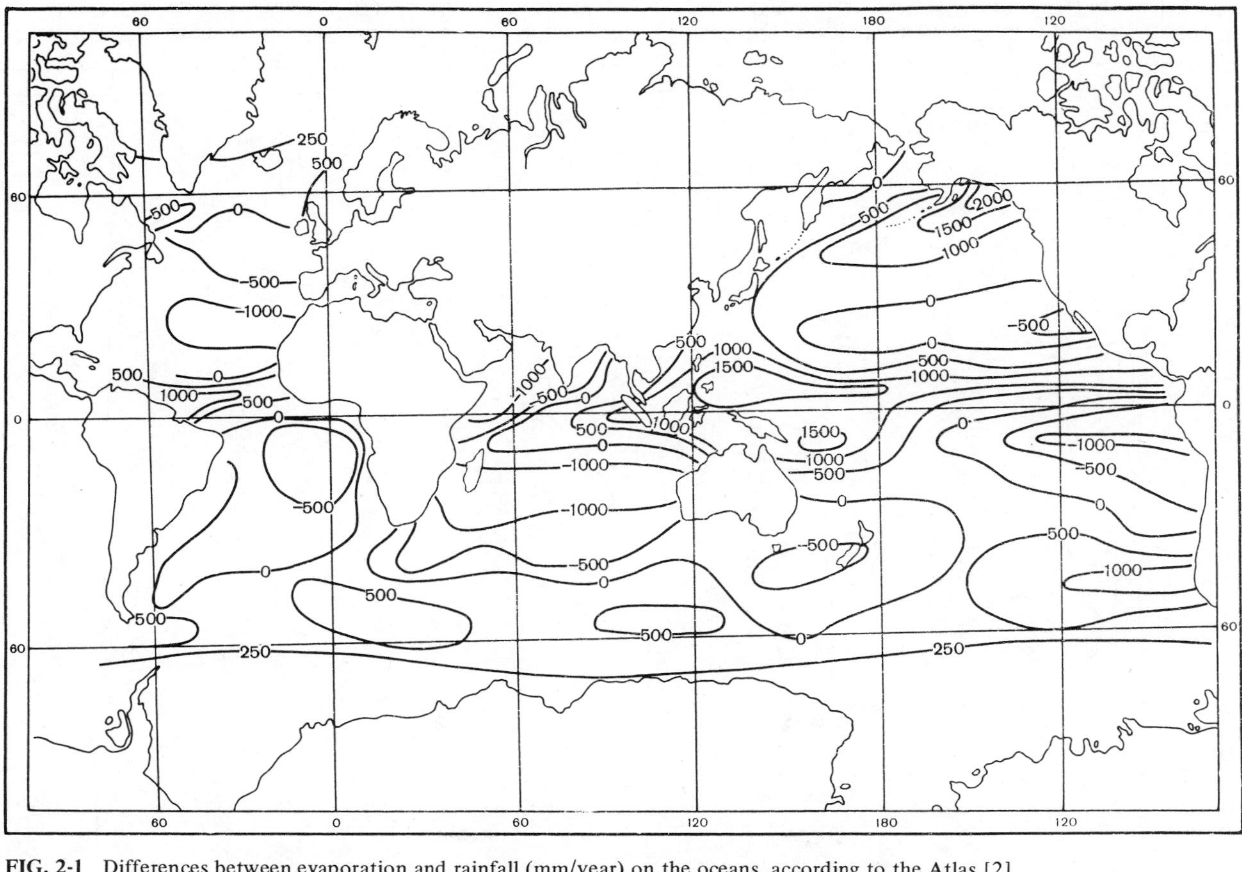

FIG. 2-1 Differences between evaporation and rainfall (mm/year) on the oceans, according to the Atlas [2].

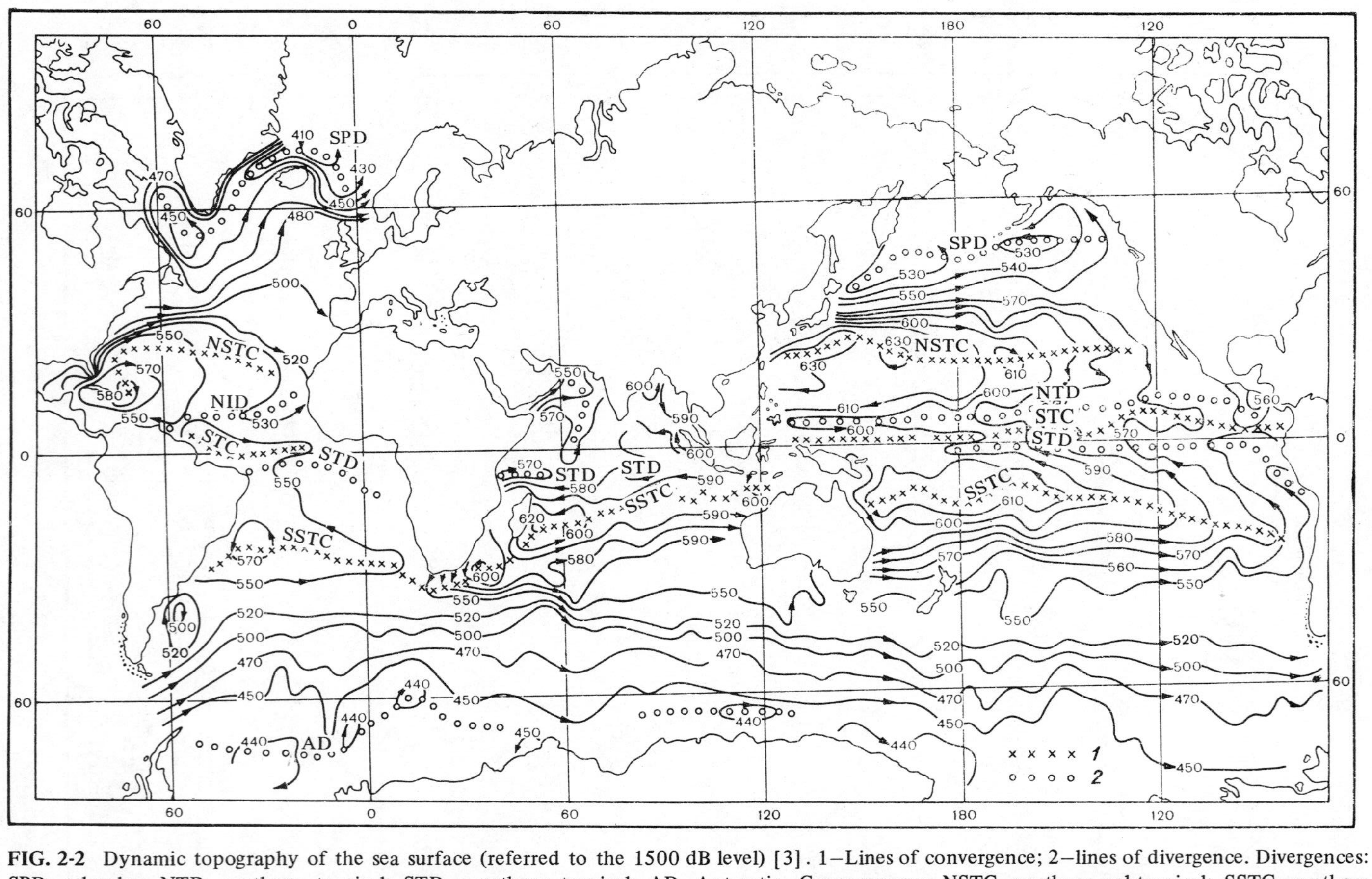

FIG. 2-2 Dynamic topography of the sea surface (referred to the 1500 dB level) [3]. 1–Lines of convergence; 2–lines of divergence. Divergences: SPD–subpolar; NTD–northern tropical; STD–sourthern tropical; AD–Antarctic. Convergences: NSTC–northern subtropical; SSTC–southern subtropical.

a great degree the averaged wind field in the surface air layer, which itself is controlled by the distribution of atmospheric pressure over the sea surface.

Thus the circulation of the upper ocean reflects both the *intertropical convergence zone* (or the *equatorial trough*, generally lying somewhat north of the equator) and the *subtropical highs*—Azores, Honolulu, South Atlantic (St. Helena high), South Pacific, and South Indian (Mauritius high), and the *trades* connecting them to the equatorial trough (the trades blow toward the southwest in the Northern Hemisphere and to the northwest in the Southern Hemisphere). In addition, the upper-ocean circulation reflects the *west-to-east transport* in middle latitudes, the Iceland and Aleutian lows, the Arctic and Antarctic highs and, in its seasonal aspect, also the *monsoons* (especially in the northern Indian Ocean).

The thermohaline circulation. An important source of kinetic energy that drives the quasistationary circulation in the ocean is conversion of its *available potential energy* in the gravitational field into kinetic energy. Specifically, the available potential energy is understood here as the sum of the potential and internal energies, less that part of this sum that would remain if the sea were adiabatically reduced to a stably stratified state with a constant pressure on any isentropic surface. The conversion to kinetic energy becomes possible by virtue of the fact that sea water is a compressible and, moreover, a *baroclinic* fluid with its density depending not only on pressure, but also on temperature and salinity. Hence *isobaric-isopycnic solenoids* can form in it and drive the circulation.

The circulation generated by factors that directly affect the temperature and salinity at the sea surface is called the *thermohaline* circulation. An example is the sinking and spreading across the sea floor of water whose density has increased by cooling from the top in the polar regions of the sea. Thermal processes, that is, heating and cooling of the ocean waters, contribute the most to the shaping of the thermohaline circulation. The primary source of heat in the atmosphere and oceans is the short-wave electromagnetic radiation of the sun (which is concentrated in the wavelength range of 0.38 to 2.5 μm). Its input rate to the top of the atmosphere averages over the year 1.952 gram-calories per square centimeter of area perpendicular to the sun rays per minute (this is the *solar constant* on the American scale). For the entire earth, the input from this energy source is $1.8 \cdot 10^{14}$ kW. A certain fraction α of this radiation is reflected by the atmosphere and surface of the earth back into space. According to recent satellite data, this fraction (the *albedo*) amounts to about 27 percent. But until quite recently climatologists assumed an α of 35 to 40 percent, so that they must now increase by 12 to 22 percent the solar heat absorbed by the earth and make the corresponding corrections in the climatological mechanisms. Also, mechanisms in which all the energy flows are not estimated independently, but are "balanced" with the aid of fitting factors, require radical revision.

The shortwave solar radiation is scattered and absorbed in the atmosphere. Only a small fraction, which differs with latitude and over the years, but never exceeds one-fourth the multiyear average, reaches the sea surface in the form of direct and scattered (the latter being generally a somewhat smaller fraction) radiation. (See Fig. 2-3, which was plotted from McLellan's data [4].) A small fraction of this insolation fraction is reflected (the albedo of the sea surface being assumed to average only 8

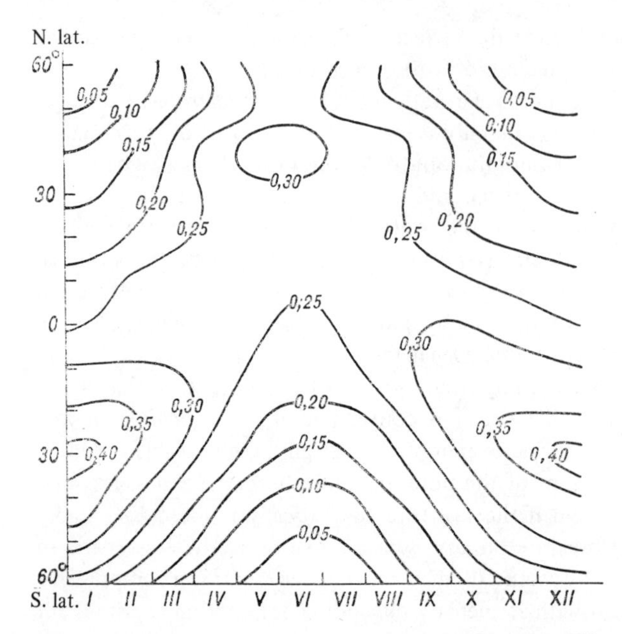

FIG. 2-3 Sums of direct and scattered solar radiation [cal/(cm² · min)] reaching the surface of the ocean at various latitudes during various months (according to McLellan [4]).

percent), but most of it is absorbed by a thin surface water layer. About 60 percent of the radiation is absorbed in the first meter, and less than 0.5 percent penetrates deeper than 100 m. The absorbed heat is then redistributed over the entire upper mixing layer of the sea. The sea surface radiates heat into the atmosphere in the form of longwave electromagnetic radiation, basically at wavelengths of 5 to 20 μm, and in turn, receives the longwave counterradiation from the atmosphere. The difference between these two radiation fluxes is called the effective radiation.

The sea loses large amounts of heat by evaporation. This heat is transferred to the atmosphere together with the water vapor in the form of latent heat. Rudloff [5] estimates the average rate of evaporation from the entire sea surface at $3.34 \cdot 10^{20}$ g/yr, that is, a layer of water about 1 meter thick is evaporated during a year. Consequently, the average residence time of a water molecule in the sea until it is transported to the atmosphere is about 4000 years. The amount of precipitation falling on the sea is estimated at $2.97 \cdot 10^{20}$ g/yr. However, it is possible that both the evaporation and precipitation figures are low. Hence, the excess of evaporation over precipitation is calculated to be $3.7 \cdot 10^{19}$ g/yr, or about 10 cm of water per year. This quantity is compensated by runoff from land. If the latent heat of evaporation is 600 cal/g, then the sea loses $2 \cdot 10^{23}$ cal/yr by evaporation, and the average rate of heat loss is 0.11 cal/(cm^2 · min). If this loss were derived from the upper 100-meter layer of the ocean and were totally uncompensated, the ocean would be cooled by 6°C each year.

The rate of *turbulent (contact) heat exchange* across the interface between the ocean and the atmosphere is an order of magnitude lower. The mechanism of this exchange involves molecular heat transfer with subsequent transport to (or from) the boundary layers of the atmosphere and the ocean by means of turbulent mixing. The heat (or, more commonly cold) flow arriving with precipitation to its surface appears to be comparable to that produced by turbulent exchange (at least in certain areas of the sea). The heat released by dissipation of kinetic energy in the sea is not significant. We do not yet have reliable estimates of the average kinetic energy dissipation rate per unit mass (ϵ) for the sea. However, comparisons can be made with the astronomical estimate of the tidal energy dissipation rate, $2.76 \cdot 10^{19}$ erg/sec $\approx 2 \cdot 10^{19}$ cal/yr. This corresponds to $\epsilon = 2 \cdot 10^{-5}$ cm^2/sec^3, obtained by multiplying by the mass of the ocean and dividing by its area, or a heat flux of only 5.5 cal/(cm^2 $\cdot$ yr) $\approx 1.1 \cdot 10^{-5}$ cal/(cm^2 $\cdot$ min). In the atmosphere, however, $\epsilon \sim 5$ cm^2/sec^3, which corresponds to a heat flux of $4 \cdot 10^3$ cal/(cm^2 $\cdot$ yr) $\approx 0.8 \cdot 10^{-2}$ cal/(cm^2 $\cdot$ min). This flux is comparable to the multiyear-average values of some of the more intense heat fluxes of variable sign. The geothermal heat flux through the sea floor, which is of the order of 10^{-4} cal/(cm^2 $\cdot$ min) is probably only of minor local importance. It also appears that the net heat effects of chemical reaction occurring in sea water are small. Finally, there are also the local thermal effects associated with ice formation and melting in polar regions.

The shortwave solar radiation heats the ocean. But, according to Albrecht [6], evaporation, effective radiation, and contact heat transport tend to cool it on the average. The ratio of these heat fluxes is 51:32:7, although the relative importance of evaporation may have been underestimated here. However, these three heat fluxes vary over the ocean and between seasons, and may sometimes even change sign. The principal heat fluxes through the sea surface in the Northern Hemisphere, averaged by longitudes and over the year, are shown in Fig. 2-4 (after Defant [7]). We see that, in sum, they heat the sea south of the twenty-fifth parallel and cool it north of that latitude. This difference must be compensated by heat transfer via a meridional water circulation.

Generally speaking, salinity variations in sea water make a smaller contribution to the thermohaline circulations than thermal processes. The average salinity is 34.72 per mille [the principal cations are Na^+ (10.76‰), Mg^{2+} (1.30‰), Ca^{2+} (0.41‰), K^+ (0.39‰), and Sr^{2+} (0.01‰), and the principal anions Cl^- (0.41‰), SO_4^{2-} (2.70‰), HCO_3^- (0.14‰), Br^- (0.07‰), and CO_3^{2-} (0.07‰)]. The most intense salinity-changing processes are salinization of the top layer of the sea by evaporation from its surface. This releases more than 10^{13} tons of salt per year (only a small fraction enters the atmosphere with spray). The other process, desalinization by rainfall and freshening by river water (which is of local nature and an order of magnitude smaller), compensates the salinization produced by the average excess of evaporation over precipitation.

Increases (or decreases) in the density of the surface waters on cooling (or heating) and salinization (or desalinization) generate buoyancy forces that cause the water to sink (or rise), thereby creating the prime mover of thermohaline circulation. The net effect of these factors can be described by the vertical mass flux at the surface (which is positive when upward, that is, increases the buoyancy):

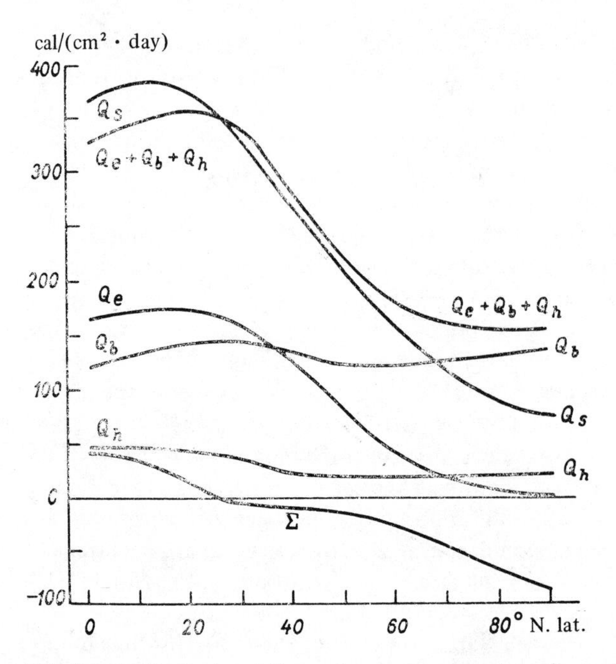

FIG. 2-4 Latitudinal variations of annual average heat fluxes (according to Defant [7]); Q_S is the flux of direct and scattered shortwave solar radiation, Q_e is the latent heat flux, Q_b is the effective radiation, Q_h is the turbulent (contact) heat flux, and Σ is the resultant heat input.

$$M=(P-E)S-\frac{\alpha}{c}(LE+Q)+\alpha P(T_p-T_w), \tag{2.1}$$

where P is the rate of precipitation; E is the rate of evaporation; S is the salinity; Q is the sum of the radiative and turbulent heat fluxes in the surface air layer (which is positive when the flux Q is upward); $\alpha \approx 2 \cdot 10^{-4}$ $(°C)^{-1}$ is the thermal expansion coefficient of the water; c is its heat capacity; L is the latent heat of vaporization; and T_p and T_w are the temperatures of the precipitation and the water surface, respectively [8]. Here gM/ρ is the kinetic energy generation rate per unit mass.

Since $\alpha L/c \approx 0.12$ and $S \approx 0.03$, the contribution from evaporation is four times larger in the second term than in the first (i.e., the rate of cooling due to evaporation is four times greater than that of salinization). The values of M are on the order of 10 g/(cm^2 · yr), which corresponds to a kinetic-energy generation rate of the order of $3 \cdot 10^{-4}$ cm^2/sec^3. Unfortunately, no charts of the values of M have yet been prepared for the sea (see Fig. 2-1 for a P-E chart).

The overall circulation. Zones of *divergence* or set-down of the surface waters (both in the open sea and along coasts where the prevailing wind blows offshore) and zones of *convergence* or set-up (in the open sea and along coasts where the prevailing wind is onshore) can be identified in the field of the wind-driven (pure drift and gradient) and thermohaline currents in the upper ocean. Upwelling of cold, deep water occurs in the divergence zones, and it replaces surface water that is pushed aside. Given the higher density of the deep waters, upwelling tends to maintain a depressed sea level in the divergence zones. On the other hand, sinking (downwelling) of colliding warm surface waters occurs in convergence zones (downwelling of a different, i.e., thermohaline nature also occurs in areas of higher density of surface waters, produced by cooling or salinization). Given the lower density of surface waters, the downwelling in convergence zones tends to maintain a higher sea level in these zones.

Thus, dynamic factors, horizontal divergence or convergence of currents, generate large-scale vertical motions in the sea which are sometimes even stronger than flows of thermohaline origin. Their typical velocity can be estimated [9] from $w \sim HD \sim H/L\ U\mathrm{Ki}$, where H is the thickness of the ocean and $D \sim U/L\ \mathrm{Ki}$ is the typical horizontal velocity divergence of the geostrophic currents (U is the typical current velocity, L is the width of the ocean, $\mathrm{Ki} = U/Lf$ is the Kibel' number, and f is the Coriolis parameter). At $H = 4$ km, $L = 10^3$ km, $U = 10$ cm/sec, and f of about 10^{-4} sec^{-1}, we obtain $\mathrm{Ki} \sim 10^{-3}$, $D \sim 10^{-10}$ sec^{-1}, and $w \sim 4 \cdot 10^{-5}$ cm/sec. The vertical velocities in flow around the sea floor rises may be 2 to 3 orders higher. It is possible that they make the main contribution to the vertical mixing of the ocean.

The pattern of the overall circulation of the surface water is seen in Fig. 2-2. In its most general outlines, the horizontal circulation of the surface waters involves gigantic *anticyclonic gyres* in the northern and southern halves of the ocean around the corresponding atmospheric subtropical highs, with intensification of the currents on the western peripheries of the gyres. At these peripheries we also find the strongest ocean currents, including the Gulf Stream and the Kuroshio. The northern and southern gyres are separated by an equatorial, or, more precisely, *north tropical* convergence, which is formed by the colliding equatorial currents. There are *subtropical* convergence zones on the quasilatitudinal axes of these gyres. If the radius of a gyre is taken as 2500 km and the average velocity of the currents around its periphery as 10 cm/sec, then the period of this circulation is 5 years (the period of the Arctic gyre has been found to be 4 years from the drift of tagged ice floes). Thus, there are natural periods on the order of a few years (say, 3 to 10 years) in the circulation of the ocean.

There are also secondary circulation elements, the largest of which are the *cyclonic gyres* around Antarctica and the Iceland and Aleutian atmospheric lows. These are separated from the giant anticyclonic gyres by *polar oceanic fronts.* There are also *tropical* divergence zones between the equatorial convergence and the gigantic anticyclonic gyres. *Polar* convergences are observed in places along the polar peripheries of the cyclonic gyres. Let us note here that Fig. 2-2 is only a simplified schematic diagram. True current charts, derived from observational data, are much more complicated.

Our knowledge of hydrology of the sea. The existing observational data on the hydrodynamics of the ocean have been accumulated by hydrologic stations at which sampling bottles bring up water from various depths. From these data the vertical temperature and salinity distributions and, consequently, the density distributions can be plotted. The latter are then used to prepare charts of the absolute and relative topography of isobaric surfaces (*dynamic charts*), from which the geostrophic currents can be estimated. Since one does not know the actual sea surface topography, dynamic charts are plotted under the assumption that a certain isobaric surface $p = p_0$ (the so-called zero surface) at a great depth in the ocean is strictly horizontal, that is, perpendicular to the local gravity vector at each of its points. It is from this surface that the heights of the other isobaric surfaces p = const are reckoned, including the height of the free sea surface, at which atmospheric pressure is constant, $p = p_a$. Clearly, the zero surface theory can hold only approximately, and the choice of this surface is somewhat arbitrary in practice. As a result, the value of the dynamic current-estimating method is limited and, in addition, it is only used to estimate the geostrophic component of the currents.

All told, about 100,000 hydrological stations have been run so far in the Atlantic Ocean, about 70,000 in the Pacific, and only about 7000 in the Indian Ocean. These stations are distributed very unevenly over the oceans. Most of them have been concentrated in the Northern Hemisphere, and even there most of them were in the areas fairly close to the coasts. Furthermore, only small depths, for example, 500 to 1000 meters, were reached at most stations. Finally, these stations were also distributed very unevenly in time, and over the seasons of the year. Only a few stations have been run in certain five-degree squares of the sea, and in some only one. Obviously such data are totally inadequate for study of the synoptic variability of the ocean and, in the case of large areas of water, even the seasonal and year-to-year variability. Still, we can infer from them the pattern of global quasistationary oceanic circulation, although, needless to say, some of its details are not very certain.

Direct instrumental current measurements (usually over 1 to 2 days) with automatic current recorders on buoys have been even more spotty: there have been only about 550 such measurements, covering a total of 1100 days. Isolated measurements have also been made with electromagnetic current meters (EMCM). Finally, some information has been derived from ship drift data and "bottle mail," and form a basis for estimates (although very rough ones) of surface currents. Given this scarcity of direct measurements and the limitations of the indirect dynamic method, it is not surprising that new ocean-current systems are still being discovered. Thus, the strong deep equatorial countercurrents which flow eastward, that is, oppositely to the surface equatorial currents, were not discovered until the 1950's and 1960's. The very long Antilles-Guiana countercurrent on the western periphery of the Sargasso Sea was discovered only quite recently. Recent measurements have revealed complex structures in the principal currents, which commonly consist of narrow jet flows in opposite directions, a phenomenon for which we do not yet have a theoretical explanation.

The stratification of the ocean. The sea is heated from above but, unlike the atmosphere, its temperature decreases with depth. Hence its thermal stratification is generally very stable.

Because of relatively intensive mixing, the upper, approximately 100 meters thick, layer of the sea is quasihomogeneous. Immediately below this layer there is a seasonal thermocline, and below it the temperature drops off appreciably with depth in a layer about 1 to 1.5 km thick that is known as the *permanent thermocline.* Still deeper, temperature decreases very slowly with increasing depth, reaching values of around 1 to 2°C near the bottom layers of the deep ocean, to which sink the surface waters that have been chilled in polar regions.

As a result, the sea is rather cold over most of its depth: the average temperature is 3.8°C and the potential temperature 3.52°C (the corresponding values are 3.7 and 3.36° in the Pacific Ocean, 3.8 and 3.72° in the Indian Ocean, and 4.0 and 3.73° in the Atlantic) [10]. Nekrasov and Stepanov [11] suggested a classification of vertical temperature profiles into five types: polar, subantarctic, subarctic Atlantic, subarctic Pacific, and temperate-tropical, with a few subtypes. The typical temperature profiles are shown in Fig. 2-5.

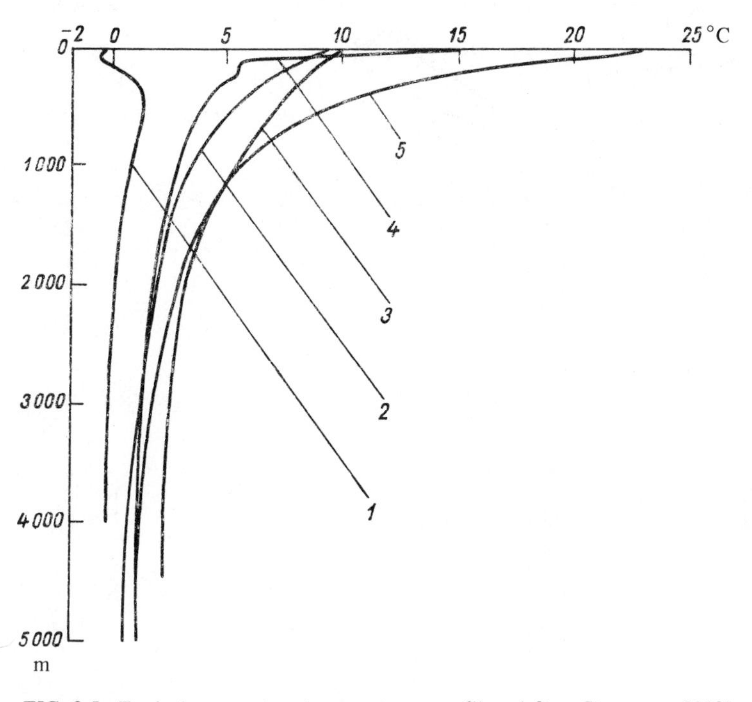

FIG. 2-5 Typical sea-water temperature profiles (after Stepanov [11]). 1–Polar; 2–subantarctic; 3–subarctic Atlantic; 4–subarctic Pacific; 5–temperate-tropical.

The oceans differ slightly in terms of salinity. The average salinities are 34.62 per mille in the Pacific Ocean, 34.76 per mille in the Indian Ocean, and 34.90 per mille in the Atlantic Ocean [12]. In polar regions, salinity increases with depth because salt water is denser, that is, heavier. It is 1.5 to 2.5 per mille higher at the bottom than at the surface. It varies little in the top, mixed layer, increases significantly below it in a layer 1 to 1.5 km thick (the *halocline*), and increases very slowly at still greater depths. However, the water density is controlled to a greater degree by temperature than by salinity, and the hydrostatic stability of density permits the formation of local low-salinity layers in the sea. Stepanov [12] identifies seven types and several subtypes of vertical salinity variations, and published a distribution chart for them. The basic types are polar and subpolar with salinity increasing with depth, the temperate-tropical type with a salinity minimum at depths of 800 to 1000 m (and a weak minimum at the bottom), and the equatorial-tropical type, which has another secondary minimum at the surface. Locally important are the North Atlantic type in which salinity decreases with depth to the bottom of the ocean, the Atlantic-Mediterranean[1] type with salinity maxima at the surface and at depths of 500 to 1000 m, and the Indo-Malayan type with a salinity maximum at 500 m. Typical salinity profiles from [12] are shown in Fig. 2-6.

[1] Ed. note: Atlantic water with a Mediterranean intermediate layer.

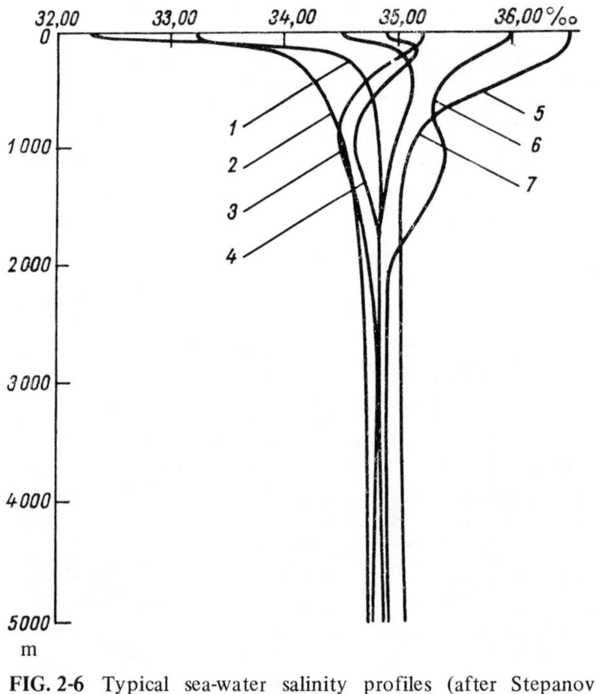

FIG. 2-6 Typical sea-water salinity profiles (after Stepanov [12]). 1–Polar; 2–subpolar; 3–temperature-tropical; 4–equatorial-tropical; 5–North Atlantic; 6–Atlantic-Mediterranean; 7–Indo-Malayan.

The thermohaline vertical structure of the ocean waters is represented conveniently by *TS* curves, on which depths are represented by points. Stepanov [13] has also classified the *TS* curves into eight types as follows: polar, subpolar, temperate-tropical, tropical, equatorial, North Atlantic, Atlantic-Mediterranean, and Indo-Malayan. His distribution chart for these is shown in Fig. 2-7, where the wide variety of *TS* curves encountered in the ocean can also be seen.

Sea water is a very slightly compressible liquid, and its density, which averages just over 1 g/cm^3, varies only by a few hundredths with the depth. It is convenient to measure the density reduced to atmospheric pressure (at T and S = const) in units of $\sigma_t = 10^3 \ (\rho - 1)$; then the density at pressure p will be equal to $\rho = 1 + 10^{-3}\sigma_t/1 - \mu p$, where $\mu \sim (4 \text{ to } 5) \cdot 10^{-5}$ atm^{-1} is the isothermal compressibility of water which decreases slightly with increasing pressure. Both the total density and σ_t increase almost everywhere with depth. Hence the density stratification is almost always hydrostatically stable. The lowest density (σ_t of about 22 to 23) is observed in the surface layers of equatorial waters. At depths greater than 3 km, σ_t exceeds 27.8 nearly everywhere. (See the distribution of σ_t on average meridional sections of the oceans, cited after Stepanov [14], in Fig. 2-8.) A major part of this vertical gradient of σ_t is accounted for by the seasonal *discontinuity layer*, which is the lower boundary of the top mixing layer of the ocean (the latter is quasihomogeneous in density) and usually occurs at depths of 50 to 100 m. Below it, σ_t increases slowly with depth over a *pycnocline* 1 to 1.5 km thick, but it varies very slowly at greater depths.

While the relative vertical variations of density are small, the absolute variations compared, for example, to the variations of air density in the atmosphere, are by no means small. Thus, at a density gradient $\delta\sigma_t = 3$ in a discontinuity layer 20 meters thick, the vertical density gradient is $\partial\rho/\partial z = 1.5 \cdot 10^{-6}$ g/cm^4. However, in a strong temperature inversion (1°C/m), the gradient in the atmospheric surface layer is only $\partial\rho_a/\partial z \approx \rho_a/T(\partial T/\partial z) \approx 4 \cdot 10^{-8}$ g/cm^4, that is, two orders smaller than in the discontinuity layer in the sea. Thus, the sea is much more stably stratified than the atmosphere, and therefore vertical mixing in it is much less intensive. On the other hand, internal waves develop much more strongly in the sea.

The density gradient $\partial\rho/\partial z$ is a more indicative characteristic of stratification of the sea than the function $\rho(z)$ itself. The density gradient can be compared to the Väisälä frequency N, with the latter written as

$$N = \left\{ \frac{g}{\rho} \left[\frac{\partial\rho}{\partial z} - \frac{g\rho}{c^2} \right] \right\}^{1/2}, \tag{2-2}$$

where c is the speed of sound (which is approximately 1.5 km/sec in sea water, so that

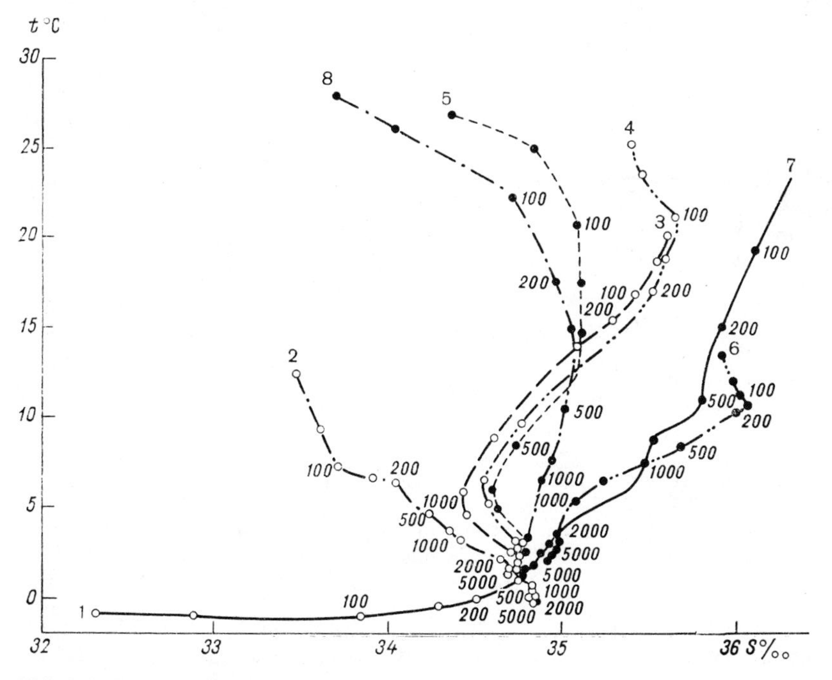

FIG. 2-7 Typical *TS* curves for ocean waters (after Stepanov [13]). 1–Polar; 2–subpolar; 3–temperate-tropical; 4–tropical; 5–equatorial; 6–North Atlantic; 7–Atlantic-Mediterranean; 8–Indo-Malayan.

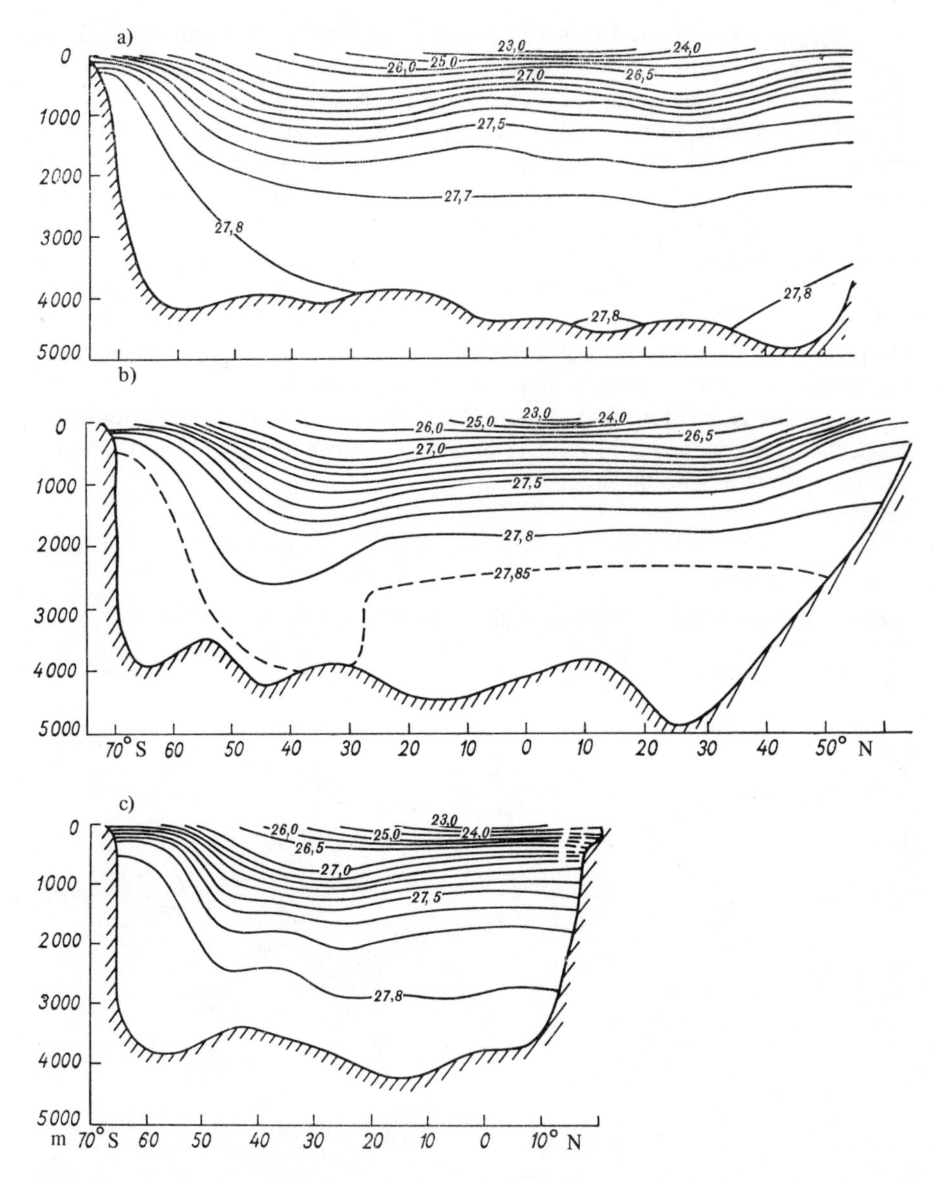

FIG. 2-8 Distribution of latitude-averaged water density δ_t on longitudinal vertical sections of the Pacific (*a*), Atlantic (*b*), and Indian (*c*) oceans (after Stepanov [14]).

$g\rho/c^2 \approx 4.4 \cdot 10^{-8}$ g/cm^4; this term is significant only deep in the ocean, where $\partial\rho/\partial z$ drops below, say, $4 \cdot 10^{-7}$ g/cm^4). The frequency N usually increases with depth from the surface to the discontinuity layer, in which the period $2\pi/N$ has a value on the order of 10 minutes. However, it drops to less than a tenth of its surface value between the discontinuity layer and the sea floor. According to Monin, Neyman, and Filyushkin [15], it drops approximately as $N = w_*/g$, where $w_* \approx 2.2$ m/sec, in the upper half of the sea, and approximately in proportion to the distance to the bottom in the lower half.

Still another characteristic of the stratification of the sea and one that is very important in hydroacoustics and various engineering problems (for example, in interpretation of echo-sounder readings), is the vertical distribution of the sonic velocity

$$c = \left\{ \left(\frac{\partial p}{\partial \rho} \right)_{\eta, S} \right\}^{1/2},$$

where η is the entropy and S is the salinity. It increases both with temperature and with pressure and salinity (although it depends much less strongly on the latter). The result is that, on the average, it decreases in the upper ocean from about 1540 m/sec at the surface to a minimum of 1480 to 1490 m/sec at depths of 700 to 1300 m (where an acoustic wave guide is formed as a result). Below that level, it rises smoothly, again reaching values around 1540 m/sec at a depth of 5 km.

The deep circulation. It is customary to subdivide the water masses that constitute the vertical structure of the ocean into a) *surface* waters (depths from 0 to an average of 250 m), which are shaped by atmospheric factors; b) *intermediate*-depth waters (with a lower boundary at depths between 1 and 2 km) with higher temperatures in polar regions and below- or above-average salinities in other areas; c) *deep* waters (with a lower boundary at about 4 km), mainly in the regions of cyclonic circulation systems of high latitudes; and d) *bottom* waters, in high latitudes [11, 17]. These water masses circulate in different ways.

The circulation of the surface waters has already been discussed, wind-driven currents being the most important factor in this circulation. It is more intense than the circulations of deeper waters.

A circulation pattern has been derived by the dynamic method for intermediate-depth waters by Burkov [18], Bylatov [19], and Burkov, Bylatov, and Neyman [3]. Both wind and thermohaline factors appear to affect this circulation, which exhibits almost complete damping of the wind-driven circulation in the tropics and some shift of the subtropical circulation systems toward the poles. The intermediate-depth circulation is responsible for the flow of intermediate waters of low salinity from polar fronts into the subtropics and tropics, and of water of relatively higher temperature into the subpolar regions. The equatorial countercurrents mentioned above, the Cromwell current in the Pacific, the Lomonosov current in the Atlantic, and the analogous current in the Indian Ocean, are parts of this circulation.

A circulation pattern for the deep and bottom waters has been derived thus far only for the Pacific Ocean [20]. This circulation is primarily in the direction opposite to that in the upper ocean. Its velocities increase slightly toward the bottom, and the streamlines tend to follow the isobaths of the sea-floor relief. The transition from the upper-ocean circulation to the reversed circulation of the deep and bottom waters occurs at depths of 1 to 2 km, where the velocities decrease accordingly.

Stepanov [16, 17] attempted to calculate longitude-averaged circulation in meridional planes of the Atlantic, Indian, and Pacific oceans, and to determine the

Current	Point of measurement	Velocities (cm/sec) at depths of				Cross section	Transport, 10^6 m^3/sec	
		0 m	100 m	250 m	500 m		w	s
North Equatorial	17° N, 120°W, w	4	7	3	2	150° E	95	85
	13° N, 150° E, s	20	23	14	1	170° E	65	65
	13° N, 170° W, s	27	26	10	6			
South Equatorial	4° N, 148° W, s	192	191	85	65	150° E	35	
	2° N, 160° E, s	65	38	20	39	170° W	35	
	5° S, 155° W, s	56	56	26	20	110° W	85	
West Wind Drift	57° S, 165° W, w	7	7	6	4	170° W	75*	
	62° S, 97° W, w	12	10	8	5	110° W	55	
	62° S, 79° W, w	8	7	7	7			
Kuroshio	31° N, 133° E, s	31	34	33	23	30° N	35	45
	34° N, 141° E, s	46	45	40	27	145° E	45	60
	36° N, 149° E, s	63	55	40	25			
Equatorial undercurrent	0°, 159° W, w	20	110	33	35	140° W	39	
	0°, 164° E, w	29	24	26	24			
North Pacific	38° N, 160° W, s	7	5	3	1	180° W	35	25
Equatorial countercurrent	5° N, 140° E, s	17	24	1	5			
	5° N, 170° W, s	70	80	2	4			
	8° N, 125° W, s	16	10	17	16			
Taiwan [18]	21° N, 122° E, s	29	34	27	9	20° N	30	30
	22° N, 122° E, s	17	20	18	5			
Kurile	46° N, 152° E, s	6	7	7	7	45° N	25	15
East Australia	33° S, 154° E, w	51	44	34	19	33° S	20	
Alaska	55° N, 152° W, s	9	8	7	5	150° W	15	15
Aleutian	48° N, 152° W, s	9	6	4	3	180° W	15	10
California	26° N, 117° W, s	6	4	3	1	30° N	12	8
	40° N, 128° W, s	8	5	3	2			
Peru	27° S, 73° W, w	1	6	5	2	30° S	10	
	35° S, 74° W, w	11	8	8	7			
South Pacific [18]	28° S, 177° W, w	3	4	4	3	170° W	5	
	30° S, 107° W, w	12	17	15	6			
	32° S, 107° W, w	8	10	10	9			

*Recent estimates of the transport in this current [21] show 218 to 234 · 10^6 m^3/sec in Drake Passage.

meridional and vertical velocity components v and w from the equations

$$v \frac{\partial \bar{T}}{\partial y} + w \frac{\partial \bar{T}}{\partial z} = K \frac{\partial^2 \bar{T}}{\partial z^2};$$
$$v \frac{\partial \bar{S}}{\partial y} + w \frac{\partial \bar{S}}{\partial z} = K \frac{\partial^2 \bar{S}}{\partial z^2}, \tag{2-3}$$

where $\bar{T}$ and $\bar{S}$ are the latitude-averaged temperature and salinity found from hydrological-station data. A value of 10 cm^2/sec was assumed for K. He found that the average meridional velocity for the entire ocean is about 2.4 cm/sec. If the same value is also used for the zonal velocity, so that the overall velocity averaged over the entire depth will be 3.5 cm/sec, a value of the order of 10^{18} J is found for the kinetic energy of the quasistationary circulation of the ocean, that is, a value three orders smaller than the kinetic energy of the atmospheric motion, which is estimated at 10^{21} J. This is probably to be expected because the ocean acquires its kinetic energy chiefly from the atmosphere, and the coupling between the two systems is fairly weak.

The principal currents. Narrow jet-type currents with velocities considerably higher than the 3.5 cm/sec average indicated above would appear to be typical of the upper ocean. The quantitative characteristics of these currents are given in the table on page 40, using the Pacific Ocean currents as an example.

In the table, the letters w and s indicate Northern Hemisphere winter and summer. The currents are given in approximate order of decreasing flow, with the latter calculated from dynamic charts referred to the 1500-decibar surface. The table shows the order of these flows to range between 10^7 and 10^8 cm^3/sec. For the equatorial Cromwell countercurrent, the transport was calculated within the limits of the 25 cm/sec isotach from the data of Knauss [22].

REFERENCES

1. Stepanov, V. N. The principal dimensions of the global ocean and of its principal parts, Okeanologiya, **1**, No. 2, 213–219 (1961).
2. Morskoy Atlas (*Sea Atlas*), **2**, Published by the Naval General Staff of the USSR Ministry of Defense (1953).
3. Burkov, V. A., R. P. Bylatov, and V. G. Neyman. Large-scale features of the oceanic circulation, Okeanologiya, **13**, No. 3, 395–403 (1973).
4. McLellan, J. H. *Elements of Physical Oceanography*, Pergamon Press, 150 (1965).
5. Rudloff, W. Wetterlotse, **14**, No. 188, 183 (1962).
6. Albrecht, F. The heat and water balance of the earth, Ann. Meteorol., **2**, No. 1, 16–31 (1949).
7. Defant, A. *Physical Oceanography*, Pergamon Press, **1**, 729 (1961).
8. Monin, A. S. Turbulent mass fluxes in the oceans, Dokl. Akad. Nauk SSSR, **193**, No. 5, 1038–1041 (1970).
9. Monin, A. S. Prognoz pogody kak problema fiziki (*Weather Forecasting as a Problem of Physics*), Nauka Press, Moscow (1969).
10. Montgomery, R. B. Water characteristics of Atlantic Ocean and of world ocean, Deep-Sea Res., **5**, No. 2, 134–148 (1958).
11. Nekrasova, V. A. and V. N. Stepanov. Types of vertical water temperature variation in the oceans, Dokl. Akad. Nauk SSSR, **143**, No. 3, 713–716 (1972).

12. Stepanov, V. N. and V. A. Shagin. Types of vertical salinity variation in the oceans, Dokl. Akad. Nauk SSSR, **136**, No. 4, 927–930 (1961).
13. Stepanov, V. N. Study of the basic water-structure relationships in the oceans. In: Okeanologicheskiye issľedovaniya (*Oceanological Research*), USSR Academy of Sciences Press, Moscow, No. 13, 9–16 (1965).
14. Stepanov, V. N. The principal specific features of the water structure, Okeanologiya, **2**, No. 1, 26–30 (1962).
15. Monin, A. S., V. G. Neyman, and B. N. Filyushkin. Density stratification in the ocean, Dokl. Akad. Nauk SSSR, **191**, No. 6, 1277–1279 (1970).
16. Stepanov, V. N. A general classification of the water masses of the ocean, their formation and transport, Okeanologiya, **9**, No. 5, 755–767 (1969).
17. Stepanov, V. N. Water circulation in meridional planes of the oceans, Okeanologiya, **9**, No. 3, 387–397 (1969).
18. Burkov, V. A. Water circulation. In: Tikhiy okean (*The Pacific Ocean*), **II**, *The Hydrology of the Pacific Ocean*, Nauka Press, Moscow, 206–289 (1968).
19. Bulatov, R. P. Circulation of Atlantic Ocean waters in various space-time scales. In: Okeanologicheskiye issledovaniya (*Oceanological Research*), No. 22, 7–93 (1971).
20. Burkov, V. A. The general circulation of Pacific Ocean waters. In: Tikhiy okean (*The Pacific Ocean*), Nauka Press, Moscow (1972).
21. Reid, J. L. and W. D. Nowlin. Transport of water through the Drake Passage, Deep-Sea Res., **18**, No. 1, 51–64 (1971).
22. Knauss, J. A. Measurements of the Cromwell Current, Deep-Sea Res., **6**, No. 4, 263–334 (1960).

3 SMALL-SCALE PHENOMENA

3-1 WIND WAVES

In discussing wind waves, and short waves on the sea surface in general, we can neglect the effect of the rotation of the earth (the Coriolis force). We can also neglect, at least initially, the effect of density stratification, and can assume that sea water is incompressible, that its density ρ is constant and its motion nondivergent, i.e., div $\mathbf{u} = 0$. Then the equation of motion for the water can be written as

$$\frac{\partial \mathbf{u}}{\partial t} + \nabla \left(\frac{p}{\rho} + gz + \frac{|\mathbf{u}|^2}{2} \right) = \mathbf{u} \times \text{curl } \mathbf{u} + \nu \Delta \mathbf{u} \tag{3-1-1}$$

where z increases upward and ν is the kinematic viscosity of the water $\nu \sim 1.5 \times 10^{-2}$ cm^2/sec.

Generally speaking, this equation must be analyzed in conjunction with a similar equation for the motion of the air above the water. Here it is convenient to separate out the buoyant force in explicit form in order to describe the dynamic role of stratification, which may be significant. The conditions that must be satisfied on a disturbed water surface $z = \zeta(x, y, t)$ include, first of all, the *kinematic* boundary condition

$$w_s = \frac{\partial \zeta}{\partial t} + \mathbf{u}_s \cdot \Delta \zeta . \tag{3-1-2}$$

(Here and throughout this section, the subscript s identifies quantities pertaining to the surface of the sea.) This equation can also be written as

$$(u_n)_s = (1 + |\nabla \zeta|^2)^{-\frac{1}{2}} \frac{\partial \zeta}{\partial t} \tag{3-1-2'}$$

where u_n is the velocity component normal to the surface. Second, *dynamic* boundary conditions must be satisfied. These mean the continuity of the velocities of the tangential components of the force acting on a unit area of this surface and a condition of pressure discontinuity, generated by surface tension on a curved (wavy) surface, the surface tension being proportional to the Gaussian curvature of the surface:

$$p_s - p_a = -\gamma\rho \operatorname{div}\left\{(1+|\nabla\zeta|^2)^{-\frac{1}{2}}\nabla\zeta\right\}. \quad (3\text{-}1\text{-}3)$$

Here p_a is the atmospheric pressure and $\gamma \approx 72.5 \text{ cm}^3/\text{sec}^2$ at $20°\text{C}$ for the water-air boundary. However, this value may vary strongly under the influence of surface films.

The complexity of the herein-stated program for fluid-mechanical description of surface waves is compounded by several factors. Thus the turbulence of the surface air layer and, most importantly, the turbulent fluctuations of the atmospheric pressure p_a at the surface, both terms in Eq. (3-1-3), play a significant role in wind-wave generation (the eddy viscosity of the water may be a significant factor in the damping of the waves). But turbulence can only be calculated by methods of *statistical fluid mechanics*. Given its complexity, the above description has still not been completely achieved. In fact, the very large classical and contemporary literature dealing with this topic presents solutions for only a few simplified problems, with somewhat tenuous relation to reality. Here we shall briefly discuss only some trends in this research.

Let us represent the velocity field as $\mathbf{u} = \nabla\varphi + \mathbf{v}$, where φ is the velocity potential and $\mathbf{v} = \text{curl } \mathbf{A}$ is the solenoidal or vortical component of the velocity field (curl $\mathbf{U} = -\Delta\mathbf{A}$). Then the nondivergence condition assumes the form $\Delta\varphi = 0$; from this it is at once evident that $\Delta\mathbf{u} = \Delta\mathbf{v}$, i.e., the viscous force is a function of the solenoidal velocity field component alone, and the velocity potential of *sinusoidal waves*, which depend on the horizontal coordinates x, y, in accordance as $e^{i\,\mathbf{k}\cdot\mathbf{x}}$, decreases with depth as e^{-kz}. Real waves appear to exhibit only small vorticity, of the order of the square of the slope imparted to them by viscous forces in thin layers at the surface and at the bottom. Thus *purely potential waves*, for which $\mathbf{u} = \nabla\varphi$, can serve as a certain approximation of reality. With this condition, the right-hand side of the equation of motion (3-1-1) vanishes and the *Bernoulli equation*

$$\frac{\partial\varphi}{\partial t} + \frac{p}{\rho} + gz + \frac{1}{2}|\nabla\varphi|^2 = 0$$

follows from it. It needs only to be used for the surface of the water $z = \zeta(x, y, t)$, eliminating the pressure p_s with the aid of Eq. (3-1-3). As a result, the potential waves are described by

$$\Delta\varphi = 0$$

$$\frac{\partial\zeta}{\partial t} = \left(\frac{\partial\varphi}{\partial z}\right)_s - (\nabla\varphi)_s\,\nabla\zeta = (1+|\nabla\zeta|^2)^{\frac{1}{2}}\left(\frac{\partial\varphi}{\partial n}\right)_s$$

$$\left(\frac{\partial\varphi}{\partial t}\right)_s + g\zeta - \gamma \operatorname{div}\left\{(1+|\nabla\zeta|^2)^{-\frac{1}{2}}\nabla\zeta\right\} + \frac{1}{2}|\nabla\varphi|_s^2 = -\frac{p_a}{\rho} \quad (3\text{-}1\text{-}4)$$

Let us deal only with the case of a flat sea floor at depth $z = -H$. Note that the second and third equations of (3-1-4) contain, in addition to p_a which must be determined from the dynamic equations of the surface air layer, three functions of the horizontal coordinates and time: $\zeta(\mathbf{x}, t)$, $\Phi(\mathbf{x}, t) = \varphi_s$ and $\Psi(\mathbf{x}, t) = \left(\frac{\partial \varphi}{\partial z}\right)_s$. However, the third of these functions can be expressed in terms of the other two upon solution of the boundary-value problem for the equation $\Delta\varphi = 0$.

Thus, for example, for a specified value of $\varphi(\mathbf{x}, z_0, t) = \int e^{i\mathbf{k}\cdot\mathbf{x}}\, dA(\mathbf{k}, t)$ at a certain depth z_0 below the troughs of the waves, the solution of this equation that satisfies the condition $\frac{\partial \varphi}{\partial z} \longrightarrow 0$ $(z \to -H)$ takes the form

$$\varphi(\mathbf{x}, z, t) = \int e^{i\mathbf{k}\cdot\mathbf{x}} \frac{\cosh k(z+H)}{\cosh k(z_0+H)}\, dA(\mathbf{k}, t), \tag{3-1-5}$$

and the relation between Ψ, Φ, and ζ can be obtained from

$$\Phi(\mathbf{x}, t) = \int e^{i\mathbf{k}\cdot\mathbf{x}} \frac{\cosh k(\zeta+H)}{\cosh k(z_0+H)}\, dA(\mathbf{k}, t);$$

$$\Psi(\mathbf{x}, t) = \int e^{i\mathbf{k}\cdot\mathbf{x}} \frac{\sinh k(\zeta+H)}{\cosh k(z_0+H)}\, k\, dA(\mathbf{k}, t). \tag{3-1-6}$$

The total energy of a vertical column of water of unit cross section equals ρE. Here

$$E = \frac{1}{2}\int_{-H}^{\zeta} |\nabla\varphi|^2\, dz + \frac{1}{2} g\zeta^2 + \gamma\left\{(1 + |\nabla\zeta|^2)^{\frac{1}{2}} - 1\right\}, \tag{3-1-7}$$

and the first term gives the kinetic energy and the others the potential energy. The horizontally averaged total energy $\overline{E}$ is a functional of two functions: $\Phi(\mathbf{x}, t)$ and $\zeta(\mathbf{x}, t)$ (the contribution of kinetic energy to $\overline{E}$ can be reduced to the form $\frac{1}{2}\overline{\Phi \frac{\partial \zeta}{\partial t}}$). It is seen with the aid of Eq. (3-1-4) that $\overline{E}$ varies in time at a rate $-\overline{\frac{p_a}{\rho}\frac{\partial \zeta}{\partial t}}$. The waves computed under the condition $p_a = \text{const}$ (this constant can be made to vanish by renormalization of φ) are called *free* waves (let us note, in passing, that Shuleykin [1] does not consider *growing* sea waves to be free because, as the wind flows over them, the pressure p_a is lowest at the crests of the waves and it is higher on the upwind wave slopes than on the lee slopes). The total energy $\overline{E}$ is constant for free waves. Zakharov [2] showed that for free waves, $\overline{E}$ is a Hamiltonian with the canonical variables ζ (the generalized coordinate) and Φ (generalized momentum). That is,

$$\frac{\partial \Phi}{\partial t} = -\frac{\delta \overline{E}}{\delta \zeta}; \quad \frac{\partial \zeta}{\partial t} = \frac{\delta \overline{E}}{\delta \Phi}. \tag{3-1-8}$$

For small-amplitude free waves, Eqs. (3-1-4) can be linearized (the conditions at the surface of the water are then written for $z = 0$, and are noted by substituting a zero for the subscript s). Equations (3-1-4) then become

$$\Delta\varphi = 0; \quad \frac{\partial \varphi_0}{\partial t} + g\zeta - \gamma\,\Delta\zeta = 0; \quad \frac{\partial \zeta}{\partial t} = \left(\frac{\partial \varphi}{\partial z}\right)_0. \tag{3-1-9}$$

Their solutions are elementary plane waves:

$$\zeta = a\cos(\mathbf{k}\cdot\mathbf{x} - \omega t);$$
$$\varphi = \frac{a\omega\cosh k(z+H)}{k\sinh kH}\sin(\mathbf{k}\cdot\mathbf{x} - \omega t);$$
$$\omega^2 = (gk + \gamma k^3)\,\text{th}\,kH. \qquad (3\text{-}1\text{-}10)$$

In this case the phase velocity $c = \omega/k$ of the waves depends on k, that is, the waves have dispersion, and wave packets should become diffuse with time. At $k \ll \left(\frac{g}{\gamma}\right)^{\frac{1}{2}}$, the surface-tension effect is negligibly small, so that the waves are pure *gravity waves*; at $k \gg \left(\frac{g}{\gamma}\right)^{\frac{1}{2}}$, on the other hand, gravity can be neglected, and the waves are *capillary*. The cutoff wavelength $2\pi\left(\frac{\gamma}{g}\right)^{\frac{1}{2}}$ is 1.7 cm for a clean water surface, the phase velocity of this wave being 23 cm/sec. At large kH (in fact, even at $kH > \pi/2$), we can put $\tanh kH \approx 1$ and $\cosh k(z+H)/\sinh kH \approx e^{kz}$ in Eq. (3-1-10); this is the case of waves over deep water. Conversely, at $kH \ll 1$ (shallow water), we can put $\tanh kH \approx kH$ and $\cosh k(z+H)/\sinh kH \approx 1/kH$; then $c \approx \sqrt{gH}$, so that long gravity waves propagate without dispersion.

If we point the x axis along the vector $\mathbf{k}$, the paths of the water particles in the waves (3-1-10) are given by

$$\frac{d\xi}{dt} = \frac{a\omega\cosh k(z+H)}{\sinh kH}\cos(kx - \omega t) = \frac{a\omega}{\sinh kH}\times$$
$$\times(\cosh\mu\cos\chi - k\xi\cosh\mu\sin\chi + k\eta\sinh\mu\cos\chi + \ldots),$$
$$\frac{d\eta}{dt} = \frac{a\omega\sinh k(z+H)}{\sinh kH}\sin(kx - \omega t) = \frac{a\omega}{\sinh kH}\times$$
$$\times(\sinh\mu\sin\chi + k\xi\sinh\mu\cos\chi + k\eta\cosh\mu\sin\chi + \ldots) \qquad (3\text{-}1\text{-}11)$$

with zero initial conditions, where $\xi = x - x_0$; $\eta = z - z_0$; x_0, z_0 are the initial coordinates of the particle; $\mu = k(z_0 + H)$; $\chi + kx_0 - \omega t$. If all but the first terms in the parentheses in the right-hand side are dropped, the computed paths are ellipses with the semiaxes $a\cosh\mu/\sinh kH$, $a\sinh\mu/\sinh kH$. (In deep water, they are circles with radius ae^{kz_0}.) However, subsequent terms also give the time-averaged drift velocity $\overline{d\xi/dt} \approx a^2k\omega\cosh 2\mu/2\sinh^2 kH$, which was observed back in 1847 by Stokes [3].

The nonlinear corrections to the waves (3-1-10) can be found by expanding the functions φ and ζ in Eq. (3-1-4) in series $\varepsilon f_1 + \varepsilon^2 f_2 + \ldots$ in powers of a small parameter ε that is of the order of the wave slope. A correction for viscosity (case of low vorticity) can be introduced similarly for the surface and near-bottom water. Note that this correction will contain an additional drift-current term that for a clean surface over deep water will be equal to the Stokes drift mentioned above (so that the latter is doubled), and an analogous induced current term at the bottom. This current is one-and-a-half times greater than the Stokes drift (see Chap. 3 of Phillips [4]).

A convenient way of describing the nonlinear interactions between waves is obtained upon canonical transformation of the small oscillations $ia^*(\mathbf{k})$ and $a(\mathbf{k})$ in the Hamiltonian $\overline{E}(\Phi, \zeta)$ to *normal* variables, using the expressions

$$\Phi(\mathbf{k}) = -i\sqrt{\frac{\omega(k)}{2k}}\,[a(\mathbf{k}) - a^*(\mathbf{k})];$$

$$\zeta(\mathbf{k}) = \sqrt{\frac{k}{2\omega(k)}}\,[a(\mathbf{k}) + a^*(\mathbf{k})], \tag{3-1-12}$$

where $\Phi(\mathbf{k})$, $\zeta(\mathbf{k})$ are the Fourier transforms of the original variables $\Phi(\mathbf{x})$ and $\zeta(\mathbf{x})$ (the time dependence is implicit). The eigenfrequencies $\omega(k)$ are given by the third equation of (3-1-10), and the asterisk indicates a complex conjugate.

Then Hamilton's equations (3-1-8) assume the form $\dfrac{\partial a(\mathbf{k})}{\partial t} = -i\,\dfrac{\delta\overline{E}}{\delta a^*(\mathbf{k})}$. The Hamiltonian $\overline{E}(a, a^*)$ can be represented as a functional power series in powers of its arguments, beginning with the quadratic term $\int \omega(k)a(\mathbf{k})a^*(\mathbf{k})d\mathbf{k}$ which makes the contribution $-i\omega(k)\,a(\mathbf{k})$ corresponding to linear oscillations, to $\dfrac{\partial a(\mathbf{k})}{\partial t}$. By setting $a(\mathbf{k}) = A(\mathbf{k})e^{-i\omega(k)t}$ we write Hamilton's equation for the wave amplitudes $A(\mathbf{k})$ in the form

$$\begin{aligned}\frac{\partial A(\mathbf{k})}{\partial t} &= -i\int V(-\mathbf{k}, \mathbf{k}_1, \mathbf{k}_2)\,A(\mathbf{k}_1)\,A(\mathbf{k}_2)\times\\ &\times e^{-i[\omega(k_1)+\omega(k_2)-\omega(k)]t}\,\delta(\mathbf{k}_1 + \mathbf{k}_2 - \mathbf{k})\,d\mathbf{k}_1 d\mathbf{k}_2 + \ldots\\ &-i\int W(-\mathbf{k} - \mathbf{k}_1, \mathbf{k}_2, \mathbf{k}_3)\,A^*(\mathbf{k}_1)\,A(\mathbf{k}_2)\,A(\mathbf{k}_3)\times\\ &\times e^{-i[\omega(k_2)+\omega(k_3)-\omega(k_1)-\omega(k)]t}\,\delta(\mathbf{k}_2 + \mathbf{k}_3 - \mathbf{k}_1 - \mathbf{k})\,d\mathbf{k}_1\,d\mathbf{k}_2\,d\mathbf{k}_3 + \ldots,\end{aligned} \tag{3-1-13}$$

where we have not written out terms with other combinations of the amplitudes $A(\mathbf{k}_1), A^*(\mathbf{k}_1), A(\mathbf{k}_2), A^*(\mathbf{k}_2), \ldots$ or higher-order terms. Functions $V, W, \ldots$, whose form is given by Eqs. (3-1-4), are the coefficients of interaction of groups of three, four, etc., waves with corresponding wave vectors. From Eq. (3-1-13) we see that the interaction between three waves with wave vectors $\mathbf{k}, \mathbf{k}_1, \mathbf{k}_2$ may generate an aperiodic variation (increase or decrease) of the amplitude $A(\mathbf{k})$ if the *resonance* conditions

$$\mathbf{k} = \mathbf{k}_1 + \mathbf{k}_2; \quad \omega(k) = \omega(k_1) + \omega(k_2) \tag{3-1-14}$$

are satisfied.

Then either waves $\mathbf{k}_1, \mathbf{k}_2$ will grow at the expense of damped-out wave $\mathbf{k}$ (i.e., the latter will break into two waves) or, on the other hand, wave $\mathbf{k}$ will grow as a result of damping of $\mathbf{k}_1$ and $\mathbf{k}_2$ (i.e., they will merge into a single wave). This resonant three-wave interaction might be called *break-up* interaction. For a more detailed description, see Kadomtsev and Karpman [5]. The resonance conditions (3-1-14) cannot be satisfied for gravity waves in deep water ($\omega^2 = gk$), but three-wave break-up interactions are possible in the case of capillary-gravity waves ($\omega^2 = gk + \gamma k^3$) (they were computed by McGoldrick [6]). In analogy to the above, the conditions

$$\begin{aligned}\mathbf{k} + \mathbf{k}_1 &= \mathbf{k}_2 + \mathbf{k}_3;\\ \omega(k) + \omega(k_1) &= \omega(k_2) + \omega(k_3),\end{aligned} \tag{3-1-15}$$

must be satisfied for resonant interaction among four waves $\mathbf{k}$, $\mathbf{k}_1$, $\mathbf{k}_2$, and $\mathbf{k}_3$. These conditions can be satisfied in the case of gravity waves in deep water. At $\mathbf{k} = \mathbf{k}_1$, these resonant interactions can also be called break-up interactions. The configuration $\mathbf{k} = \mathbf{k}_1 \perp \mathbf{k}_2$ is particularly useful for reproduction of these interactions in experiment. They were calculated by Phillips [7] and in [2, 8–14, 15]. The effects predicted by the calculation were observed experimentally (see [16, 17] and Feir [15]; one example from [17] appears in Fig. 3-1-1). At sea states with a developed spectrum, each wave vector can be a term in very many resonant quadruplets, so that one can only describe the interaction effects statistically. A mathematical formalism for such description was developed by Hasselmann [10, 11]. Among other things, he proved that only very close wave numbers interact significantly, with the apparent result that in real waves energy is redistributed very slowly over the spectrum compared to the energy input from the outside. Thus a statistical equilibrium between the influx and spectral energy transfer over the entire wave spectrum is hardly possible.

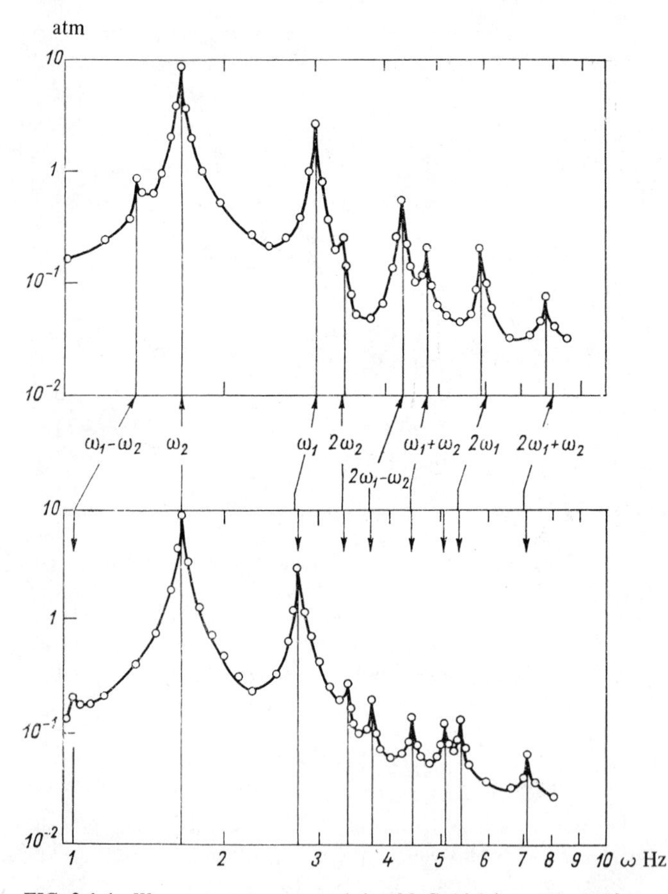

FIG. 3-1-1 Wave spectra measured by McGoldrick et al. [17] in a channel 107 cm deep. The top spectrum corresponds to the resonant ratio of the primary-wave frequencies, $\omega_1/\omega_2 = 1.77$ and the peak at the frequency $2\omega_1 - \omega_2$ is quite pronounced. The lower spectrum corresponds to the nonresonant ratio $\omega_1/\omega_2 = 1.60$, and the peak at $2\omega_1 - \omega_2$ is indistinct. The smaller peaks are those of harmonics of the primary waves.

On the other hand, nonlinear interactions between the elementary waves composing a wave packet can, in theory, compensate for the tendency of the packet to disintegrate because of dispersion of its components. In other words, solutions of Eqs. (3-1-4) may exist in the form of *steady-state* (i.e., propagating without shape change) free waves of finite amplitude.

Steady-state plane waves propagating along the x axis at a constant velocity c are described conveniently in a coordinate frame that moves with them. Then the second equation of (3-1-4) becomes

$$g\zeta-\gamma\frac{\partial}{\partial x}\left\{\left(1+\left|\frac{\partial\zeta}{\partial x}\right|^2\right)^{-\frac{1}{2}}\frac{\partial\zeta}{\partial x}\right\}+\frac{1}{2}|\mathbf{u}|_s^2=\frac{c^2}{2}. \tag{3-1-16}$$

For two-dimensional irrotational flows, we can introduce, along with the velocity potential φ, the stream function ψ ($u=\partial\psi/\partial z$, $\omega=-\partial\psi/\partial x$), which will also satisfy the Laplace equation $\Delta\psi = 0$ (absence of vorticity). In steady-state flow, the streamlines ψ = const will coincide with the paths of fluid parcels. One of them, say, $\psi = 0$, will coincide with the liquid surface $z=\zeta(x)$ [this condition replaces the third equation of (3-1-4)]. Steady-state periodic plane waves are described by the functions $\psi(\chi, z)$ and $\zeta(\chi)$, $\chi = kx$, which, together with the velocity c, must be determined from the equations listed above.

Gravity waves of this type are known as *Stokes waves.* Stokes attempted to define the characteristics of steady-state waves (for deep water in [3], and for water of finite depth in [18]) in the form of series in powers of a parameter on the order of their slope and found, to within terms of power of 4,

$$\zeta=-a\cos\chi+\left(\frac{1}{2}ka^2+\frac{17}{24}k^3a^4\right)\cos 2\chi-$$

$$-\frac{3}{8}k^2a^3\cos 3\chi+\frac{1}{3}k^3a^4\cos 4\chi+\ldots;$$

$$c^2=\frac{g}{k}\left(1+a^2k^2+\frac{5}{4}a^4k^4+\ldots\right). \tag{3-1-17}$$

The expression for ζ differs from the corresponding expansion of a trochoid only in the term $\frac{2}{3}k^2a^4\cos 2\chi$. However, the convergence of these series was not proven.

A rigorous proof of the existence of steady waves of finite amplitude was given by Nekrasov [19, 20]. He made use of the fact that the complex potential $W=\varphi+i\psi$ is an analytic function of the complex variable $Z=x+iz$, whereby $dW/dZ=u-iv=|\mathbf{u}|e^{i\theta}$ acts as a complex velocity. By assuming the wave to be symmetrical (this property was proven by Levi-Civita [21]) in the case of deep water, we map the entire region occupied by water between wave troughs in the Z plane onto a unit circle with a radial cut in the plane $Z_1=|Z_1|\,e^{i\chi}$, setting $\frac{dZ}{dZ_1}=\frac{i\lambda}{2\pi}\frac{f(Z_1)}{Z_1}$. Here λ is the wavelength and $f(Z_1)$ is a power series beginning at unity.

The complex potential can be taken in the form $W=(ic\lambda/2\pi)\ln Z_1$, from which we obtain $f(e^{i\chi})=\frac{c}{|u|_s}e^{i\theta_s}=R(\chi)e^{i\theta(\chi)}$ for the water surface. By taking logarithms in this formula and using the power series for f, we can obtain a relation between $R(\chi)$ and $\theta(\chi)$ that reduces to

$$\theta(\chi) = -\frac{1}{2\pi}\int_0^{2\pi} \ln\left|\frac{\sin\frac{\chi'+\chi}{2}}{\sin\frac{\chi'-\chi}{2}}\right| \frac{d\ln R(\chi')}{d\chi'}\, d\chi'.$$

A second relation between R and θ is obtained by differentiating Eq. (3-1-16) with respect to χ for $\gamma = 0$, using the expression $\frac{dZ}{d\chi} = \frac{dZ}{dZ_1}\frac{dZ_1}{d\chi}$, which gives $\frac{d\zeta}{d\chi} = -\frac{\lambda}{2\pi} R\sin\theta$ for the water surface; it takes the form $\frac{d}{d\chi}\frac{1}{R^2} = \frac{g\lambda}{\pi c^2} \times R\sin\theta$. Using this to express R in terms of χ, we reduce the preceding relation to the form

$$\theta(\chi) = \frac{\mu}{3\pi}\int_0^{2\pi} \frac{\sin\theta(\chi')}{1+\mu\int_0^{\chi'}\sin\theta(\chi'')\,d(\chi'')} \ln\left|\frac{\sin\frac{\chi'+\chi}{2}}{\sin\frac{\chi'-\chi}{2}}\right| d\chi', \qquad (3\text{-}1\text{-}18)$$

where $\mu = 3g\lambda c/2\pi|\mathbf{u}|_0^3$ and the subscript "0" pertains to the crest of the wave.

Nekrasov proved the existence of nontrivial solutions of this equation at $\mu > 3$. He has analyzed the case of water of finite depth in a similar fashion.

Given the possibility of resonant four-wave interactions, Stokes waves are inherently unstable [2, 14, 15]; moreover, waves with excessively sharp crests will break. It would seem that the limiting form of stable waves would be defined solely by the single parameter g, so that in polar coordinates r, θ (with the origin on the crest and the axis $\theta = 0$ pointing vertically downward) we would have, from dimensional considerations, $\psi(r, \theta) \sim \sqrt{gr^3}\Psi(\theta)$ near the crest the equation $\Delta\psi = 0$ giving $\Psi(\theta) = \cos 3\theta/2$. The water surface $\psi = 0$ will correspond to angles $\theta = \pm 60°$, so that the sharpest crests have vertex angles of 120°. This result was established by Stokes himself.

An exact theory of steady-state *capillary* waves of finite amplitude was derived, using different methods, by Slezkin [22] and Crapper [23]. In Crapper's method, conversion to dimensionless variables with the aid of the units of length γ/c^2 and velocity c and the assumption that $|\mathbf{u}| = e^{\tau}$, so that $\ln(dW/dZ) = \tau - i\theta$ will be an analytic function of W, allow one to reduce the problem to solution of the Laplace equation $\frac{\partial^2\tau}{\partial\varphi^2} + \frac{\partial^2\tau}{\partial\psi^2} = 0$ with the boundary condition $\partial\tau/\partial\psi = \sinh\tau$ at $\psi = 0$, which is obtained from Eq. (3-1-18) at $g = 0$. It is easily verified that the solution is contained in the relation $\frac{dW}{dZ} \left(\frac{1+Ae^{ikW}}{1-Ae^{ikW}}\right)^2$ where $A^2 = \frac{k-1}{k+1}$. The height h of these waves equals $8A/k(1 - A^2)$, and their velocity c is $\left(k\gamma\frac{1-A^2}{1+A^2}\right)^{\frac{1}{2}}$. Figure 3-1-2 shows the streamlines for various slopes $\delta = kh/2\pi$. With increasing slope, the troughs of these waves become sharper and the zero streamline crosses itself, indicating the formation of air bubbles under the sharp troughs.

In 1802, Gerstner [24] produced an exact solution of the Lagrangian equations of fluid dynamics for steady-state plane *rotational* free waves of finite amplitude on the surface of an ideal liquid of infinite depth. In this solution, the coordinates x, z of a

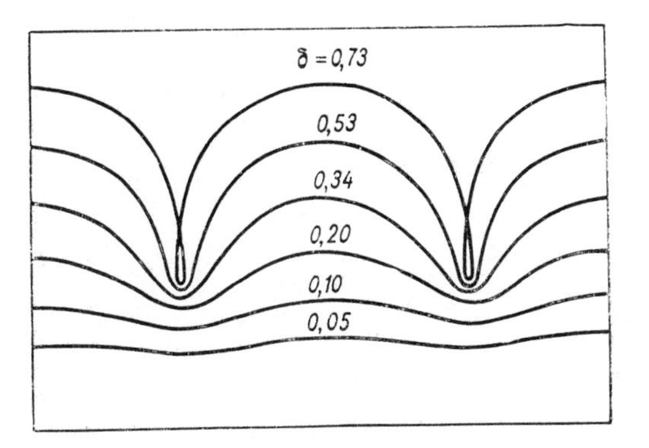

FIG. 3-1-2 Streamlines in steady-state plane capillary waves of finite amplitude at various slopes δ.

fluid parcel that in the absence of a wave was situated at the point (x_0, z_0), changes in the wave as a function of time:

$$x = x_0 - ae^{kz_0} \sin k(x_0 - ct);$$
$$z = z_0 + ae^{kz_0} \cos k(x_0 - ct), \qquad (3\text{-}1\text{-}19)$$

where k is an arbitrary wave number, $c = \sqrt{g/k}$, a is an arbitrary amplitude of fluctuations of the water surface $z_0 = 0$. Here the water parcels move along circles with radii ae^{kz_0} that decrease exponentially with the depth $-z_0$. In a coordinate system that moves with the wave, the streamlines are trochoids:

$$x = \frac{\chi}{k} - ae^{kz_0} \sin \chi;$$
$$z = z_0 + ae^{kz_0} \cos \chi; \quad \chi = kx.$$

As $a \to 1/k - 0$, the surface of the water assumes the limiting shape of a cycloid with tapered crests (see Fig. 3-1-3). Note that, according to Shuleykin [1], real wind waves have sharper vertices and shallower troughs than trochoids with the same slope. The profiles of such waves can be described by the equations $x = \chi/k - an \sin \chi$, $z = a \cos \chi$, $n > 1$. The pressure in the fluid parcels of Gerstner waves is constant, and is identical in parcels with the same z_0; the curl $\partial u/\partial z - \partial w/\partial x$ equals $-\dfrac{2ca^2k^3e^{2kz_0}}{1 - a^2k^2e^{2kz_0}}$. A more general theory of rotational waves was developed by Dubreil-Jacotin [25], Gouyon [26], and Moiseyev [27].

Let us now consider real ocean waves. In the case of developed sea states, it is helpful to resort to a statistical description, treating $\zeta(\mathbf{x}, t)$ as a *random* moving surface. Here we examine the *probability distributions* for the values of ζ on finite sets of points of the space-time continuum $(\mathbf{x}_n, t_n)$. Wave recordings indicate that the probability distribution for ζ at a fixed point $\mathbf{x}$ is near-Gaussian for developed sea states. The joint probability distribution for the slopes of the sea surface $\partial\zeta/\partial x$, $\partial\zeta/\partial y$

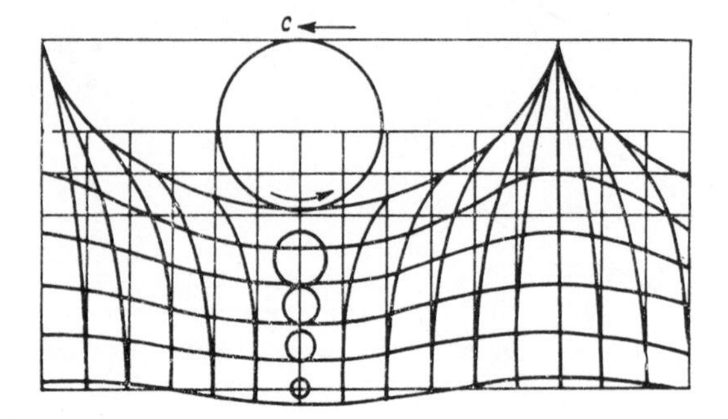

FIG. 3-1-3 Streamlines (lines of constant pressure) in a Gerstner wave.

measured by Cox and Munk [28, 29] on aerial photographs of sun highlights on the sea surface, proved to be approximately Gaussian. However, the marginal distributions for $\partial\zeta/\partial x$ and $\partial\zeta/\partial y$ were sharper than Gaussian, and the distribution for slopes in the wind direction was slightly skewed. Therefore the model of a Gaussian random surface $\zeta(\mathbf{x}, t)$ analyzed by Longuet-Higgins [30, 31] may prove adequate for many purposes.

One of the most important statistical characteristics of the random function $\zeta(\mathbf{x}, t)$ is its two-point correlation function

$$B(\mathbf{r}, \tau) = \langle \zeta(\mathbf{x}, t)\zeta(\mathbf{x}+\mathbf{r}, t+\tau)\rangle, \tag{3-1-20}$$

where the angle brackets indicate statistical averaging. If horizontal statistical homogeneity is lacking, as near coasts, B will also depend on $\mathbf{x}$; if statistical steady state in time is lacking, as in the case of developing waves, B will also depend on t. The Fourier transform with respect to $\mathbf{r}$ and τ is called the wave spectrum:

$$E(\mathbf{k}, \omega) = (2\pi)^{-3} \int e^{i(\mathbf{k}\cdot\mathbf{r} - \omega\tau)} B(\mathbf{r}, \tau)\, d\mathbf{r}\, d\tau. \tag{3-1-21}$$

A statistically horizontally homogeneous and stationary random function $\zeta(\mathbf{x}, t)$ can always be represented as a superposition of harmonic plane waves $e^{i(\mathbf{k}\cdot\mathbf{x} - \omega t)}$ with random uncorrelated amplitudes $dZ(\mathbf{k}, \omega)$, and the meaning of the spectrum is given by the formula $E(\mathbf{k}, \omega)\, d\mathbf{k}\, d\omega = \langle |dZ(\mathbf{k}, \omega)|^2\rangle$. The integral of $E(\mathbf{k}, \omega)$ over all $(\mathbf{k}, \omega)$ gives $\langle\zeta^2\rangle$; the integral over all $\mathbf{k}$ gives the *frequency spectrum* $E(\omega)$ of the waves $\zeta(\mathbf{x}, t)$ at a fixed point $\mathbf{x}$; the integral over all ω gives the *spatial* (two-dimensional) spectrum $E(\mathbf{k})$ of the waves at a fixed time t. By setting $k_x = k \cos\alpha$ and $k_y = k \sin\alpha$ and integrating $kE(k_x, k_y)$ over all α, we obtain the wave-number spectrum $E(\mathbf{k})$.

Each plane sinusoidal wave is presented on the $E(\mathbf{k})$ plot by a point, carrying a certain weight, in the $\mathbf{k}$ plane; ripples consisting of plane waves in the same direction are represented by points on a single straight line passing through the origin; developed wind waves are represented by points in a certain region, which will probably be elongated in the direction of the prevailing wind.

If the dispersion relation $\omega^2 = gk$ is satisfied for the spectral components $e^{i(\mathbf{k}\cdot\mathbf{x} - \omega t)}\, dZ(\mathbf{k}, \omega)$ in the case of gravity waves in deep water (actually, it may be violated because of transport of short wave by long waves and because of three-wave interactions), then we have $E(\mathbf{k}, \omega) = E(\mathbf{k})\delta(\omega - \sqrt{gk})$. We then obtain

$$E(k_x, k_y)dk_x dk_y = \left[\frac{2\omega^3}{g^2} E\left(\frac{\omega^2}{g}\cos\alpha, \frac{\omega^2}{g}\sin\alpha\right)\right] d\alpha d\omega,$$

and the expression in brackets becomes the *direction and frequency spectrum* $E(\alpha, \omega)$ [of which the integral of $kE(\mathbf{k}, \omega)$ over all k is the exact definition; the integral of $E(\alpha, \omega)$ over all α again gives the frequency spectrum].

The frequency spectrum of the waves is easily derived from the traces produced by a single wave recorder. The typical example in Fig. 3-1-4 indicates that the wave spectrum is quite narrow: the main contribution to $\langle\zeta^2\rangle$ is made by a frequency band of only three octaves. A direction- and frequency spectrum was estimated by Cote et al. [33] from measurements of the fluctuations and tilts of a free-floating buoy.

A two-dimensional spectrum was estimated in [34] from aerial stereophotographs of the ocean surface. The use of aircraft and satellite-borne interferometric altimeters and radars offers good prospects for measurements of the three-dimensional spectra of wave states.

Calculation of wave spectra is a problem of statistical fluid mechanics. Generally speaking, the evolution of the spatial spectrum $E(\mathbf{k})$ is described by the equation [4]

$$\left[\frac{\partial}{\partial t} + (C_\alpha + U_\alpha)\frac{\partial}{\partial x_\alpha}\right] E(\mathbf{k}) + \frac{S_{\alpha\beta}(\mathbf{k})}{\rho g}\frac{\partial U_\alpha}{\partial x_\beta} = \Gamma_+(\mathbf{k}) - \Gamma_-(\mathbf{k}) + T(\mathbf{k}), \tag{3-1-22}$$

where C_α and U_α are the components of the wave group velocity and the current velocity, respectively, $S_{\alpha\beta}(\mathbf{k})$ is the spectral contribution to the excess momentum flux

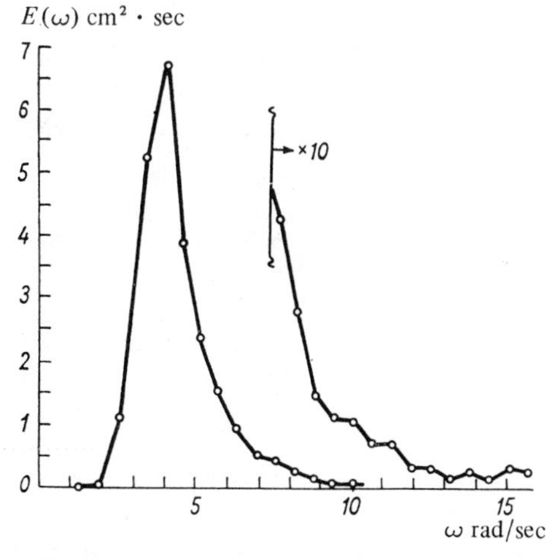

FIG. 3-1-4 Frequency spectrum of waves with short wind fetch (~ 2 km), measured by Kinsman [32]. The right-hand curve represents values of $E(\omega)$ for $\omega > 8$ rad/sec, enlarged tenfold.

tensor due to the wave motions, Γ_+ is the rate of energy input from the wind, Γ_- is the wave energy dissipation rate, and T is the rate of energy input from other wave components. In the case of horizontal homogeneity, the left-hand side equals $\partial E(\mathbf{k})/\partial t$. Our present concepts of the qualitative behavior of the terms in this equation are represented in Fig. 3-1-5. The atmospheric excitation $\Gamma_+(k)$ exhibits a fairly broad spectrum with a peak near the maximum of the wave spectrum $E(k)$, the dissipation $\Gamma_-(k)$ is appreciable only at large k, and the wave interaction $T(k)$ transfers energy from short to long waves. The exact forms of these terms have not yet been defined.

As we have already noted, the slowness of the redistribution of energy over the spectrum via the interaction between waves prevents establishment of statistical equilibrium between this transfer and the atmospheric excitation in the neighborhood of the $E(k)$ maximum. If the wind action persists, then the waves grow: the maximum of $E(k)$ increases and shifts to smaller k (the mechanism of this wave growth was described by Shuleykin [1]). However, the growth of waves at fixed k is limited by their stability condition, that is, the requirements that the local accelerations at their crests not exceed a certain fraction of g (i.e., $g/2$ for Stokes waves); crossing this limit results in breaking of the waves. In the range $k \gg k_0$, where k_0 corresponds to the maximum of $E(k)$, the spectrum cannot grow and a statistical equilibrium is established. At $k \ll (g/\gamma)^{1/2}$, it is defined by the parameter g alone. The shape of the spectrum in the equilibrium range is determined with the aid of dimensional analysis [35]:

$$E(k)=Bk^{-4}; \quad E(\omega)=\beta g^2\omega^{-5}, \tag{3-1-23}$$

where B and β are numerical constants.

Data collected by Kitaygorodskiy [36] indicate that $\beta \sim 6 \cdot 10^{-3}$, when ω is measured in rad/sec; if $\omega^2 = gk$, we obtain $B \sim \beta/2$. Figure 3-1-6 shows a set of wave

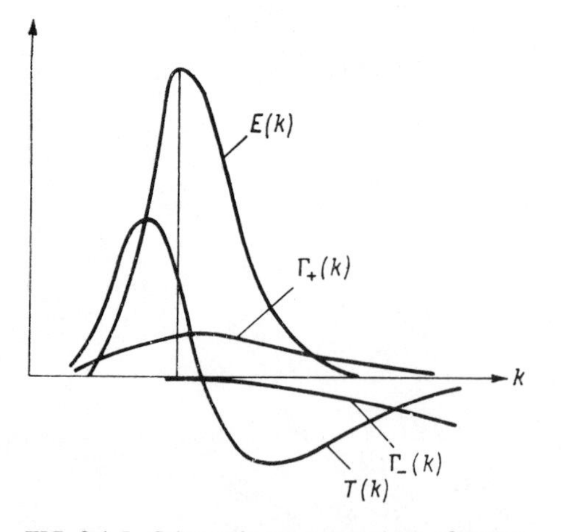

FIG. 3-1-5 Schematic representations of terms in the spatial-spectrum equation (3-1-22).

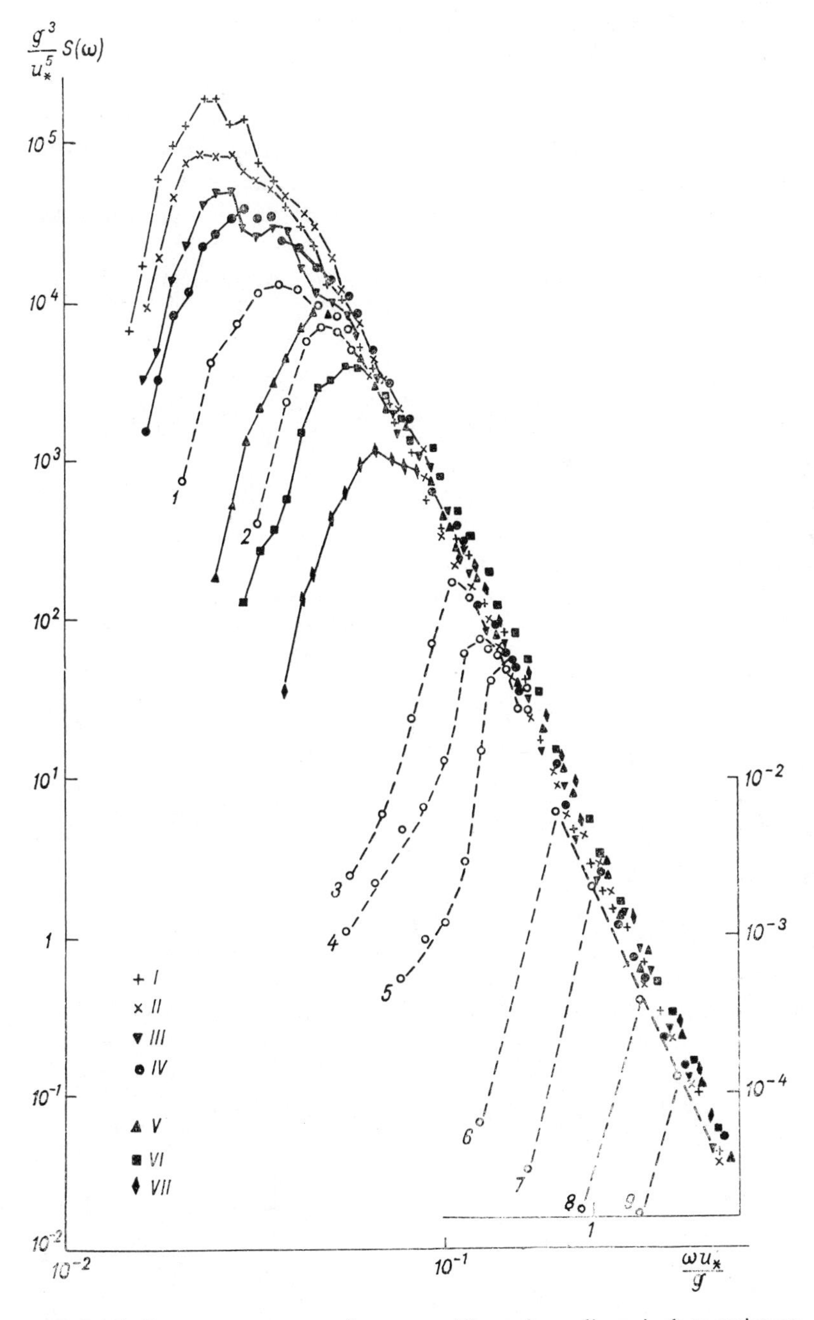

FIG. 3-1-6 Frequency spectra of waves with various dimensionless variances $g^2/u_*^4 \langle \zeta^2 \rangle$ (after Volkov [37]). I-VII–from 5730 to 116; 1-9–from 570 to $2.1 \cdot 10^{-3}$. The scales have been modified for curves 6-9 (laboratory experiments) [see the coordinate system in the lower right-hand corner of figure].

frequency spectra with various dimensionless variances $g^2/u_*^4\langle\zeta^2\rangle$ (u_* is the friction velocity), as measured by Volkov [37]; all the high-frequency branches lie on the same straight line that is a plot of Phillips' law (3-1-23). Note that corrections to Phillips' law (for example, a certain variability of β) may appear in a more detailed theory because of the effect exerted on the equilibrium waves by local accelerations generated by longer waves, and perhaps also because of stratification effects in the upper ocean [38]. Note further that in the equilibrium range, the spectral tensor of the sea-surface slopes $\partial\zeta/\partial x_i$, should have the form $C_{ij}\omega^{-1}$ with $C_{12} = C_{21} = 0$ in the case of symmetry about the wind direction x_1; this prediction was confirmed by the measurements of [33], which gave $C_{11} + C_{22} \approx 8 \cdot 10^{-3}$. Since the growth of capillary waves is also limited by their stability limit, above which air bubbles form under sharp troughs, we may also expect their spectra to exhibit an equilibrium range, defined this time only by parameter γ. In this range, $E(\omega) \sim \gamma^{\frac{2}{3}}\omega^{-\frac{7}{3}}$, $E(k) \sim k^{-4}$ and the spectral slope tensor must have the form $C'_{ij}\omega^{-1}$; this last result was confirmed by the laboratory measurements of Cox [39], who found $C'_{11} + C'_{22} \approx 1 \cdot 10^{-2}$.

The nonequilibrium part of the spectrum in the neighborhood of its maximum, and hence also the total variance $\langle\zeta^2\rangle$ may depend primarily on the wind action time t and the wind fetch x. This dependence can be calculated for the total wave energy $E = \rho g\langle\zeta^2\rangle$ with the aid of the energy transport equation obtained by integrating Eq. (3-1-22) over all $\mathbf{k}$ [the integral of $T(\mathbf{k})$ will then vanish].

Shuleykin [1] represented the terms in this equation (with $U_\alpha = 0$) by semiempirical expressions. He assumed that $C \sim \zeta^{1/2}$ (i.e., that the wave lengths increase in proportion to wave heights), the atmospheric excitation was proportional to ζ and the dissipation to ζ^2. He then obtained in the appropriate dimensionless variables,

$$\frac{\partial\zeta}{\partial t} + \zeta^{1/2}\frac{\partial\zeta}{\partial x} = 1 - \zeta.$$

In 1962, Kitaygorodskiy [36] suggested a similarity theory for waves above a certain minimum length in the nonequilibrium spectrum and for t and x that are not too small, in which the spectrum is defined by only two parameters, namely, g and $u_* = \left(\frac{\tau}{\rho_a}\right)^{\frac{1}{2}}$ (where τ is the wind stress on the water surface). Then

$$E(\omega) = \frac{u_*^5}{g^3} f_1\left(\frac{u_*\omega}{g}, \frac{gt}{u_*}, \frac{gx}{u_*^2}\right);$$

$$\langle\zeta^2\rangle = \frac{u_*^4}{g^2} f_2\left(\frac{gt}{u_*}, \frac{gx}{u_*^2}\right), \tag{3-1-24}$$

where the dependence on t vanishes for x below a certain value $\frac{u_*^2}{g} f_3\left(\frac{gt}{u_*}\right)$ (which corresponds to the so-called, steady sea-state front), and where there is no dependence on x for x higher than this limit (beyond this limit, the dependence on t may also weaken). These predictions appear to agree satisfactorily with experimental data (of which Fig. 3-1-6 may serve as one example) [36]. For the description of the spectrum of the hypothetical asymptotic, "fully developed" sea state (the existence of which is doubtful in the light of what we have said concerning the nonequilibrium nature of the spectra), Pierson and Moskovitz [40] chose the formula $\frac{g^3}{U^5} E(\omega) = 4 \cdot 10^{-3} f^{-5} \exp(-0.74 f^{-4})$, where $f = U\omega/g$ and U is the wind velocity. Here $\langle \zeta^2 \rangle^{1/2} = 0.05\, U^2/g$ and the frequency $\omega = 0.88\, g/U$ corresponds to the spectral maximum. This conclusion may serve as a point of reference in estimating the heights and lengths of the dominant waves in strong winds.

At small t in the initial stage of wind-wave growth, the entire spectrum $E(\mathbf{k}, t)$ is of the nonequilibrium type. It can be calculated by means of the linearized equations (3-1-9) by replacing the right-hand side of the second equation with $-p_a/\rho$ and representing the flucutations of the atmospheric pressure p_a as the sum of statistically stationary fluctuations p_1, generated by turbulence in the surface air layer over the sea *not perturbed by waves*, and of fluctuations p_2 induced by the waves. The spatial Fourier amplitudes of the latter will be proportional to the Fourier amplitudes $dZ(\mathbf{k}, t)$ of the wave field $\zeta(\mathbf{x}, t)$, so that they can be written in the form $\alpha\rho c^2 k dZ(\mathbf{k}, t)$. Here α is a small complex number, and we shall denote its imaginary part by μ. Now, using the above-linearized equations to express $dz(\mathbf{k}, t)$ in terms of the Fourier amplitudes of the function $p_1(\mathbf{x}, t)$, we obtain Eq. (3-1-25) for $\theta(\mathbf{k}) \ll t \ll 1/\omega$ (here $(\mathbf{k})$ is the correlation time scale of the $\mathbf{k}$-component of the p_1 field in a reference system that moves at the velocity $\mathbf{u}(\mathbf{k})$ at which this component is transported by the wind) [4]

$$E(\mathbf{k}, t) = \frac{\pi}{\rho^2 c^2} P(\mathbf{k}, \omega) M(t);$$

$$M(t) = \frac{\operatorname{sh} \mu\omega t}{\mu\omega}, \tag{3-1-25}$$

where $P(\mathbf{k}, \omega)$ is the spatial-frequency spectrum of the field $p_1(\mathbf{x}, t)$ and ω and c are defined by the dispersion relation for deep water. For small $\mu\omega t$ we obtain $M(t) \approx t$; this is the *resonant* stage of wave growth that was discovered by Phillips [41]. By the definition of $\mathbf{u}(\mathbf{k})$, the spectrum $P(\mathbf{k}, \omega)$ has a peak at $\omega = \pm\, \mathbf{k} \times \mathbf{u}(\mathbf{k})$. Thus the response of the water surface to the action of the pressure field p_1 is strongest when the dominant frequency of the wind-transported $\mathbf{k}$-components of the pressure p_1 coincides with the frequency of free waves with the same wave vector. The pressure fluctuations induced by the waves are insignificant at this stage.

At $\mu\omega t > 1$, the growth of the spectrum (3-1-25) with time becomes exponential. The wave-induced pressure fluctuations p_2 (or at least the fraction governed by the parameter μ, which is in phase with the slope of the wave surface) become significant in this case, that is, feedback is introduced in the wind-and-wave system. This *exponential* stage of wave growth was first calculated by Miles [42] in the solution of the classical problem of hydrodynamic instability of the velocity profile in an air flow above a free water surface. (This problem reduces to the eigenvalue problem for the corresponding Orr-Sommerfeld equation. Here the air flow was assumed to be quasilaminar, but the logarithmic velocity profile was typical for turbulent flows.)

The difficulty in this theory centers around calculation of the parameter $\mu = 2\tau_2/\rho c^2 k^2 a^2$ (where τ_2 is the wave-induced frictional stress on the surface of the water and a is the wave amplitude $\zeta = a \cos kx$), which depends both on k and on the wind velocity profile. A calculation of this kind was described by Phillips [4]. His results for the wave-amplification factor M as a function of the dimensionless frequency $u_* \omega/g$ and the dimensionless time gt/u_* for the case

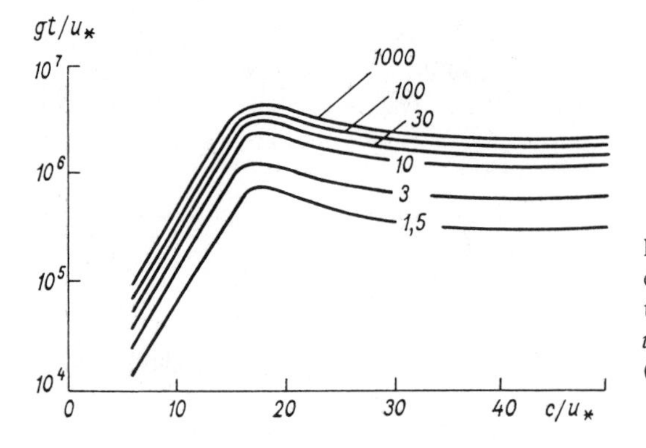

FIG. 3-1-7 Wave amplification coefficient M in Eq. (3-1-25) as a function of the dimensionless frequency $u_* \omega/g$ and dimensionless time gt/u_* (after Phillips [4]).

k||u(k) appears in Fig. 3-1-7. Note, however, that certain very recent data, which were summarized by Pond [43], show the wave growth to be considerably faster than that indicated by the above calculations. As a result, the latter may require a review.

REFERENCES

1. Shuleykin, V. V. The theory of ocean waves, Trudy Morskogo gidrofizicheskogo Instituta Akad. Nauk SSSR, **9** (1956).
2. Zakharov, V. Ye. Stability of periodic waves of finite amplitude on the surface of a deep liquid, Prikl. Mekh. i Tekhn. Fiz., No. 2, 86–94 (1968).
3. Stokes, G. G. On the theory of oscillatory waves, Trans. Cambr. Phil. Soc., **8**, 441–455 (1847).
4. Phillips, O. *The Dynamics of the Upper Ocean*, Cambridge University Press (1966).
5. Kadomtsev, B. B. and V. I. Karpman. Nonlinear waves, Uspekhi Fizicheskikh Nauk, **103**, No. 2, 193–211 (1971).
6. McGoldrick, L. F. Resonant interactions among capillary-gravity waves, J. Fluid Mech., **21**, Part 2, 305–332 (1965).
7. Phillips, O. M. On the dynamics of unsteady waves of finite amplitude, J. Fluid Mech., **9**, Part 1, 193–217 (1960).
8. Longuet-Higgins, M. S. Resonant interactions between two trains of gravity waves, J. Fluid Mech., **12**, Part 2, 321–332 (1962).
9. Longuet-Higgins, M. S. and O. M. Phillips. Phase velocity effect of wave interactions, J. Fluid Mech., **12**, Part 2, 333–336 (1962).
10. Hasselmann, K. On the non-linear energy transfer in a gravity wave spectrum, J. Fluid Mech., **12**, Part 3, 481–500 (1962).
11. Hasselmann, K. On the non-linear energy transfer in a gravity wave spectrum, J. Fluid Mech., **15**, Part 2, 273–281 (1963).
12. Benney, D. J. Non-linear gravity wave interaction, J. Fluid Mech., **14**, Part 4, 577–584 (1962).
13. Bretherton, F. P. Resonant interactions between waves, J. Fluid Mech., **20**, Part 4, 457–480 (1964).
14. Benjamin, T. B. and J. E. Feir. The disintegration of wave trains on deep water, J. Fluid Mech., **27**, Part 3, 417–430 (1967).
15. The Nonlinear Theory of Wave Propagation, Translated from the English, Mir Press, Moscow (1970), 231 pp.
16. Longuet-Higgins, M. S. and N. D. Smith. An Experiment on Third-Order Resonant Wave Interaction, J. Fluid Mech., **25**, Part 3, 417–435 (1966).
17. McGoldrick, F., O. M. Phillips, N. Huang, and T. Hodgson. Measurements of Third-Order Resonant Wave Interactions, J. Fluid Mech., **25**, Part 3, 437–456 (1966).
18. Stokes, G. G. Supplement to a paper on the Theory of Oscillatory Waves, Math. and Phys. Pap., Cambr. Univ. Press, **1**, 197–229 (1880).

19. Nekrasov, A. I. Waves of Steady-State Form, Part 1, Izv. Ivanovo-Voznesenskogo politekhn. Inst., No. 3, 52–65 (1921).
20. Nekrasov, A. I. Nonlinear Integral Equations, Part II, Izv. Ivanovo-Voznesenskogo politekhn. Inst., No. 6, 3–19 (1922).
21. Levi-Civita, T. Rigorous Determination of Finite-Amplitude Steady-State Waves, Math. Ann., **93**, No. 4, 264–314 (1925).
22. Slezkin, N. A. Steady-State Capillary Waves, Uch. zapiski Moskov. Gosud. Univ., No. 7, 71–102 (1937).
23. Crapper, G. D. An exact solution for progressive capillary waves of arbitrary amplitude, J. Fluid Mech., **2**, Part 6, 532–540 (1957).
24. Gerstner, F. J. The theory of waves, Abh. koenigi. Bohm. Ges. Wissensch., **3**, 1, 1408, Reprinted in Ann. Physik, **32**, 412–445 (1809).
25. Dubreil-Jacotin, L. Rigorous determination of periodic steady-state waves of finite amplitude, J. Math. Pure Appliq., **13**, No. 3, 217–291 (1934).
26. Gouyon, R. Plane swells at infinite depth, Compt. Rend. Acad. Sci., **247**, Nos. 1 (33–35), 2 (180–182), 3 (266–269) (1958).
27. Moiseyev, N. N. Existence and uniqueness theorems for turbulent waves of the periodic type, Prikl. Matem. i Mekh., **24**, No. 4, 711–714 (1969).
28. Cox, C. S. and W. H. Munk. Measurement of the roughness of the sea surface from photographs of the Sun's glitter, J. Opt. Soc. Am., **44**, No. 11, 838–850 (1954).
29. Cox, C. S. and W. H. Munk. Statistics of the sea surface derived from Sun glitter, J. Marine Res., **13**, No. 2, 198–227 (1954).
30. Longuet-Higgins, M. S. Statistical analysis of a random moving surface. In: *Wind Waves*, Translated from the English, Foreign Literature Press, Moscow, 125–218 (1962).
31. Longuet-Higgins, M. S. Statistical geometry of random surfaces. In: *Hydrodynamic Instability*, Translated from the English, Mir Press, Moscow, 124–167 (1964).
32. Kinsman, B. *Wind Waves, Their Generation and Propagation on the Ocean Surface*, New York (1975).
33. Longuet-Higgins, M. S., D. E. Cartwright, and N. D. Smith. Observations of the directional spectrum of sea waves using the motions of a floating buoy. In: *Ocean Wave Spectra*, Englewood Cliffs, N.Y., 111–136 (1963).
34. Cote, L. I., I. O. Davis, W. Marks, R. J. McGough, E. Mehr, W. J. Pierson, I. F. Ropek, G. Stephenson, and R. C. Vetter. The directional spectrum of a wind-generated sea determined by the Stereo Wave-Observation Project, Meteorol. Pap. New York Univ. Coll. of Engineering, **2**, No. 6 (1960).
35. Phillips, O. M. The equilibrium range in the spectrum of wind-generated waves, J. Fluid Mech., **4**, 426–434 (1958).
36. Kitaygoroskiy, S. A. Fizika vzaimodeystviya atmosfery i okeana (*The Physics of Sea-Air Interaction*), Girdometeoizdat Press, Leningrad (1970).
37. Volkov, Yu. A. Analysis of the spectra of sea waves development under the action of a turbulent wind, Izv. Akad. Nauk SSSR, Fizika atm. i. okeana, **4**, No. 9, 968–987 (1968).
38. Zaslavskiy, M. M. and S. A. Kitaygorodskiy. The equilibrium range in the spectrum off wind-generated surface gravity waves, Izv. Akad. Nauk SSSR, Fizika atm. i okeana, **7**, No. 5, 565–575 (1971).
39. Cox, C. S. Measurements of slopes of high-frequency wind waves, J. Marine Res., **16**, No. 3, 199–225 (1958).
40. Pierson, W. I. and L. A. Moskovitz. A proposed spectral form for fully developed seas based on the similarity theory of Kitaigorodsky, S. A., J. Geoph. Res., **69**, No. 24, 5181–5190 (1964).
41. Phillips, O. M. On the generation of waves by turbulent wind, J. Fluid Mech., **2**, Part 5, 417–445 (1957).
42. Miles, J. W. On the generation of surface waves by shear flows, J. Fluid Mech., **3**, Part 2, 185–204 (1957).
43. Pond, S. Air-sea interaction, Trans. Amer. Geophys. Union, **52**, No. 6, 389–394 (1971).

3-2 INTERNAL WAVES

Just like the surface waves, internal waves are a practically ubiquitous phenomenon in the sea. They sometimes reach truly immense size (Fig. 3-2-1 shows an example from Bocket [1], whose observations of vertical salinity profile variations in the Strait of Gibraltar indicated an internal wave at a depth of 150 m with an amplitude of about 100 m and a period of about half a day).

We shall discuss only small-scale (short and short-period) internal waves, in the description of which one can neglect the effects of the spherical curvature of the earth and its rotation (Coriolis force). We then write the equations of motion of the water as

$$\frac{d\mathbf{u}}{dt} = -\frac{\nabla p}{\rho} + \mathbf{g} + \nu \Delta \mathbf{u}. \tag{3-2-1}$$

The main difference between the theories of surface and internal waves is that in the latter case one must allow for the *inhomogeneity of the water*, that is, the variations of its density in space and time. Since the water velocities in waves are very low compared to the velocity of sound, this motion can be assumed nondivergent, that is, we can set div $\mathbf{u} = 0$; this condition filters out sound waves from the solutions of the hydrodynamic equations.

In describing the density fluctuations, we use the entropy balance $\rho(d\eta/dt) = \epsilon$, where η is the specific entropy and ϵ is the change in the entropy of a unit volume of

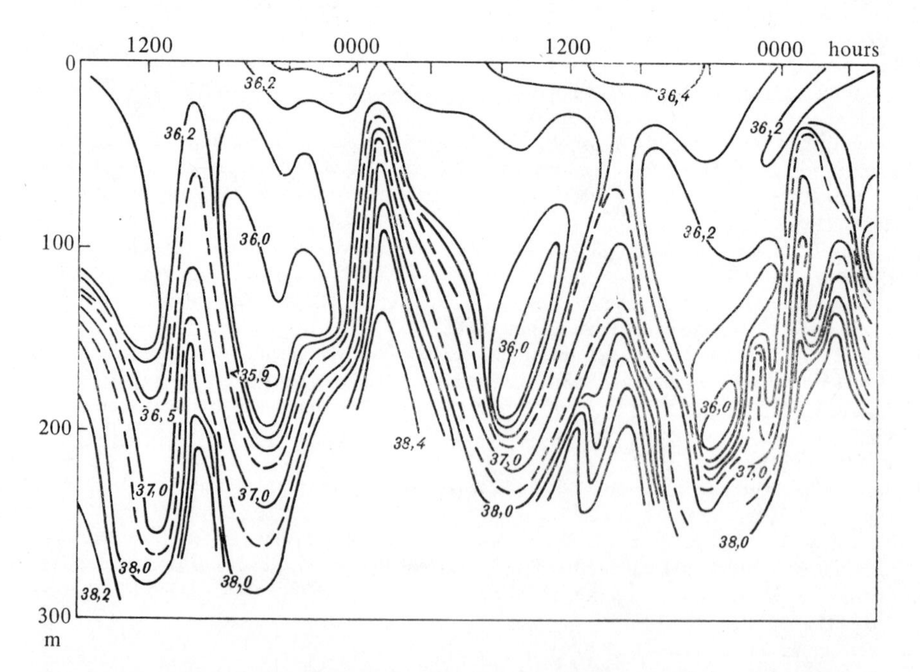

FIG. 3-2-1 Salinity fluctuations (in‰) in the Strait of Gibraltar (35°54.6′N, 5°44.4′W) on 16–18 May 1961 (after Bocket [1]).

water due to molecular effects and radiative heat transfer. Generally speaking, the entropy balance alone is insufficient for description, since the entropy of sea water varies not only with the pressure and density, but also with the salinity s; it is therefore necessary to include a salt diffusion equation $\rho(ds/dt) = \epsilon_1$, where ϵ_1 is the salt input to a unit volume of water due to molecular effects. But if we neglect here the very slow damping of the waves due to molecular effects or the fine structure of the hydrodynamic fields at the smallest scales of inhomogeneity of these fields, where only molecular effects are material, then the motion of water in the waves can be regarded as isentropic and isohaline, that is, they can be described by $d\eta/dt = 0$, $ds/dt = 0$, which can be reduced to

$$\frac{dp}{dt} = c_0^2 \frac{d\rho}{dt}, \tag{3-2-2}$$

where $c_0 = [(\partial p/\partial \rho)_{\eta, s}]^{1/2}$ is the speed of sound. If we consider the latter to be given (or a known function of p and ρ), then Eqs. (3-2-1) and (3-2-2), together with the condition div $\mathbf{u} = 0$, form a closed set in the functions $\mathbf{u}, p$, and ρ.

The boundary condition at the sea floor $z = H(x, y)$ will be the vanishing of the velocity $\mathbf{u}$ [or, if viscous forces are neglected in the equations of motion (3-2-1), the condition that the velocity component normal to the sea floor, that is, the vertical component w in the case of a flat floor, be zero]. Generally speaking, in the case of internal waves the same kinematic and dynamic boundary conditions must be satisfied on the sea surface $z = \zeta(x, y, t)$ as in the case of surface waves. Then both surface and internal waves and their possible interactions will be accounted for by the resulting expressions. In calculating internal waves, however, it appears reasonable to neglect capillary effects at the sea surface, and to write the dynamic boundary condition as $dp/dt = dp_a/dt$ at $z = \zeta(x, y, t)$.

In studies of the properties of internal waves, the surface waves are often filtered out, that is, the oscillations of the ocean surface are neglected, the surface is assumed to coincide with the plane $z = 0$, and the kinematic boundary condition $w = 0$ is imposed upon it; the consequence of the somewhat artificial nature of this statement is that the dynamic boundary condition $p = p_a$ is not, strictly speaking, satisfied. But, even though they are small, the sea-surface oscillations caused by internal waves are nevertheless nonvanishing, because seismograms recorded on the Arctic ice show oscillations with periods characteristic of internal waves. Therefore, in a more exact theory of internal waves, it is prudent to employ at least a linearized kinematic and dynamic boundary condition such as

$$w = \frac{\partial \zeta}{\partial t}; \quad \frac{\partial p}{\partial t} + g\rho_0 w = \frac{\partial p_a}{\partial t}, \tag{3-2-3}$$

referring them approximately to the level $z = 0$ (here ρ_0 is the quasistationary part of the density).

Let us first consider free (p_a = const) internal waves of *small amplitude* in a sea of constant depth H (a sea which would be in a state of rest in the absence of the waves), in which the pressure $p = p_0(z)$, the density $\rho = \rho_0(z)$, and the speed of sound $c_0 = c_0(z)$ depends only on depth, with $dp_0/dz = g\rho_0$ and $p_0(0) = p_a$. Assuming that $p = p_0 + p'$ and $\rho = \rho_0 + \rho'$ in the presence of waves (where p' and ρ' are small fluctuations generated by the waves), we obtain in the linear approximation

$$-\frac{\nabla p}{\rho}+\mathbf{g}\approx-\frac{\nabla p'}{\rho_0}+\mathbf{g}\frac{\rho'}{\rho_0}. \tag{3-2-4}$$

This expression is known as the *Boussinesq approximation* (let us emphasize that it is much too approximate for purposes of a more exact theory that accounts for nonlinear interactions between waves [2]); the second term in its right-hand side is the acceleration due to *buoyant forces.*

With Eq. (3-2-4), the linearized hydrodynamic equations reduce to

$$\frac{\partial \mathbf{u}}{dt}=-\frac{\nabla p'}{\rho_0}+\mathbf{g}\frac{\rho'}{\rho_0}; \quad \operatorname{div}\mathbf{u}=0;$$

$$\frac{\partial p'}{\partial t}+g\rho_0 w=c_0^2\left(\frac{\partial \rho'}{\partial t}+w\frac{d\rho_0}{dz}\right). \tag{3-2-5}$$

They must be solved with the homogeneous boundary conditions $\partial p'/\partial t + g\rho_0 w = 0$ at $z = 0$, and $w = 0$ at $z = H$. The sea-surface oscillations will be described by the values taken by $\zeta = -p'/g\rho_0$ at $z = 0$.

From these equations, the vertical projection of the velocity curl $\partial v/\partial x - \partial u/\partial y$ is time-invariant; it should be set equal to zero for internal waves, so that the horizontal motions in such waves are nonvortical (but this does not mean that internal waves are potential waves!). We shall consider only the case in which the stratification of the sea is everywhere hydrostatically stable, so that the Brunt-Väisälä frequency N is real and positive. We shall employ the *potential density* ρ_p defined from $d \ln \rho_p/dz = N^2/g$ under the condition $\rho_p = \rho_0$ at $z = 0$. Using ρ_p we introduce the energy integral

$$E=\frac{1}{2}\rho_s g\zeta^2+\int_0^H\left[\frac{u^2+v^2+w^2}{2}+\frac{g^2}{2}\left(\frac{p'-c_0^2\rho'}{\rho_0 c_0^2 N}\right)^2\right]\rho_p\,dz. \tag{3-2-6}$$

Employing here Eq. (3-2-5), one can prove that the horizontally averaged quantity $\bar{E}$ does not vary with time [the first term in Eq. (3-2-6) vanishes under the simplified boundary condition $w = 0$ at $z = 0$].

The following equations and boundary conditions for the vertical velocity are easily derived from Eq. (3-2-5):

$$\frac{\partial^2}{\partial t^2}\left(\Delta w+\frac{N^2}{g}\frac{\partial w}{\partial z}\right)+N^2\Delta_h w=0;$$

$$\frac{\partial^3 w}{\partial t^2 \partial z}+g\Delta_h w=0 \quad \text{at} \quad z=0; \quad w=0 \quad \text{at} \quad z=H. \tag{3-2-7}$$

Thus, the Brunt-Väisälä frequency proves to be the only characteristic of ocean stratification that controls the behavior of internal waves. This equation, in somewhat different form, is found in the books of Phillips [3] and Krauss [4], where it is derived by inaccurate use of the continuity equation.

In the case of a flat sea floor (H = const), Eqs. (3-2-5) have elementary solutions in the form of plane waves that depend on the horizontal coordinates x, y and the time t in accordance with $e^{i(k_x x + k_y y - \omega t)}$ and whose amplitudes are functions of the depth z; they are related by

$$\zeta = \frac{iw_0}{\omega}; \quad u = \frac{ik_1}{k^2}\frac{\partial w}{\partial z}; \quad v = \frac{ik_2}{k^2}\frac{\partial w}{\partial z};$$

$$p' = \frac{i\omega\rho_0}{k^2}\frac{\partial w}{\partial z}; \quad \rho' = \frac{i\omega\rho_0}{k^2 c_0^2}\left(\frac{\partial w}{\partial z} - \frac{k^2 c_0^2 N^2}{g\omega^2} w\right), \tag{3-2-8}$$

where the possible frequencies ω and the amplitudes $w(z)$ that correspond to them must be determined from

$$\frac{\partial^2 w}{\partial z^2} + \frac{N^2}{g}\frac{\partial w}{\partial z} + \frac{N^2 - \omega^2}{\omega^2} k^2 w = 0;$$

$$\omega^2 \frac{\partial w}{\partial z} + gk^2 w = 0 \quad \text{at} \quad z = 0; \quad w = 0 \quad \text{at} \quad z = H. \tag{3-2-9}$$

We note that *internal waves* prove to be rotational: if the x axis is pointed in the direction of propagation of the wave, we obtain for the amplitude of the velocity curl $\partial u/\partial z - \partial w/\partial x$ the expression $\frac{i}{k}\left(\frac{\partial^2 w}{\partial z^2} - k^2 w\right)$, which vanishes at the sea surface, but is generally nonzero at depth.

The general theory of the eigenfrequency spectrum and the vertical structure of small oscillations in the ocean is discussed in Chap. 4 (Sec. 4-1). We note that owing to the presence of the energy integral E = const, all frequencies ω are real. The spectrum of the frequencies ω and the shape of the eigenfunctions $w(z)$ depend on the form of the function $N(z)$. Typically, the values of $N(z)$ for the upper mixed layer (whose thickness we shall denote by h) and below the pycnocline (whose thickness we shall denote by δ) are small. A peak of $N(z)$ occurs in the density discontinuity layer in the pycnocline. The net result is that a three-layer model in which N vanishes outside of the pycnocline and is constant in it will give a qualitatively correct picture. In this model, assuming an ocean of infinite depth, the everywhere continuous eigenfunctions $w(z)$ with continuous derivatives at $z = h + \delta$ have the form

$$w(z) = \begin{cases} w(h)\dfrac{\omega^2 \cosh kz - gk \sinh kz}{\omega^2 \cosh kh - gk \sinh kh}, \quad 0 \leqslant z \leqslant h, \\[2ex] w(h)\, e^{-\frac{N^2}{2g}(z-h)} \dfrac{l \cos l(h+\delta-z) + k\left(1 - \dfrac{N^2}{2gk}\right)\sin l(h+\delta-z)}{l \cos l\delta + k\left(1 - \dfrac{N^2}{2gk}\right)\sin l\delta}, \\ \qquad h \leqslant z \leqslant h+\delta, \\[2ex] w(h)\, e^{-\frac{N^2}{2g}\delta} \dfrac{l}{l\cos l\delta + k\left(1 - \dfrac{N^2}{2gk}\right)\sin l\delta} e^{-k(z-h-\delta)}, \quad z \geqslant h+\delta, \end{cases} \tag{3-2-10}$$

where l is the vertical wave number, which is related to the frequency ω by

$$\omega^2 = N^2\left(1 + \frac{l^2}{k^2} + \frac{N^4}{4g^2k^2}\right)^{-1}. \tag{3-2-11}$$

At $z = h$, the condition for continuity of $\partial w/\partial z$ is satisfied for $\omega^2 = gk$ by all N and k [here $w(z) = w(0)e^{-kz}$ for all z i.e., there are surface waves that are not distorted by the stratification of the ocean]. But if $\omega^2 \neq gk$ we can derive from that continuity condition the following equation for determination of the possible vertical wave numbers l [or, by virtue of Eq. (3-2-11), of frequencies ω] of the internal waves:

$$\tan l\delta = f(l); \quad f(l) = \frac{\frac{l}{k}(1 + \coth kh)}{\frac{l^2}{k^2} - \left(1 + \frac{N^2}{2gk}\right)\left(\coth kh - \frac{N^2}{2gk}\right)}, \tag{3-2-12}$$

which has a denumerable set of real positive roots $l_0 < l_1 < \ldots \to \infty$, which lie at the intersections of the tangent curves $y = \tan l\delta$ with the algebraic curve $y = f(l)$ and approach the points $l_n = n\pi/\delta$ from above at large wave numbers. The frequencies corresponding to these roots satisfy the inequalities $N > \omega_0 > \omega_1 > \ldots \to 0$ and the corresponding modes of $w_n(z)$ have n nodes (i.e., depths where w changes sign) each in the pycnocline at wave numbers that are not too small.

It is not difficult to prove that the lowest mode l_0 has the properties

$$l_0 < \frac{\pi}{2\delta} \quad \text{at} \quad \frac{\tanh kh}{kh} \geq \max\left[\frac{1 + \varepsilon + \frac{\delta}{h}kh}{\varepsilon^2\frac{h}{\delta} - (1 - \varepsilon)kh}, \frac{1}{\varepsilon}\frac{\delta}{h}\right];$$

$$\frac{\pi}{\delta} < l_0 < \frac{3\pi}{2\delta} \quad \text{at} \quad \frac{1 + \varepsilon\frac{\delta}{h}kh}{\varepsilon^2\frac{h}{\delta} - (1 - \varepsilon)kh} \geq \frac{\tanh kh}{kh} > \frac{1}{\varepsilon}\frac{\delta}{h};$$

$$l_0 < \frac{\pi}{2\delta} \quad \text{at} \quad \frac{1}{\varepsilon}\frac{\delta}{h} \geq \frac{\tanh kh}{kh} > \frac{1}{\varepsilon}\frac{\delta}{h}\left[1 + \frac{\pi^2}{4\varepsilon(\varepsilon + k\delta)}\right]^{-1};$$

$$\frac{\pi}{2\delta} < l_0 < \frac{\pi}{\delta} \quad \text{at} \quad \frac{\tanh kh}{kh} < \frac{1}{\varepsilon}\frac{\delta}{h}\left[1 + \frac{\pi^2}{4\varepsilon(\varepsilon + k\delta)}\right]^{-1}, \tag{3-2-13}$$

where $\epsilon = N^2\delta/2g \approx \Delta\rho/2\rho$ is half the relative density change in the pycnocline, that is, a small quantity on the order of 10^{-3}.

The first of these inequalities for k can be satisfied only at $\delta < \frac{\varepsilon^2}{1+\varepsilon}h$, that is, with a very thin pycnocline (an "almost two-layer" model of the oceanic stratification), and at $k < \frac{\varepsilon^2}{1+\varepsilon}\frac{1}{\delta}$, that is, for waves that are not too short. With an "equals" sign in the first condition of (3-2-13) for k we also have a "zeroth mode" $l = 0$ [in which there is a linear decrease of $w(z)e^{-N^2z/2g}$ with depth in the pycnocline] that can occur at only one specific value of k. With an inequality sign, on the other hand, this condition corresponds to a range of not too large k at which there exists another "imaginary mode" $l = i\hat{l}$ in which $\hat{l} < \frac{1}{\delta}(\varepsilon + k\delta)^{-\frac{1}{2}}\left(\varepsilon - \frac{k\delta}{\operatorname{th} kh}\right)^{-\frac{1}{2}}$. The frequency ω corresponding to the "imaginary mode" is higher than the frequencies of all real modes with the same k, but is still smaller than N.

For waves that are long compared to the pycnocline thickness (i.e., at $k\delta \ll 1$), the function $f(l)$ in Eq. (3-2-12) can be approximated by $(1 + \coth kh/l\delta)k\delta$, so that $\tan l\delta$ will also be small, and we can set $\tan l_0\delta \approx l_0\delta$ for the lowest mode. Then Eq. (3-2-12) will assume the form $(l_0\delta)^2 \approx k\delta(1 + \coth kh)$, and we obtain for the square of the corresponding frequency $\omega_0^2 \approx N^2 \frac{k^2}{l_0^2}$ [3]

$$\omega_0^2 \approx \frac{\Delta\rho}{\rho} \frac{gk}{1 + \coth kh}. \tag{3-2-14}$$

Since $l_0\delta$ is small, the variations of w over the depth of the pycnocline will also be small, that is, it will oscillate as a single entity in the lowest long-wave mode. Then the second term in Eq. (3-2-9) can be neglected for the pycnocline and the longitudinal-velocity gradient in the wave will take the form $\frac{\partial u}{\partial z} = \frac{i}{k}\frac{\partial^2 w}{\partial z^2} \approx ik\frac{\omega_0^2 - N^2}{\omega_0^2} w(h)$. Assuming $w(h) = -i\omega_0\zeta_h$ (where ζ_h is the amplitude of oscillations of the pycnocline), we obtain the Richardson number for the lowest mode in the pycnocline [3]:

$$\mathrm{Ri} = N^2\left(\frac{\partial u}{\partial z}\right)^{-2} \approx \left(\frac{\omega_0}{N} - \frac{N}{\omega_0}\right)^{-2} k^{-2}\zeta_h^{-2}. \tag{3-2-15}$$

Finally, using Eq. (3-2-8) and the first relation in Eq. (3-2-10), we obtain $u_s \approx \omega_0\zeta_h/\sinh kh$ for the longitudinal velocity at the ocean surface. The maximum convergence of u_s will occur above the nodes of the wave in the pycnocline, where bands of foam perpendicular to the direction of wave propagation are sometimes observed (for one description, see LaFond [5]).

If $l\delta \gg \varepsilon$, as is the case for the higher modes and often also for the lowest mode, relation (3-2-11) can be approximated by $\omega \approx \pm N \frac{k}{\sqrt{k^2 + l^2}} = \pm N\cos\theta$, where θ is the angle between the three-dimensional wave vector $\varkappa = (k_1, k_2, l)$, that is, the direction of wave propagation, and the horizontal plane. Thus, waves with frequencies approaching N propagate almost horizontally, but waves with very small ω/N (higher modes) propagate almost vertically (i.e., different layers of water periodically shift in horizontal directions relative to one another, the periods of this motion being large). This frequency dependence of the wave-propagation direction is a manifestation of wave *dispersion.* Their phase velocity $c = \omega/\varkappa \approx Nk/k^2 + l^2$ is always smaller than

$N/k = L/\tau$ ($L = 2\pi/k$ is the wavelength and $\tau = 2\pi/N$) and decreases with increasing mode number (e.g., at τ of about 10 min, waves with lengths in the kilometer range propagate at velocities of no more than a few meters per second), and the group velocity $\partial\omega/\partial\varkappa$ is perpendicular to $\varkappa$ and exceeds c by a factor of $\tan\theta$.

Because of the slowness of propagation of internal waves, their behavior may be strongly influenced by currents, and especially by the presence of a vertical current velocity gradient $\Gamma = \partial\bar{u}/\partial z$. Let the current be dynamically stable, that is, let the Richardson number $\mathrm{Ri} = (N/\Gamma)^2$ be large (Miles and Howard [6], state the stability condition as $\mathrm{Ri} > 1/4$, i.e., $|\Gamma| < 2N$). Internal waves with large vertical wave numbers will be sufficiently well characterized in this case by a local description in which N and Γ are assumed to be quasiconstant, whereas Eq. (3-2-7) will hold for description of internal waves in a system that moves with the current at the reference depth provided one substitutes $\frac{d}{dt} = \frac{\partial}{\partial t} + \Gamma z \frac{\partial}{\partial x}$ for $\partial/\partial t$ and adds the term $\frac{\Gamma N^2}{g}\frac{d}{dt}\frac{\partial w}{\partial x}$ in the right-hand side. The modified equation will then have solutions of the form $w = w(t)e^{-\frac{N^2 z}{2g} + ik_x(x-\Gamma z t) + ik_y y}$, that is, they will be waves with vertical wave numbers $l = -k_x\Gamma t$ that increase in time and, consequently, with a three-dimensional wave vector $\varkappa = (k_x, k_y, -k_x\Gamma t)$ that gradually rotates from an initially horizontal to a vertical position (since a current with a velocity gradient rotates the constant-phase planes of the waves toward the horizontal plane and tends to bunch them closer together). The frequency $\omega(t)$ of the wave decreases in time, and in the approximation considered above, in which $\omega \approx N\cos\theta$, the modification of Eq. (3-2-7) indicated here yields (see [3]) $w(t) \sim \omega^{3/2} e^{i\int\omega dt}$, so that $|w|^2 \sim \omega^3$ also decreases with time. In this case the kinetic energy of the wave is transferred to the current.

If there exists a stationary source of internal waves in a current with a velocity gradient, the above-described evolution of wave frequencies $\omega(t)$ and the squared amplitudes $|w(t)|^2 \sim \omega^3(t)$ produces a stationary spectrum of fluctuations of energy of vertical velocity $E_w(\omega)$, whereby waves that exhibit the frequency range $(\omega, \omega + d\omega)$ during a time dt will contribute to the fluctuation energy $\overline{w^2}$. Here $E_w(\omega)d\omega$ is proportional to ω^3 and dt. Since $d\omega = -\frac{k_x\Gamma}{kN^2}(N^2 - \omega^2)^{1/2}\omega^2 dt$, the spectrum is obtained in the form

$$E_w(\omega) \sim (N^2 - \omega^2)^{-1/2}\omega, \tag{3-2-16}$$

and the spectra of the vertical displacement ζ, of the temperature fluctuations $T' \approx \frac{\partial\overline{T}}{\partial z}\zeta$, and of the potential and kinetic energies are obtained from Eq. (3-2-16) by dividing by ω^2 (these results differ from those given by Phillips [3]; Frankignoul [7] derived them with allowance for the earth's rotation). Needless to say, spectral formulas of this kind cease to be useful at very small ω, when the vertical velocity gradients in the wave become very large and may generate turbulence (and, in addition, the effects of the earth's rotation may become significant). Figure 3-2-2 shows temperature fluctuation spectra at a depth of 70 m in the thermocline at Γ of about $4 \cdot 10^{-3}$ sec^{-1} and N of about $3 \cdot 10^{-3}$ sec^{-1}. These were calculated, assuming an inertial frequency of about 10^{-4} sec^{-1}, by Kitaygorodskiy et al. [8] in the tropical area of the Atlantic on the second voyage of the R/V *Dmitriy Mendeleyev* by means of an array of photoelectric thermographs on three anchored buoy stations about 5 miles apart. In the high-frequency range, these spectra obey $E_T(\omega) \sim \omega^{-1}$, which follows from Eq. (3-2-16).

Nonlinear interactions may be a very important factor in the dynamics of internal waves, even more important than in the case of gravity waves on the surface of a deep ocean. This is because the most efficient *resonant* interactions can occur even among

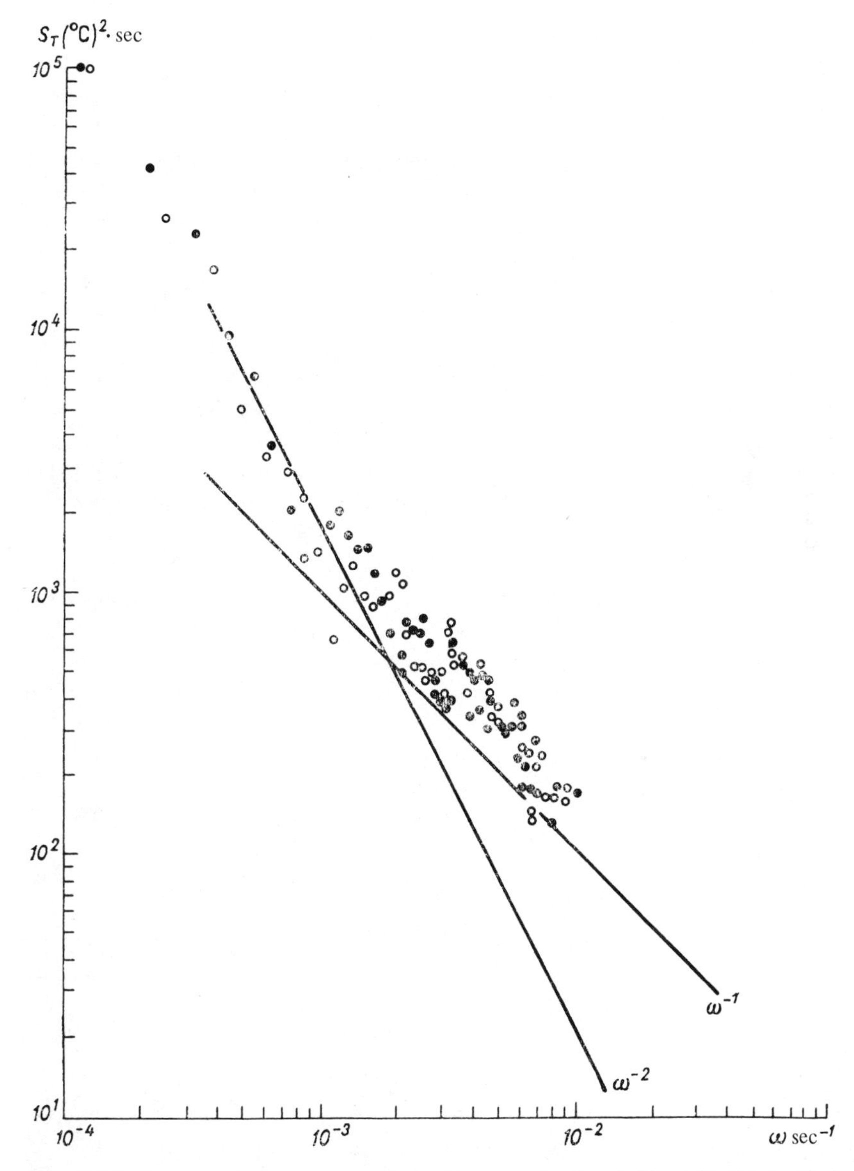

FIG. 3-2-2 Spectra of temperature fluctuations at depth of 70 m in the tropical Atlantic at $\Gamma \sim 4 \cdot 10^{-3}$ sec^{-1} and $N \sim 3 \cdot 10^{-3}$ sec^{-1}. From measurements made with photoelectric thermographs at three neighboring buoy stations (after Kitaygorodskiy, Miropol'skiy, and Filyushkin [8]). Different symbols denote measurements made at different stations.

wave triplets. Thus the dispersion relation $\omega = \omega(\varkappa)$ for internal waves, which is given by Eq. (3-2-11) or by the simplified expression $\omega = \pm N \cos\theta$, admits satisfaction of the resonance conditions

$$\omega(\varkappa) = \omega(\varkappa_1) \pm \omega(\varkappa_2) \quad \text{at} \quad \varkappa = \varkappa_1 \pm \varkappa_2. \tag{3-2-17}$$

Also possible are three-wave resonant interactions between surface and internal waves. The three-wave interactions were calculated by Thorpe [9], Phillips [3, 10], and Brekhovskikh et al. [2]. It was found, among others, that the typical resonant interaction time is on the order of $\tau \sim [(\varkappa_1 w_1)(\varkappa_2 w_2)]^{-1/2}$ and is large compared to $2\pi/N$ (since the periods $2\pi/\omega$ of the oscillations are large). There are also nonresonant interactions between internal waves. In this case each pair of waves $\boldsymbol{\varkappa}_1$, $\boldsymbol{\varkappa}_2$ generates "forced modes," that is, waves $\boldsymbol{\varkappa}=\boldsymbol{\varkappa}_1 \pm \boldsymbol{\varkappa}_2$ with frequencies $\omega=\omega(\boldsymbol{\varkappa}_1) \pm \omega(\boldsymbol{\varkappa}_2)$ that do not satisfy the dispersion relation (i.e., $\omega \neq \omega(\boldsymbol{\varkappa})$. If $\tau \gg N^{-1}$ the amplitudes of the forced modes are small, but when $\tau \sim N^{-1}$ they become comparable with the amplitudes of the original waves and perhaps larger, and the interactions of these forced modes with one another and with free wave modes will generate a spectrum of vortical oscillations that do not satisfy any specific dispersion relation, that is, turbulence. If $E(k)$ is the spectral density of the kinetic energy of the oscillations (per unit mass), we can set $\tau=[k^3 E(k)]^{-1/2}$, and treat $E(k)$ as the spectrum of the interacting internal waves at $\tau \gg N^{-1}$ or as the turbulence spectrum at $\tau \not> N^{-1}$ [11].

As in the case of surface waves, nonlinear interactions between internal waves can, in theory, compensate the tendency of waves packets to lose identity because of dispersion of their constituent elementary harmonic waves, that is, there may exist *steady-state* internal waves of finite amplitude. A limiting specific case of these waves is a harmonic wave in the pycnocline ($N \approx$ const) that depends on the spatial coordinates as $e^{i\,\boldsymbol{\varkappa}\cdot\mathbf{x}}$. At any amplitude, this wave is the exact solution of the *nonlinear* dynamic equations, since at $\mathbf{u} \sim e^{i\,\boldsymbol{\varkappa}\cdot\mathbf{x}}$ the nondivergence condition for the velocity field assumes the form $\boldsymbol{\varkappa}\cdot\mathbf{u}=0$ and all nonlinear terms vanish in the equations, which in this case have the form $(\mathbf{u}\cdot\nabla)e^{i\,\boldsymbol{\varkappa}\cdot\mathbf{x}}$.

Magaard [12] calculated steady-state internal waves of finite amplitude for arbitrary stratification. For any hydrodynamic characteristic $\eta(x-ct,\ z)$ of steady-state plane waves propagating along the x axis in the plane x, z at velocity c we obtain $\dfrac{d\eta}{dt}=\dfrac{\partial(\eta,\psi)}{\partial(x,z)}$, where $\psi(x-ct,\ z)$ is the stream function, which is introduced by the relations $u-c=\dfrac{\partial\psi}{\partial z}$, $w=-\dfrac{\partial\psi}{\partial x}$, so that any characteristic conserved in the motions of the fluid parcels (i.e., satisfying the equation $\dfrac{\partial\eta}{\partial t}=0$) is an arbitrary function of ψ. Treating the motion as isopycnic (which, incidentally, is not rigorous treatment: it would be more correct to consider them isentropic), that is, setting $\rho=\rho(\psi)$ and introducing the new variable $\varphi=\int\rho^{1/2}d\psi$ we reduce the equations of motion, after eliminating pressure terms (see [4, Sec. 174]), to the form $\Delta\varphi+F(\varphi)=gZ\dfrac{d\rho}{d\varphi}$ where $F(\varphi)$ is an arbitrary function. If, for example, we set $F(\varphi)=a^2\varphi$ and $\rho(\varphi)=\rho_0+b\varphi$, where a, b and ρ_0 are constant, it is easy to obtain wave solutions for φ. Thus, with the boundary conditions $w=0$ at $z=0, H$ the functions

$$\varphi_n = A_n \sin\frac{\pi n z}{H}\sin k_n(x-ct)+\frac{g b_n z}{a_n^2};\quad k_n^2=a_n^2-\frac{\pi^2 n^2}{H^2} \tag{3-2-18}$$

will be the solutions.

In the upper layer of the ocean, the corresponding isopycnic lines at $n=1$ (which were calculated by Magaard) are waves with flat crests and sharp troughs (and vice versa in the bottom

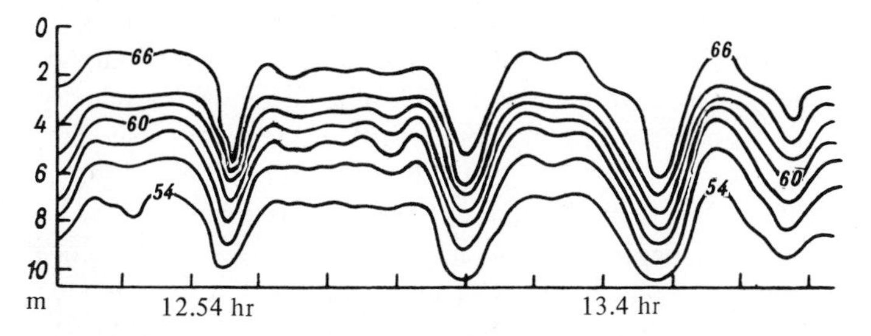

FIG. 3-2-3 Steady-state internal waves of finite amplitude in the shallow thermocline near San Diego, as observed by LaFond [7]. Isotherms in degrees Fahrenheit.

layer); Thorpe [9] obtained a similar result by an approximate method for the case of a shallow pycnocline. This prediction of the shape of the internal waves is confirmed by LaFond's data [5], which are represented in Fig. 3-2-3.

We can postulate several possible mechanisms for generating internal waves. Thus, tide-producing forces generate internal waves with tidal periods. Flows over roughnesses on the sea floor give rise to the so-called obstacle waves, which are standing or, in the case of a periodically varying current (tidal or created by long surface waves), traveling; they correspond to the boundary condition at the sea floor $w = \mathbf{u} \cdot \nabla H$ at $z = H(x, y)$. The theory of these waves was developed by Long [13] and Yih [14, 15]. Internal waves can be generated by time-dependent variations of the atmospheric pressure at the sea surface. In the linear theory, this latter effect is described by adding the term $\frac{1}{\rho_0} \Delta_h \frac{\partial p_a}{\partial t}$ to the right-hand side of the boundary condition (3-2-7) for the sea surface. This mechanism is analogous to the resonant excitation of surface waves by atmospheric-pressure fluctuations. (For calculations, see Bryant's paper [16].) Internal waves may also be excited by wind friction at the sea surface. In calculating such waves, it is best to allow for eddy viscosity in the equations of motion (see [4, Sec. 165]).

A significant factor may be generation of internal waves by nonlinear resonant interaction between pairs of surface waves $\mathbf{k}_1$, $\mathbf{k}_2$ with internal waves $\mathbf{k} = \mathbf{k}_1 - \mathbf{k}_2$ the latter decay with depth z as $e^{-|\mathbf{k}_1 - \mathbf{k}_2| z}$, that is, very slowly at $|\mathbf{k}_1 - \mathbf{k}_2| h \ll 1$ (h is the depth of the pycnocline). Such interactions were calculated in [2, 3, 7, 9]. According to [9], the initial rate of increase of the amplitude a of the internal wave with the resonant frequency $\omega(\mathbf{k}) = \omega(\mathbf{k}_1) - \omega(\mathbf{k}_2)$ is

$$\frac{\partial a}{\partial t} = \frac{\omega(\mathbf{k})}{2\sqrt{k_1 k_2}} \left(1 + \frac{\mathbf{k}_1 \cdot \mathbf{k}_2}{k_1 k_2}\right) (a_1 k_1)(a_2 k_2) e^{-|\mathbf{k}_1 - \mathbf{k}_2| \cdot h}. \quad (3\text{-}2\text{-}19)$$

Internal waves may also degenerate via several different mechanisms. Their damping by the molecular viscosity of the water, which is described by $e^{-\nu\varkappa^2 t}$, where ν is the viscosity, proves to be very slow, since the wave numbers $\varkappa$ of internal waves are small (but damping becomes much faster in turbulent zones where the effective viscosity is the eddy viscosity rather than ν). Internal waves may overturn when the local accelerations $\omega^2 a$ in them reach the order of g, but this is almost impossible because of their usually small frequencies ω. However, in the presence of a current with a velocity gradient Γ, overturning may be produced by horizontal drift of the wave crests. This "convective instability" was examined by Orlanski and Bryan [17], who found that it can occur at $\mathrm{Ri} \leqslant 1 + k^2/l^2$. The hydrodynamic instability of internal waves is an efficient mechanism for producing degeneracy (the net result is turbulence) in regions in which the Richardson number in the waves is smaller than 1/4. For lowest-mode waves in the pycnocline at $\omega \ll N$, the stability limit $\mathrm{Ri} = 1/4$ reduces, according to Eq. (3-2-15), to the form $\zeta_h^2 = \dfrac{4\omega^2}{k^2 N^2}$. The two-dimensional spectrum of the pycnocline oscillations, which is bounded by this limit, is determined from the relation $E_\zeta(k) \sim \zeta^2/k^2$, which can be reduced, with the aid of expression (3-2-14) for the frequency ω, to the form [3]

$$E_\zeta(k) \sim \delta (1 + \coth kh)^{-1} k^{-3}. \tag{3-2-20}$$

For short waves ($kh \gg 1$, and ω^2 proportional to k), this spectrum is proportional to k^{-3}; for long waves ($kh \ll 1$, and ω^2 proportional to k^2), it is proportional to k^{-2}. The one-dimensional spectrum is obtained from these results by multiplying by k. The frequency spectrum $E_\zeta(\omega) \sim \zeta^2/\omega$ is proportional to ω^{-3} for short waves and to ω^{-1} for long waves. It is possible that the temperature fluctuation spectra $E_T(\omega) \sim \omega^{-3}$ that are often measured in the sea are generated by short internal waves that have reached the hydrodynamic stability limit indicated here.

It is very difficult to measure internal waves in the ocean. The most effective method now known is to measure them in the temperature field with the aid of thermistor chains, that is, chains of temperature sensors that are suspended from anchor buoys or towed with a sinker behind a moving ship [18]. If the instruments are towed, the measurements must be corrected for the Doppler effect created by the motion of the ship [19, 20]. Thus, the correlation function of random internal waves in the temperature field at a fixed depth has the following form, when calculated from towing measurements:

$$B(\tau) = \sum_n \int \exp\left[i\tau \mathbf{v} \cdot \left(\mathbf{k} - \omega_n(k) \frac{\mathbf{v}}{v^2} \right) \right] E_n(\mathbf{k})\, d\mathbf{k}, \tag{3-2-21}$$

where τ is the time shift, $\mathbf{v}$ is the towing speed, and $\omega_n(k)$ and $E_n(\mathbf{k})$ are the frequency and the two-dimensional spatial spectra of the internal waves of mode $\mathbf{n}$. By determining the value of this function at several different towing speeds $\mathbf{v}$ and by extrapolating to $\mathbf{v} = 0$, one obtains the true (undistorted by the Doppler effect) correlation function $B_0(\tau)$. Let $E_0(\omega)$ be its Fourier transform with respect to τ, that is, the true frequency spectrum. Further, regarding $B(\tau)$ as a function of $\mathbf{r} = \mathbf{v}$ and v and extrapolating it to the value $v = \infty$ for a fixed towing direction $\mathbf{v}/v$, one obtains the one-dimensional spatial correlation function $B\left(r, \frac{\mathbf{v}}{v}\right)$ whose Fourier transform with respect to r is the sum of the one-dimensional spectra of all modes, $\sum_n E_n\left(k, \frac{\mathbf{v}}{v}\right)$. When only the lowest internal-wave mode is significant, one obtains from Eq. (3-2-21)

$$\int_0^{\omega_1(k)} E_0(\omega)\, d\omega = \pi \int_0^k \widetilde{E}_1(k)\, k\, dk, \tag{3-2-22}$$

where $\widetilde{E}_1(k)$ is the two-dimensional spectrum $E_1(\mathbf{k})$, averaged over all directions $\mathbf{k}/k = \mathbf{v}/\mathrm{v}$ of the wave vector $\mathbf{k}$. This relation can be used for empirical determination of the dispersion relation for the lowest mode $\omega = \omega_1(k)$ by using the above procedure to measure the functions $E_0(\omega)$ and $\widetilde{E}_1(k)$. An attempt at applying this procedure was made by Miropol'skiy and Filyushkin [11], who obtained an empirical dispersion relation $\omega_1(k)$ that agrees well with the prediction obtained for the three-layer model (3-2-10). In this case the spatial spectrum was approximately isotropic and generally decreased more rapidly with increasing k than in Phillip's law (3-2-20), that is, it was not "saturated" (see the example in Fig. 3-2-4). If not only the lowest mode, but other modes are also significant, then one must work with measurements made simultaneously at several depths. No complete calculations of this type have yet been made.

A useful statistical characteristic of the random wave field is the cross-frequency spectrum $E_{\xi\eta}(\omega)$ of the fluctuations $\xi(t)$ and $\eta(t)$ of a pair of hydrodynamic variables of this field (velocity components, temperature, salinity, etc.), that is, the Fourier transform of their cross-correlation function $B_{\xi\eta}(\tau) = \langle \xi(t)\eta(t+\tau) \rangle$ (here ξ and η can be understood as the values of the same variable at different points in space, or of different variables at the same or different points). The real part $C_{\xi\eta}(\omega)$ of the function $E_{\xi\eta}(\omega)$, that is, the Fourier transform of the even part $\frac{1}{2}\left[B_{\xi\eta}(\tau) + B_{\xi\eta}(-\tau)\right]$ of the cross-correlation function, is known as the cospectrum, while the imaginary part $Q_{\xi\eta}(\omega)$, that is, the Fourier transform of the odd part $-\frac{1}{2}[B_{\xi\eta}(\tau) - B_{\xi\eta}(-\tau)$ is called the quadrature spectrum of the fluctuations $\xi(t)$ and $\eta(t)$. If we set

$$E_{\xi\eta}(\omega)\left[E_{\xi\xi}(\omega)E_{\eta\eta}(\omega)\right]^{-1/2} = \mathrm{Co}_{\xi\eta}(\omega)\, e^{i\Phi_{\xi\eta}(\omega)}, \tag{3-2-23}$$

the absolute value of the left-hand side $\mathrm{Co}_{\xi\eta}(\omega)$ is called the coherence, and $\Phi_{\xi\eta}(\omega)$ is the spectrum of the phase shift between $\xi(t)$ and $\eta(t)$. Thus, for example, according to Eq. (3-2-8), the phases of fluctuations of the variables u, v, ζ, p, ρ, and T in a field of small-amplitude internal waves are shifted, at all frequencies, by $\pi/2$ with respect to the fluctuations of w at the same point in space. This constitutes the sharp distinction between internal waves and turbulence.

A summary of data on the spectra and vertical and horizontal coherence of internal waves is given by Garrett and Munk [21] (see also [11] and Navrotskiy's paper [22]); the data on the frequency spectra $E_u(\omega)$ and the one-dimensional spatial spectra $E_\zeta(k_1)$ are given in Fig. 3-2-5. On the basis of this summary, Garrett and Munk proposed a simplified model for the spectral energy density $E(\mathbf{k}, \omega)$ of internal waves, assuming them, first of all, to be multimodal (since the one-mode model cannot explain the observed decay of coherence at vertical distances on the order of 10^1 to 10^2 m) and replacing the discrete set of modes $\omega_n(k)$ by an equivalent continuum $f \leqslant \omega \leqslant \omega_1(k)$. Here $\omega_1(k)$ corresponds to the lowest mode and f is the inertial frequency (Coriolis parameter). Further, recognizing that resonant three-wave interactions generate a tendency to isotropy, the spectrum was assumed isotropic, that is, $E(\mathbf{k}, \omega) = \dfrac{E(k, \omega)}{2\pi k}$.

Calculation of the vertical coherence indicates that the width of the band of wave numbers k of internal waves with a fixed frequency ω is a function of the frequency $\mu(\omega)$. Garrett and Munk assumed self-preservation of the spectrum $E(k, \omega) \sim \dfrac{1}{\mu} A\left(\dfrac{k}{\mu}\right) E(\omega)$ and made $A(\chi)$ an elementary function that equals unity at $0 \leqslant \chi \leqslant 1$ and zero at $\chi > 1$ (although, strictly speaking, it need not

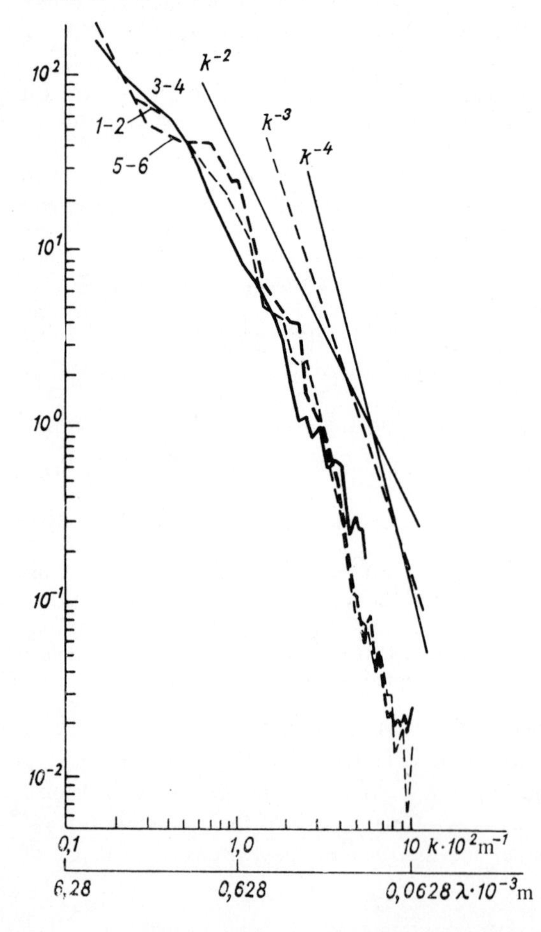

FIG. 3-2-4 One-dimensional spatial functions $E\left(k, \dfrac{\mathbf{v}}{v}\right)$ of temperature fluctuations in the thermocline of the tropical Atlantic, measured by towing a thermistor chain on three tacks of the ship [tacks 1-2 and 3-4 were perpendicular to tack 5-6] (after Miropol'skiy and Filyushkin [11]).

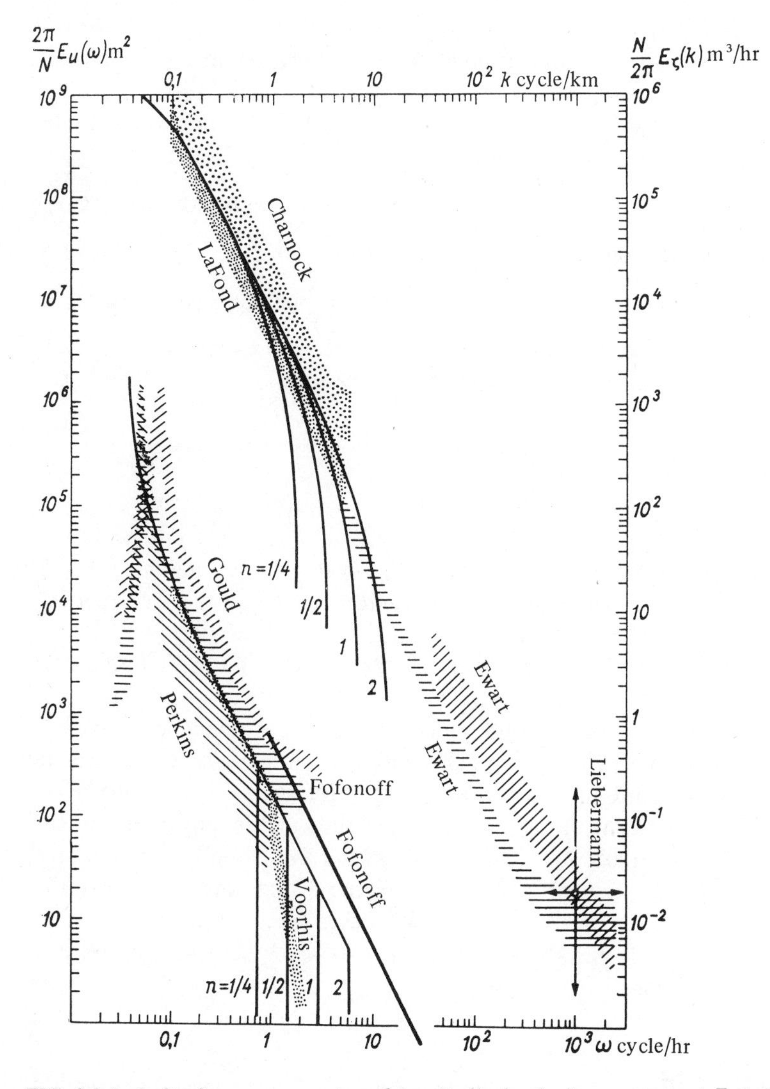

FIG. 3-2-5 Left: frequency spectra of longitudinal velocity component $E_u(\omega)$ measured from anchored buoys (Voorhis' spectra were measured from drifting buoys of neutral buoyancy); right: one-dimensional spatial spectra $E_\zeta(k)$ of the vertical displacements of isotherms, measured with towed thermistor chains (after Garrett and Munk [21]). The curves were plotted from the theoretical expressions, assuming the values of parameter n indicated on the curves; these expressions are easily obtained from Eq. (3-2-24). For details of the calculations and a description of the observations reported, see [21]. The workers who obtained the spectra are indicated on the figure.

necessarily be constant at $0 \leqslant \chi \leqslant 1$ and must vanish only at $k > k_1(\omega) > \mu(\omega)$, where $k_1(\omega)$ corresponds to the lowest mode. They assumed the power-law formula $E(\omega) \sim \omega^{-p+2s}(\omega^2 - f^2)^{-s}$ for the frequency factor, in which case $0 < s < 1$ to achieve convergence of the energy integral; the value $s = 1/2$ was chosen for the sake of argument. A similar power-law formula was selected for $\mu(\omega)$ $\sim j\pi\left(\frac{\omega}{f}\right)^{r-1}(\omega^2 - f^2)^{1/2}$; then j becomes the equivalent number of modes at the inertial frequency, and the one-dimensional spatial spectrum $E(k_1)$ turns out to be approximately proportional to $k_1^{-\frac{p+r-1}{r}}$. From data on spectra, such as Fig. 3-2-5, Garrett and Munk selected the values $p = 2$ and $r = 1$ for the exponent on k. Thus, their model of the dimensionless energy spectrum has the form

$$E(k, \omega) = \frac{2Ef}{j\pi^2}\omega^{-1}(\omega^2 - f^2)^{-1};$$

$$f \leqslant \omega \leqslant N; \quad 0 \leqslant k \leqslant j\pi(\omega^2 - f^2)^{1/2}, \tag{3-2-24}$$

where the frequencies are measured in units of N_m cycles per second and the wave numbers in units of M cycles per second, so that the total energy of the waves per unit area of sea surface equals $\rho E/2\pi M^3 N_m$.

Garrett and Munk recommend the values $E = 2\pi \cdot 10^{-5}$ and $j = 20$. In this case, at $M = 1.22 \cdot 10^{-6}$ cycles $\cdot$ cm^{-1} and $N_m = 0.83 \cdot 10^{-3}$ cycles $\cdot$ sec^{-1}, the total energy is 0.4 J/cm^2. In [23], Garrett and Munk used their spectral model to estimate the mixing that results from hydrodynamic instability of internal waves. The method involved finding the Richardson number $\mathrm{Ri} = 2/\pi^2 j^2 E(N/N_m)^{-1}$ from the mean-square vertical velocity gradient in the waves and assuming that mixing occurs at $\mathrm{Ri} < 1/4$ and, according to Thorpe's data [24], results in the formation of a mixed layer with an average Richardson number of 0.4 ± 0.1. We note that under this condition, the kinetic energy dissipation rate of the internal waves (i.e., the rate of turbulent energy generation) equals $\frac{5}{2}K(2\pi N)^2$, where K is the mixing coefficient and N is measured in cycles per second.

REFERENCES

1. Bockel, M. Oceanographic studies of the *Origny* at Gibraltar, Cahiers Oceanographiques, **14**, No. 4, 325–329 (1962).
2. Brekhovskikh, L. M., V. V. Goncharov, V. M. Kutepov, and K. A. Naugol'nykh. Resonant excitation of internal waves in nonlinear interaction of surface waves, Izv. Akad. Nauk SSSR, Fizika atm. i okeana, 8, No. 2, 192–203 (1972).
3. Phillips, O. *The Dynamics of the Upper Ocean*, Cambridge University Press (1966).
4. Krauss, W. *Internal Waves*, Translated from the German, Gidrometeoizdat Press, Leningrad (1968).
5. LaFond, E. C. Internal Waves. In: *The Sea*, **1**, 731–763 (1962).
6. Miles, J. W. On the stability of heterogeneous shear flows, J. Fluid Mech., **10**, Part 3, 496–508 (1961). Howard, L. N. Note on a paper of John W. Miles, J. Fluid Mech., **10**, Part 3, 509–510 (1961).
7. Frankignoul, C. J. The effect of weak shear and rotation on internal waves, Tellus, **22**, No. 2, 194–204 (1970).

8. Kitaygorodskiy, S. A., Yu. Z. Miropol'skiy, and B. N. Filyushkin. Use of data on sea temperature fluctuations to distinguish internal waves from turbulence, Izv. Akad. Nauk SSSR, Fizika atm. i okeana, **9**, No. 3, 272–292 (1973).
9. Thorpe, S. L. On wave interactions in a stratified fluid, J. Fluid Mech., **24**, Part 4, 737–752 (1966).
10. Phillips, O. M. Theoretical and Experimental Studies of the Interactions of Gravity Waves. In: *The Nonlinear Theory of Wave Progagation*, Translated from the English, Mir Press, Moscow, 141–160 (1970).
11. Miropol'skiy, Yu. Z. and B. N. Filyushkin. An investigation of temperature fluctuations on the scale of internal gravity waves in the upper layer of the ocean, Izv. Akad. Nauk SSSR, Fizika atm. i okeana, 7, No. 7, 778–798 (1971).
12. Magaard, L. Toward a theory of two-dimensional nonlinear internal waves in stably stratified media, Kiel. Meeresforsch, **21**, No. 1, 22–32 (1965).
13. Long, R. R. Some aspects of the flow of stratified fluids, I. A theoretical investigation, III. Continuous density gradients, Tellus, **5** No. 1, 42–58 (1953); 7, No. 3, 341–357 (1955).
14. Yih, C. S. Gravity waves in a stratified fluid, J. Fluid Mech., **8**, Part 3, 481–508 (1960).
15. Yih, C. S. Exact solutions for steady two-dimensional flow of a stratified fluid, J. Fluid Mech., **9**, Part 1, 161–174 (1960).
16. Bryant, P. T. Wind Generation of Water Waves, Ph.D. Dissertation, Univ. Cambridge (1965).
17. Orlanski, I. and K. Bryan. Formation of the thermocline step structure by large-amplitude internal gravity waves, J. Geophys. Res., **74**, No. 28, 6975–6983 (1969).
18. Paka, V. T. A study of the temperature field in the upper layers of the ocean by continuous temperature registration along the path of a ship. In collection: *Use of Radiophysical Methods in Oceanographic and Ice Studies*, Trudy AANII, 50–62 (1964).
19. Laykhtman, D. L., A. I. Leonov, and Yu. Z. Miropol'skiy. Interpretation of measurements of the statistical characteristics of the scalar fields in the ocean in the presence of internal gravity waves, Izv. Akad. Nauk SSSR, Fizika atm. i okeana, 7, No. 4, 447–445 (1971).
20. Laykhtman, D. L., A. I. Leonov, and Yu. Z. Miropol'skiy. Determination of the two-dimensional statistical characteristics of a scalar field in the ocean in the presence of internal gravity waves, Izv. Akad. Nauk SSSR, Fizika atm. i okeana, 7, No. 6, 638–648 (1971).
21. Garrett, C. and W. Munk. Space-time scales of internal waves, Geophys. Fluid Dynam., **3**, No. 3, 225–264 (1972).
22. Navrotskiy, V. V. A statistical analysis of spatial temperature fluctuations in the surface layer of the ocean, Izv. Akad. Nauk SSSR, Fizika atm. i okeana, **5**, No. 1, 94–110 (1969).
23. Garrett, C. and W. Munk. Oceanic mixing by breaking internal waves, Deep-Sea Res., **19**, No. 12, 823–832 (1972).
24. Thorpe, A. Experiments on the instability of stratified shear flows: Visible fluids, J. Fluid Mech., **46**, Part 2, 299–320 (1971).

3.3 OCEAN TURBULENCE

The extensive measurements of the characteristics of ocean turbulence and the fine vertical structure of the hydrophysical fields in it (temperature, electrical conductivity, sound velocity, refractive index, current velocities) that have been made in recent years have yielded a number of unexpected results.

It now appears that the sea is stratified practically everywhere and at all times into quasihomogeneous layers ranging in thickness from tens of meters to decimeters and centimeters. These are separated by very thin partings, within which the hydrophysical variables exhibit sharp (stepwise) vertical changes. Such layers have quite long lifetimes, at least in the tens of minutes and ranging up to hours. The turbulence is usually weak and incapable of breaking up this stratification; it develops only within quasihomogeneous layers, is local in nature (does not depend directly on depth), and is characterized by small Reynolds numbers.

Apart from their direct importance (for understanding of the nature and properties of short-period fluctuations of the hydrophysical fields in the sea), these results appear to have broad implications for oceanology in general. Thus they significantly modify past ideas on the vertical structure of the ocean and its natural vertical-mixing processes, and therefore the dispersion of various impurities in it (dissolved salts and gases, suspended mineral matter, plankton, radioactive materials).

Further, these results must be fitted into a new theory of the physical nature of small-scale internal motions in very stably stratified fluids, in which buoyant forces suppress turbulence, so that the latter can develop only locally, in regions with locally sharp velocity gradients. Such gradients appear to arise primarily in the internal gravity waves that develop intensively in stably stratified fluids. Under these conditions, the relative importance of molecular momentum, heat, and impurity transport probably increases, and the differences in the diffusion of these properties may become significant.

It is very difficult to record turbulent fluctuations of the current velocity components, temperature, conductivity, sound velocity, refractive index, or the other hydrothermodynamic variables in the sea. This requires highly sensitive, quick-response sea turbulence meters. A complicating factor is that when these instruments are towed behind a moving ship, the recorded traces of the natural fluctuations in the sea are distorted by oscillations of the instrument due to the pitching of the ship, yawing of the pods containing the turbulence meters, and the vibration of the tow cables, and, at high frequencies, by electrical noise. Because of these difficulties, serious studies of sea turbulence were not undertaken until the last decade, and very few turbulence observations have been accumulated so far. Reviews of the available information will be found in [1-5].

There is another difficulty in the fact that the frequency ranges of turbulent fluctuations and surface and internal waves overlap to a considerable extent. In general, therefore, the evaluation of the characteristics of the turbulence as such (defined as the *part of the natural fluctuations that is noncoherent with the waves*) requires that not only the aforementioned noise of mechanical and electrical origin, but also the fluctuations created by waves be filtered out of the trace recorded by the turbulence meter.

Fluctuations created by *surface* waves can be filtered out if some sort of wave recorder is used to record the oscillations of the sea surface $\xi(t)$ [or the oscillations of pressure at a certain depth] in synchronism with the total natural fluctuations $\zeta(t)$. This procedure was first used by Bowden and White [6], and Benilov and Filyushkin [7] developed it in detail, using the general method for linear filtration of stationary random processes that was set forth in Yaglom's paper [8]. Benilov and Filyushkin showed that the fluctuations $\widehat{\xi}(t)$ created by surface waves can be approximated in mean-square by the finite summation $\widehat{\xi_n}(t) = \sum_{k=1}^{n} \beta_k^{(n)} \zeta\left(t - t_k^{(n)}\right)$, where the coefficients $\beta_k^{(n)}$ are determined from equations

$$B_{\xi\zeta}\left(t_l^{(n)}\right)=\sum_{k=1}^{n}\beta_k^{(n)}B_{\zeta\zeta}\left(t_l^{(n)}-t_k^{(n)}\right),\quad l=1,2,\ldots,n. \tag{3-3-1}$$

Here $B_{\zeta\zeta}(\tau)$ is the correlation function of the waves $\zeta(t)$ and $B_{\xi\zeta}(\tau)$ is the cross-correlation function of the natural fluctuations and the waves. The mean-square filtration error $\sigma_n^2=\overline{[\hat{\xi}(t)-\xi_n(t)]^2}$ for this case is given by

$$\sigma_n^2=\int_{-\infty}^{\infty}\left|\frac{f_{\xi\zeta}(\omega)}{f_{\zeta\zeta}(\omega)}-\sum_{k=1}^{n}\beta_k^{(n)}e^{-it_k^{(n)}\omega}\right|^2 f_{\zeta\zeta}(\omega)\,d\omega, \tag{3-3-2}$$

where $f_{\zeta\zeta}(\omega)$ and $f_{\xi\zeta}(\omega)$ are the Fourier transforms of the correlation functions $B_{\zeta\zeta}(t)$ and $B_{\xi\zeta}(\tau)$, respectively.

Figure 3-3-1 shows an example of filtration from [9], where use was made of synchronous records of the natural fluctuations of temperature $\xi(t)$ at a depth of 0.5 m and of the sea-surface height $\zeta(t)$, measured with a Froude spar in the Mediterranean Sea. In this example, the "wave noise" $\hat{\xi}(t)$ was highest (on the order of 45%) at frequencies $\omega\sim 1.1$ to 1.4 rad/sec, which correspond to the maximum of the wave spectrum, and decreased rapidly with increasing frequency. Other statistical characteristics of the natural temperature fluctuations and the "filtered" fluctuations $\xi(t)-\hat{\xi}(t)$ (the part of the natural fluctuations that is noncoherent with the wave) were also compared in [9]. Figure 3-3-2 shows the probability distributions of these fluctuations [with 65 terms in the finite sum approximating $\hat{\xi}(t)$, which corresponded to a relative filtration error $\sigma_n^2/\sigma_{\hat{\xi}}^2$ of about 1%]; we note that the probability distributions were approximately Gaussian here for the wave heights $\zeta(t)$ and the "wave noise" $\hat{\xi}(t)$.

One cannot filter out the fluctuations due to *internal* waves in the same way, since the latter are not recorded separately. To determine whether it is at all possible to decompose a horizontally homogeneous random vector field $\mathbf{u}(x, y, z, t)$ (for example, a velocity field) into turbulent and wave components, Kolmogorov[1] recommends analysis of the horizontal-plane spectral representation of this field, defined by the vectorial random spectral measure $Z(M)$ (the M are sets in the plane of *horizontal* wave vectors $\mathbf{k}$; the measure Z also depends on the vertical coordinate z and the time t). In this procedure, the component $\mathbf{Z}^0 = 1/k^2\,([\mathbf{n}\cdot\mathbf{k}]\cdot\mathbf{Z})\,[\mathbf{n}\cdot\mathbf{k}]$ of measure $Z(M)$ in the horizontal plane orthogonal to $\mathbf{k}$ ($\mathbf{n}$ is the unit vector in the vertical direction) is separated at each fixed $\mathbf{k}$ from the component $\mathbf{Z}^1 = 1/k^2\,(\mathbf{k}\cdot\mathbf{Z})\mathbf{k}+(\mathbf{n}\cdot\mathbf{Z})\mathbf{n}$ that lies in the vertical plane containing $\mathbf{k}$. Then the field $\mathbf{u}^0=\int e^{i(k_x x+k_y y)}\,\mathbf{Z}^0(d\mathbf{k})$ will describe the *horizontal turbulence* (we shall return to its properties at the end of this section), and the field $\mathbf{u}^1=\int e^{i(k_x x+k_y y)}\,\mathbf{Z}^1(d\mathbf{k})$ will contain both the turbulence and the waves, which are clearly distinguishable only on nonintersecting segments of their frequency spectra. Here $\overline{|\mathbf{u}^2|}=\overline{|\mathbf{u}^0|^2}+\overline{|\mathbf{u}^1|^2}$ (the overbar denotes the mean value), but the components $\mathbf{u}^0$ and $\mathbf{u}^1$ can, in general, be correlated (even though this correlation would apparently decay with time in the approximation of linear dynamics).

Our best hope of separating turbulence and internal waves appears to lie in the use of phase relations (phase-shift spectra) between the fluctuations of the various spatial velocity components and the scalar fields, which are fixed in the case of internal waves, but are arbitrary in the case of turbulence. The development of an algorithm for filtration of internal waves from recorded traces of natural fluctuations is a task for the future.

---

[1] Personal communication.

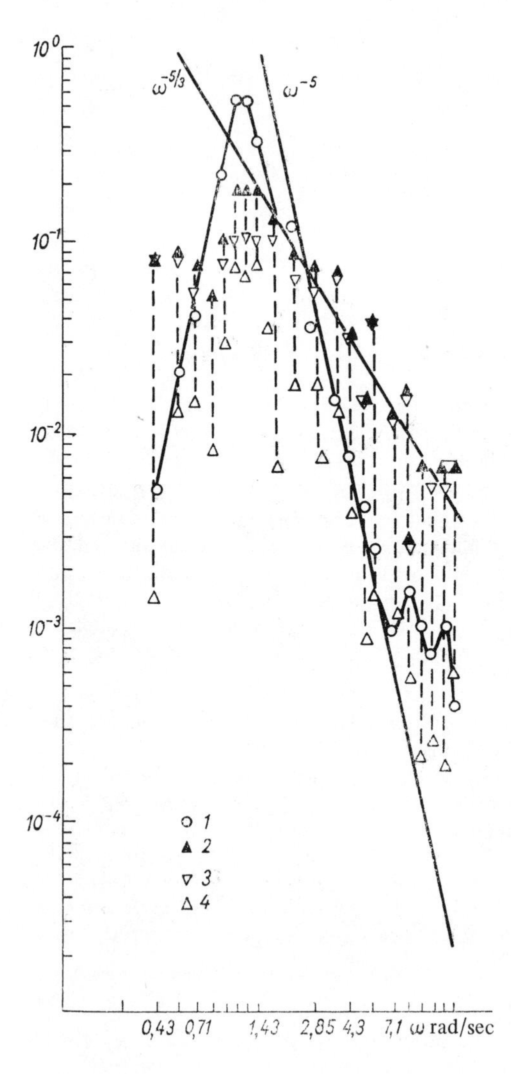

FIG. 3-3-1 Examples of normalized wave-height spectral densities $f_{\zeta\zeta}(\omega)/\sigma_{\zeta}^2$ (1), natural temperature fluctuations at depth of 0.5 m $f_{\xi\xi}(\omega)/\sigma_{\xi}^2$ (2), filtered fluctuations $f_{\xi\xi}(\omega)-f_{\hat{\xi}\hat{\xi}}(\omega)/\sigma_{\xi}^2$ (3), and "wave noise" $f_{\hat{\xi}\hat{\xi}}(\omega)/\sigma_{\xi}^2$ (4) (after Benilov [9]). The dashed arrows indicate values of the spectral density of the turbulent fluctuations at the particular frequency.

There are several possible mechanisms for generating small-scale turbulence in the sea [5]. Among them are hydrodynamic instability of the *horizontal* velocity gradients in mesoscale quasihorizontal motions (the instability being controlled by their Reynolds number), the instability of the *vertical* velocity gradients in the major (usually geostrophic) ocean currents, in drift currents in the upper ocean, in the bottom boundary layer (for example, in tidal currents), and in the internal-wave field (the latter mechanism is apparently the most important one throughout most of the depth of the ocean). Other small-scale turbulence generating processes are breaking of surface and internal waves; convection in layers with unstable density stratification (the latter resulting, among others, from cooling of the sea surface during the cold seasons and sometimes perhaps from the accumulation of salts in surface waters during periods of rapid evaporation).

In the case of stable density stratification $\partial\rho/\partial z > 0$ that is usual in the ocean (strictly speaking, ρ here is the *potential* density ρ_p), the vertical velocity gradients $\partial u/\partial z$ are unstable (and generate turbulence) if the *Richardson criterion* is satisfied:

$$\mathrm{Ri} = \frac{g}{\rho}\frac{\partial\rho}{\partial z}\left(\frac{\partial u}{\partial z}\right)^{-2} = \left(\frac{N}{\partial u/\partial z}\right)^2 < \mathrm{Ri}_{\mathrm{cr}}. \tag{3-3-3}$$

In geostrophic currents, $\dfrac{\partial V}{\partial z} \sim \dfrac{g}{f\rho}\dfrac{\partial\rho}{\partial r} \sim \dfrac{KN^2}{f}$ (r is the coordinate along a current with velocity V, and K is the slope of the isopycs), and the Richardson criterion assumes the form $KN > (\mathrm{Ri}_{\mathrm{cr}})^{-1/2}f$. In internal waves in the pycnocline, the minimum local Richardson number is $\mathrm{Ri} = a^{-2}k^{-2}\left(\dfrac{N_m}{\omega} - \dfrac{\omega}{N_m}\right)^{-2}$, where a is the amplitude of the wave and N_m is the peak $N(z)$. We have indicated in Sec. 3-2 that the condition $\mathrm{Ri} > 1/4$ is a sufficient stability criterion, so that when $ak > 2\left|\dfrac{N_m}{\omega} - \dfrac{\omega}{N_m}\right|^{-1}$ the waves can be unstable near their crests and troughs. It is at crests and troughs that patches of turbulence appear to originate within the main body of the sea.

On the basis of its turbulence, the sea can be divided into three layers, to wit: 1) an upper mixed layer (above the layer of the density discontinuity) with a thickness of

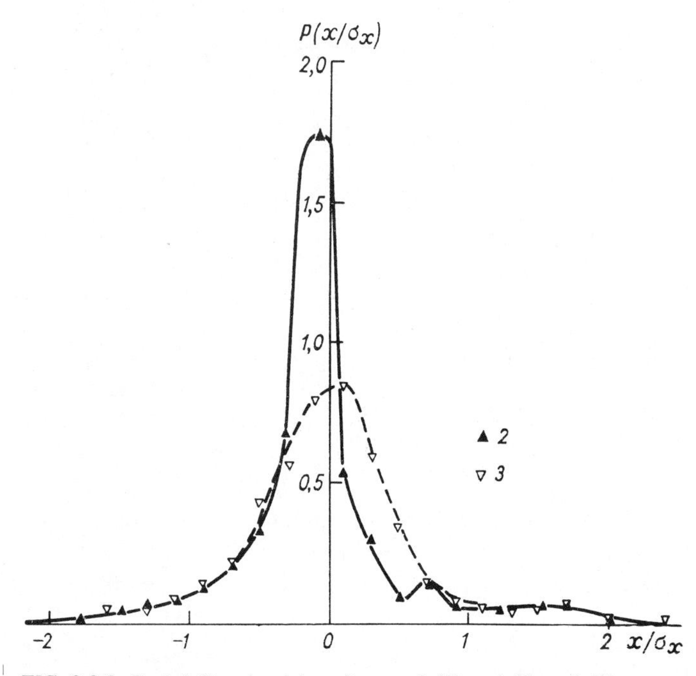

FIG. 3-3-2 Probability densities of natural (2) and filtered (3) temperature fluctuations (normalized to σ_ξ and $\sigma_{\xi-\hat{\xi}}$, respectively) for the example of Fig. 3-3-1.

the order of 100 m, which is continuously filled with turbulence generated continuously by atmospheric factors working through the breaking of surface waves, drift currents, and convection; 2) an internal layer (practically the entire thickness of the sea), in which only intermittent turbulence appears to exist in the form of isolated patches or "blini,"[2] probably formed in zones where internal waves are hydrodynamically unstable; 3) a bottom layer about 10 meters thick, which is apparently analogous to the boundary layer of the atmosphere and is at all times filled with turbulence.

The top and bottom layers are probably separated from the internal layer by distinct boundaries of irregular shape, generated by large-scale turbulent eddies (with scales on the order of the thickness of the layer) and by internal waves.

In the upper mixing layer, the root-mean-square velocity fluctuations are usually of the order of 1 cm/sec and decrease rapidly with increasing depth. The intensity of the velocity fluctuations can also be judged from the turbulent energy dissipation rate ε, which is, at large Reynolds numbers, the only parameter of the turbulent kinetic energy spectrum in the inertial subrange of wave numbers $\varkappa$ (which is sometimes observed in turbulence spectra in the upper mixing layer of the ocean), where the three-dimensional spectrum $E(\varkappa)$ is described by the Kolmogorov-Obukhov "five-thirds" law:

$$E(\varkappa) = C_1 \varepsilon^{2/3} \varkappa^{-5/3}; \tag{3-3-4}$$

Here C_1 is a numerical constant with a measured value approaching 1.4 (see [10, Secs. 21, 23]). In this case the frequency spectrum of the longitudinal velocity component, as derived from the traces recorded by turbulence meters towed at velocity U, has the form $E_1(\omega) = C_2 (\varepsilon U)^{2/3} \omega^{-5/3}$ ($C_2 \approx 0.48$). At the sea surface, ε is usually of the order of 1 to 10^{-1} cm^2/sec^3 and, on the average, decreases with depth to values of the order of 10^{-3} to 10^{-4} cm^2/sec^3, observed in the discontinuity layer.

The behavior of the root-mean-square temperature fluctuations is different from that of the velocity fluctuations. They may at first decrease with depth, but they appear to exhibit a maximum (usually on the order of 10^{-1} °C) in the thermocline, where the vertical temperature gradients are very large. An inertial-convective subrange of wave numbers is commonly exhibited by the temperature-fluctuation spectra. In this subrange, the spectra are described by the Obukhov-Corrsin "five-thirds" law

$$E_T(\varkappa) = B_1 \varepsilon_T \varepsilon^{-1/3} \varkappa^{-5/3}, \tag{3-3-5}$$

where ε_T is the rate of smoothing of temperature-field inhomogeneities and B_1 is an empirical constant of about 1.1 (see again [10, Secs. 21, 23]). The temperature-fluctuation frequency spectrum, derived from traces recorded by a towed turbulence

[2] Ed. note: A Russian pancake; the term was suggested by O. M. Phillips at the Colloquium on Atmospheric Turbulence and Radio Wave Propagation, Moscow, 1965.

meter, has the form $E_T(\omega) = B_2\varepsilon_T\varepsilon^{-1/3}U^{2/3}\omega^{-5/3}$ ($B_2 \approx 0.7$). The values of ε_T appear to vary in the range 10^{-3} to 10^{-8} $(°C)^2$/sec.

In the range of very large wave numbers (or frequencies), the turbulence spectra fall off steeply due to the effect of molecular forces. According to the Kolmogorov similarity hypotheses for locally isotropic turbulence, the falloff of the velocity fluctuation spectra is described by the universal dimensionless function $\varphi(\varkappa\eta) = (\varepsilon\nu^5)^{-1/4}E(\varkappa)$, where ν is the kinematic molecular viscosity and $\eta = (\nu^3/\varepsilon)^{1/4}$ is the Kolmogorov turbulence microscale; at small $\varkappa\eta$, the function φ is proportional to $(\varkappa\eta)^{-5/3}$, that is, the "five-thirds" law (3-3-4) holds. These predictions were confirmed by Grant, Stewart, and Moilliet [11-13], who measured ocean-turbulence spectra in a tidal current with a very large Reynolds number ($3 \cdot 10^8$). The universal function φ that they obtained is plotted in Fig. 3-3-3 (the argument plotted on the logarithmic scale on the abscissa is proportional to $\varkappa_1\eta$). The logarithmic scale for φ may tend to conceal the scatter, but the latter is in fact small, as is seen from Fig. 3-3-4, which shows, this time in the natural scale, the same function φ multiplied by $(\varkappa_1\eta)^2$ to obtain the dimensionless energy dissipation spectrum. The peak of the energy dissipation spectrum occurs at the wave number $\varkappa_1 \approx 1/8\eta$. Significant deviations of the one-dimensional longitudinal spectrum from the "five-thirds" law begin at the same point. Nasmyth [14] observed that, in the dissipation range, the ocean-turbulence spectra sometimes lie higher than the "universal curve" of Fig. 3-3-3. He explains this as the effect of buoyant forces (in statistical equilibrium with molecular forces) in the presence of density fluctuations in the medium (these latter fluctuations may be caused, among others, by salinity fluctuations arising after molecular smoothing of originally compensated temperature fluctuations).

The temperature and salinity fluctuation spectra behave somewhat differently in the range of large wave numbers. In this range the universal function $\varphi_T(\varkappa\eta, \mathrm{Pr}) = \frac{\varepsilon}{\varepsilon_T}(\varepsilon\chi^5)^{-1/4}E_T(\varkappa)$, which is proportional to $(\varkappa\eta)^{-5/3}$ at small $\varkappa\eta$ [the

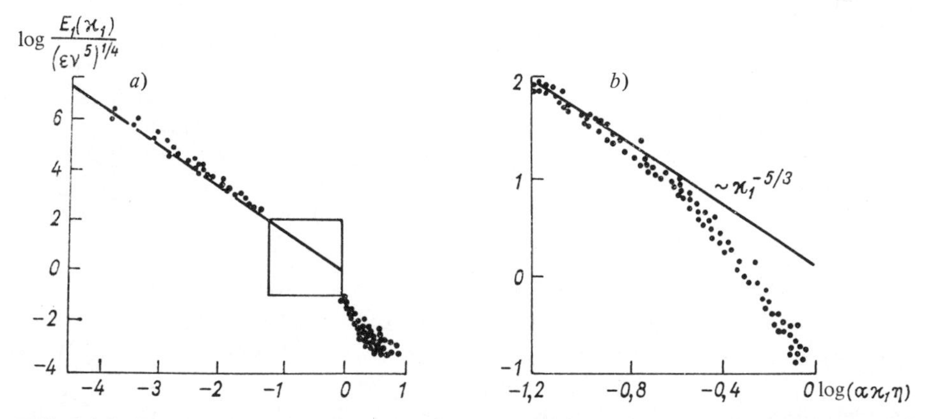

FIG. 3-3-3 Dimensionless longitudinal one-dimensional velocity fluctuation spectrum in a tidal current in the sea (*a*) (after Grant, Stewart, and Moilliet [11]); see [10, Part II, Sec. 12 for formulas for conversion from one-dimensional to three-dimensional spectra]; *b*–data inside square are given on an enlarged scale.

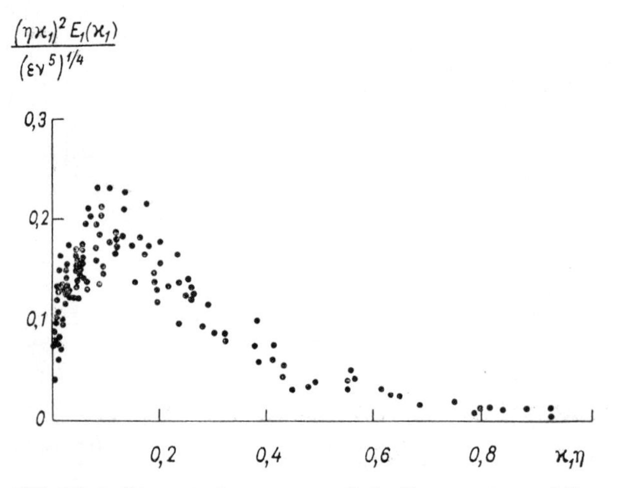

FIG. 3-3-4 Dimensionless energy dissipation spectrum (after [11]).

law (3-2-5)], depends on the Prandtl numbers $\mathrm{Pr} = \nu/\chi$ (where χ is the kinematic coefficient of molecular diffusion for heat or salts). In Batchelor's theory [15], at large Pr (in sea water Pr equals 7 for heat and 700 for salts) and $\varkappa\eta \gg 1$, the dependence of the spectrum $E_T(\varkappa)$ on ε would not be direct (since the main energy dissipation is concentrated at smaller wave numbers). Rather, $E_T(\varkappa)$ would depend on ε only through the quantity $\tau_\eta^{-1} = (\varepsilon/\nu)^{1/2}$, that is, through the typical rate of deformations that produce convective mixing by rotation and crowding of isothermal (or isohaline) surfaces. Hence $\varphi_T \sim (\mathrm{Pr})^{3/4}\, \Phi[\varkappa\eta(\mathrm{Pr})^{-1/2}]$. As long as the argument of the function Φ is small (i.e., in the *viscous-convective* subrange of the spectrum $1 \ll \varkappa\eta \ll (\mathrm{Pr})^{1/2}$), the diffusion coefficient χ should not influence $E_T(\varkappa)$, and we would necessarily have $\Phi(n) \sim n^{-1}$.

Thus, in the inertial-convective subrange of the spectrum $E_T(\varkappa) \sim \varkappa^{-5/3}$ and in the viscous-convective subrange $E_T(\varkappa) \sim \varkappa^{-1}$. Only then does the spectrum $E_T(\varkappa)$ drop off sharply, in the *viscous-diffusive* subrange $\varkappa\eta \gg (\mathrm{Pr})^{1/2}$, due to the smoothing effect of molecular diffusion. These predictions were confirmed by temperature-fluctuation spectra measured in the ocean by Grant, Hughes, Vogel, and Moilliet [16]. One example of their $E_T(\varkappa)$ spectra appears in Fig. 3-3-5, where it is compared with $E(\varkappa)$.

In the range of small wave numbers (outside of the dissipation range), the turbulence spectra of the stratified ocean may deviate from the "five-thirds" law because of the effects of buoyant forces on the turbulence. According to the Obukhov-Bolgiano similarity theory (see [10, Sec. 21.7]), this effect becomes significant at large scales $L \gtrsim L_* = \varepsilon^{5/4}\varepsilon_T^{-3/4}(\alpha g)^{-3/2}$, where $\alpha \sim 2 \cdot 10^{-4}$ deg^{-1} is the thermal expansion coefficient of the water. At $\varkappa \lesssim 1/L_*$, the factors C_1 and B_1 in (3-3-4) and (3-3-5) become functions of $\varkappa L_*$. These functions were calculated semiempirically in [17]. In Bolgiano's hypothesis, which is supported by Monin's calculations [17], at a stable density stratification, the rate of viscous energy dissipation ε in the range of small $\varkappa$ should be much smaller than the rate of energy transfer across the spectrum because of large energy losses on work against buoyant forces. Therefore, ε ceases to affect the shape of the spectrum at small $\varkappa$. This requires that $C_1 \sim (\varkappa L_*)^{-8/15}$ and $B_1 \sim (\varkappa L_*)^{4/15}$, so that

$$E(\varkappa) \sim \varepsilon_T^{2/5} (\alpha g)^{4/5} \varkappa^{-11/5}; \quad E_T(\varkappa) \sim \varepsilon_T^{4/5} (\alpha g)^{-2/5} \varkappa^{-7/5}. \tag{3-3-6}$$

This subrange may, of course, appear in the spectrum if L_* is smaller than the integral turbulence scale. On the other hand, if L_* is small, on the order of η, the subrange obeying (3-3-6) will completely replace the inertial subrange. Figure 3-3-6 shows an example of a velocity fluctuation spectrum in the upper ocean [which obeys (3-3-6)], obtained on the second voyage of the *Akademik Kurchatov* in the Atlantic [here, however, it is not possible to estimate the contribution of internal waves, which often have a spectrum $E(\varkappa) \sim \varkappa^{-3}$; this also holds for the other spectra $E(\varkappa) \sim \varkappa^{-m}$ with $m > 5/3$ that are commonly observed in the ocean].

We note, however, that for not too small scales the universal-shape spectra (3-3-4), (3-3-5), or (3-3-6) are predicted by the similarity theory only in the case of turbulence with very large Reynolds numbers. The real ocean turbulence, on the other hand, develops (and this appears to be the rule) only within the quasihomogeneous layers of the vertical microstructure with thicknesses of the order of meters, in which the

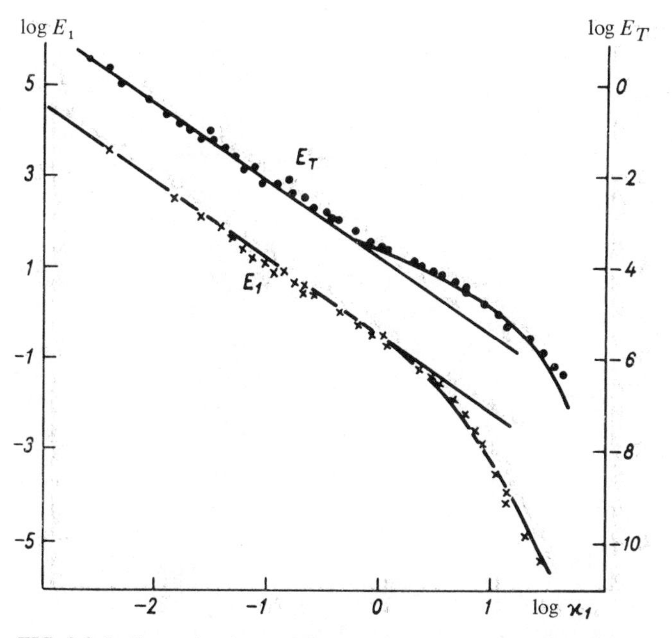

FIG. 3-3-5 Example of one-dimensional spectrum of temperature fluctuations in the ocean (upper curve, right-hand scale) compared with one-dimensional velocity fluctuation spectrum (lower curve, left-hand scale) (according to Grant, Hughes, Vogel, and Moilliet [16]).

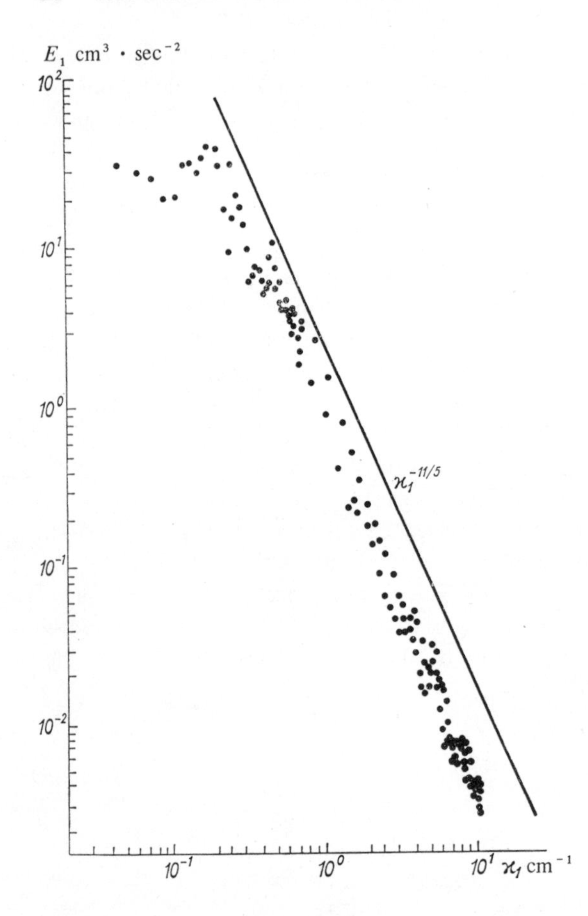

FIG. 3-3-6 Example of one-dimensional longitudinal velocity fluctuation spectrum in the ocean that obeys Eq. (3-3-6); obtained on the second voyage of the *Akademik Kurchatov* in the Atlantic.

vertical velocity differences are of the order of a few cm/sec and the Reynolds numbers of the order of 10^4 to 10^5, that is, only 1 to 2 orders higher than the typical critical values $\mathrm{Re}_{cr} \sim 3000$. At these small Re, wave-number subranges obeying the universal laws (3-3-4) to (3-3-6) no longer exist. But in scales comparable with the thicknesses of the quasihomogeneous layers, the turbulence spectra will not be defined by the parameter ε alone, but by a large number of large-scale parameters. Their calculation requires development of a specific hydrodynamic theory of the quasi-homogeneous layers in the ocean (or in very stably stratified fluids in general), and this remains a task for the future.

It is also clear that turbulence developing within a given quasihomogeneous layer will be controlled by the fluid-dynamic characteristics of that layer, rather than the depth at which this layer is situated, that is, it will not be a smooth function of depth (in a medium with a random distribution of quasihomogeneous layers, the turbulence will vary with depth *in the mean* if the properties of the layers in the microstructure vary with depth *in the mean*).

One specific property of the turbulence in the internal layer of the sea appears to be intermittency. It can be characterized by an intermittency coefficient $P(z)$, defined as the average fraction of the area occupied by the turbulence at depth z. In the measurements made by Grant, Moilliet, and Vogel [18] from a submarine, the coefficient $P(z)$ was equal to unity in an upper, 50 m-thick mixing layer of the sea, decreased to 0.05 at a depth of 100 m, and then varied little down to 300 m [ε was $2.5 \cdot 10^{-2}$ cm^2/sec^3 at $z = 15$ m and $1.5 \cdot 10^{-4}$ at $z = 90$ m; ε_T was $5.6 \cdot 10^{-7}$ ($^\circ$C)^2/sec at $z = 15$ m, reached a maximum of $6.7 \cdot 10^{-6}$ in the density discontinuity layer at $z = 43$ m, and then decreased to $7.2 \cdot 10^{-8}$ at $z = 90$ m]. It appears that typical values of $P(z)$ in the internal layer of the ocean are of the order of 10^{-2}.

However, a description of the spatial intermittency of the small-scale fluctuations with the aid of the coefficient $P(z)$ is not detailed enough. This is because spatial intermittency generally takes the form of alternation of regions with fluctuations of different intensity, and does not merely imply the presence and absence of fluctuations. Empirical probability distributions (histograms) of fluctuation intensities would give a more detailed description. Kolmogorov[3] suggested an analysis of intermittency based on more frequent (say, every 3 sec) calculations of the structure functions or spectra of the fluctuations, derived from traces recorded by towed turbulence meters over many minutes. An attempt at such calculation, based on records of electrical-conductivity fluctuations, measured on the seventh voyage of the *Dmitriy Mendeleyev*, showed that the three-second structure functions vary with periods of the order of several minutes, that is, of the order of the Brunt-Väisälä period, an apparent indication of a relation between the intermittency of the small-scale fluctuations and internal waves, for which this period is typical. It would be even more interesting to make such calculations from records of current-velocity fluctuations.

We still have almost no quantitative measurements of turbulence characteristics at great depths. Some preliminary estimates of turbulence characteristics could perhaps be extracted from information on the density stratification which, of course, is affected by vertical turbulent exchange.

Monin et al. [19] have proved that over much of the lower part of the ocean's internal layer, it is often possible to approximate the vertical profile of the Brunt-Väisälä frequency N with a function Ah, where h is the height above the sea floor (see one of the examples in Fig. 3-3-7). The coefficient A, measured at different hydrological stations, varies from 10^{-7} to 10^{-6} m^{-1} sec^{-1}. If we apply the similarity theory for turbulence in a stably stratified medium (see [10, Sec. 73]) to the internal layer of the ocean, then the relation $N = Ah$ will describe a condition in which the ratio of the exchange coefficients for heat and momentum decreases with height as h^{-2}. The coefficient A will be defined by the formula $A = \frac{\varkappa_K}{R}\left(\frac{g}{\rho}M\right)^2 u_*^{-5}$, where $\varkappa_K \approx 0.4$ is Karman's constant, $R \sim 10^{-1}$ is the maximum value of the dynamic Richardson number, u_* is the friction velocity, and $M = \overline{\rho' w'} = P\langle\rho' w'\rangle$ is the vertical turbulent mass flux (the primes identify the turbulent fluctuations, the overbar indicates averaging over the area, and the angle brackets averaging over the turbulent patches only). Then at $u_* \sim 1$ cm/sec we obtain $M \sim 10^{-8}$ g/cm^2 sec. On the other hand, the dependence of N on the depth z in the upper part of the internal layer can often be approximated by the relation $N = w_*/z$, where $w_* \approx 2.2$ m/sec varies little from station to station [19]. Equating the two expressions for N at the mid-depth of the ocean $z = H/2$, we obtain an estimate of the relative density fluctuations in the turbulent patches [20]: $\frac{\rho'}{\rho} \sim B\frac{u_*^2}{gH}$,

[3] Personal communication.

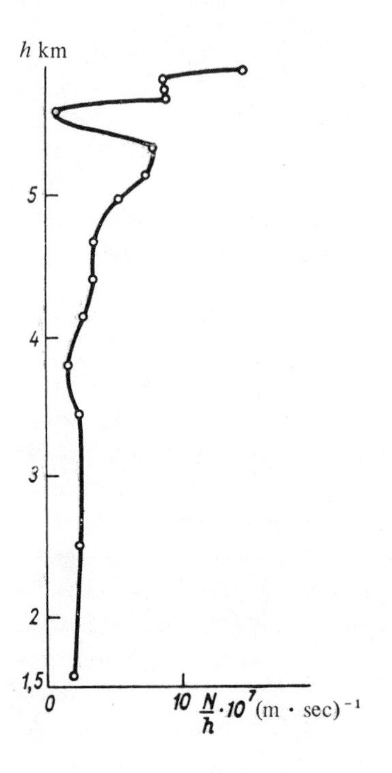

FIG. 3-3-7 Brunt-Väisälä frequency N vs. height h above the bottom at *Vityaz'* hydrological station No. 4371 in the Pacific Ocean (27°07′N, 153°45′E; depth 6 km) (see [19]).

where $B = \dfrac{1}{\varkappa_K r_{\rho w}} \left(\dfrac{4\varkappa_K R}{P} r_{uw} \dfrac{w_*}{u_*} \right)^{1/2}$, and $r_{\rho w}$ and r_{uw} are the correlation coefficients between the corresponding fluctuations, which appear to be of the order of 10^{-1}. Therefore $\rho'/\rho \sim 10^{-6}$ and $T \sim \rho'/\alpha\rho \sim 10^{-2}$ deg (needless to say, these estimates depend on the value assumed for $u_* \sim$ 1 cm/sec). Finally, we see from the relation $u_*^2 \sim P r_{uw} w'^2$ at $P \sim 10^{-2}$ that the fluctuations w' in the turbulent patches may be at least an order larger than u_*.

There are almost no measurements of turbulence in the bottom layer of the sea. The exceptions are the low-frequency fluctuations (with frequencies of the order of 10^{-1} to 10^2 cycles/hr) that were measured in the bottom layer of a deep ocean by Wimbush and Munk [21]. They were primarily interested in estimating the ability of the bottom layer to transmit geothermal heat upward, and were able to fit their results into the framework of the similarity theory for turbulence in unstable stratification (see [10, Sec. 7]).

Turbulence in the sea greatly accelerates the diffusion of various impurities in it. Study of this phenomenon is becoming particularly urgent because of the growing threat of pollution of the oceans by oil, radioactive substances, DDT, lead, and other harmful substances. It must be remembered that horizontal diffusion, generated by mesoscale horizontal turbulence, occurs much more rapidly, especially in the presence of the velocity gradients of the large-scale currents, than vertical diffusion, which is due to small-scale turbulence. The latter is characterized by vertically varying (and strongly stratification-dependent) eddy diffusivities D_z (which are of the order of 10 cm^2/sec in the upper ocean); the rms displacements of impurity particles then increase with time as $(2D_z t)^{1/2}$. Horizontal turbulence in the ocean is usually thought to be the result of breakup of large eddies into small ones. It is defined by the rate ε of energy transfer across the scale spectrum and is subject to the "five-thirds" law (3-3-4) to (3-3-5) and to the similar "four-thirds" law of Richardson, $D_h \sim \varepsilon^{1/3} L^{4/3}$, for the horizontal diffusion coefficient as a function of the size L of the diffusing patch of impurity. The rms displacements of impurity particles are then proportional to $(\varepsilon t^3)^{1/2}$.

This theory is presented by Monin [22] and Ozmidov [23]. It is confirmed, among others, by measurements of inhomogeneities of the sea-surface temperature field [24] (whose structure functions on scales from 10^0 to 10^1 km obey in some cases the "two-thirds" law $\overline{(\delta_r T)^2} \sim \varepsilon_T \varepsilon^{-1/3} r^{2/3}$, which is equivalent to (3-3-5) and in which r is distance) and by a direct empirical test of the "four-thirds" law and its corollaries on data from diffusion experiments in the ocean [25, 26] (estimates of 10^{-4} cm^2/sec^3 on scales from 10^1 to 10^3 mm and 10^{-5} cm^2/sec^3 on scales from 10^1 to 10^3 km were obtained for ε in [25]). The combined effect of vertical diffusion (with various models for the diffusion coefficient D_z) and horizontal diffusion, the latter obeying the "four-thirds" law, was analyzed in [27].

Let us emphasize that diffusion obeying the law $(\varepsilon t^3)^{1/2}$ should not be confused with ordinary diffusion obeying the law $(2D_z t)^{1/2}$ which, in the presence of a velocity gradient Γ of a large-scale current, is accelerated in the direction of this current, and then obeys the law $\left(\frac{2}{3}\Gamma^2 D_z t^3\right)^{1/2}$ (see [10, Sec. 10.4]). This effect may be responsible for the frequently observed elongation of the patches of the diffusing substance in the direction of the current.

Finally, let us mention another theory of horizontal turbulence that starts from the fact that a two-dimensional flow conserves not only kinetic energy, but also vorticity. Therefore, the structure of two-dimensional turbulence in the inertial subrange of the spectrum may be governed not only by the rate of degeneration of the energy ε, but also by the rate of degeneration of the enstrophy (i.e., the mean square of vorticity) ε_1 [28, 29]. This gives rise to a length scale $L_1 = (\varepsilon/\varepsilon_1)^{1/2}$, and the constants C_1 and B_1 in the "five-thirds" laws (3-3-4) and (3-3-5) become functions of kL_1.

If it is assumed that only the parameter ε is significant at one end of the inertial subrange, and that only ε_1 is significant at the other end, the "five-thirds" laws will be supplanted at this second end by

$$E(k) \sim \varepsilon_1^{2/3} k^{-3}; \quad E_T(k) \sim \varepsilon_T \varepsilon_1^{-1/3} k^{-1}, \tag{3-3-7}$$

and $D_h \sim \varepsilon^{1/3} L^2$ will apply instead of the Richardson "four-thirds" law. The law (3-3-7) for $E(k)$ has been confirmed to some degree in numerical experiments on two-dimensional turbulence and in the statistics of large-scale atmospheric motions, but we do not yet have the corresponding data for horizontal turbulence in the ocean.

REFERENCES

1. Bowden, K. F. Turbulence. In: Oceanogr. and Marine Biol. Annual Rev., London, **2**, 11–30 (1964); 8, 11–32 (1970).
2. Phillips, O. M. *The Dynamics of the Upper Ocean*, Cambridge University Press (1966).
3. Monin, A. S. Ocean turbulence, Izv. Akad. Nauk SSSR, Fizika atm. i okeana, **5**, No. 2, 218–225 (1969).

4. Benilov, A. Yu. Observational data on small-scale ocean turbulence, Izv. Akad. Nauk SSSR, Fizika atm. i okeana, **5**, No. 5, 513–532 (1969).
5. Monin, A. S. The basic features of ocean turbulence, Okeanologiya, **10**, No. 5, 240–249 (1970).
6. Bowden, K. F. and R. A. White. Measurements of the orbital velocities of the sea waves and their use in determining the directional spectrum, J. Roy. Astronom. Soc., **12**, No. 1, 33–54 (1966).
7. Benilov, A. Yu. and B. N. Filyushkin. Application of linear filtration methods to analysis of fluctuations in the surface layer of the ocean, Izv. Akad. Nauk SSSR, Fizika atm. i okeana, **6**, No. 8, 810–820 (1970).
8. Yaglom, A. M. Introduction to the theory of stationary random functions, Uspekhi matem. nauk, **7**, No. 5 (51), 1–168 (1952).
9. Benilov, A. Yu. Estimation of statistical characteristics of random hydrophysical fields in the upper layer of the ocean. In: Issledovaniye okeanicheskoy turbulentnosti (*Studies of Ocean Turbulence*), Nauka Press, Moscow, 49–63 (1973).
10. Monin, A. S. and A. M. Yaglom. Statistical Fluid Mechanics, Nauka Press, Moscow, Part I (1965); Part II (1967). [English translation, MIT Press.]
11. Grant, H. L., R. W. Stewart, and A. Moilliet. Turbulence spectra from a tidal channel, J. Fluid Mech., **12**, Part 2, 241–268 (1962).
12. Stewart, R. W. and H. L. Grant. Determination of the rate of dissipation of turbulent energy near the sea surface in the presence of waves, J. Geophys. Res., **67**, No. 8, 3177–3180 (1962).
13. Grant, H. L. and A. Moilliet. A spectrum of a cross-stream component of a turbulence in a tidal stream, J. Fluid. Mech., **13**, Part 2, 237–240 (1962).
14. Nasmyth, P. W. Some observations on turbulence in the upper layers of the ocean, Rapports et Proces-Verbaux des Reunions, **162**, 19–24 (1972).
15. Batchelor, G. K. Small-scale variation of convected quantities like temperature in turbulent fluid, Part I, General discussion and the case of small conductivity, J. Fluid Mech., **5**, Part 1, 134–139 (1959).
16. Grant, H. L., B. A. Hughes, W. M. Vogel, and A. Moilliet. The spectrum of temperature fluctuations in turbulent flow, J. Fluid Mech., **34**, Part 3, 423–442 (1968).
17. Monin, A. S. Turbulence spectrum in an atmosphere with inhomogeneous temperature, Izv. Akad. Nauk SSSR, Ser. geofiz., No. 3, 397–407 (1962).
18. Grant, H. L., A. Moilliet, and W. M. Vogel. Some observations of the occurrence of turbulence in and above the thermocline, J. Fluid Mech., **34**, Part 3, 443–448 (1968).
19. Monin, A. S., V. G. Neyman, and B. N. Filyushkin. Density stratification in the ocean, Doklady Akad. Nauk SSSR, **191**, No. 6, 1277–1279 (1970).
20. Monin, A. S. Turbulent mass fluxes in the oceans, Doklady Akad. Nauk SSSR, **193**, No. 5, 1038–1041 (1970).
21. Wimbush, M. and W. Munk. The Benthic Boundary Layer. In: *The Sea*, **4**, Part 1, Wiley-Interscience, 731–658 (1970).
22. Monin, A. S. Horizontal mixing in the atmosphere, Izv. Akad. Nauk SSSR, Ser. geofiz., No. 3, 327–345 (1959).
23. Ozmidov, R. V. Gorizontal'naya turbulentnost' i turbulentnyy obmen v okeane (*Horizontal Turbulence and Turbulent Exchange in the Ocean*), Nauka Press, Moscow (1968).
24. Ivanov, V. N., A. S. Monin, and V. T. Paka. Structure of the ocean-surface temperature field, Doklady Akad. Nauk SSSR, **183**, No. 6, 1304–1307 (1968).
25. Okubo, A. and R. V. Ozmidov. Empirical dependence of horizontal eddy diffusivity in the ocean on the scale of the phenomenon, Izv. Akad. Nauk SSSR, Fizika atm. i okeana, **6**, No. 5, 534–537 (1970).
26. Ozmidov, R. V., V. K. Astok, A. N. Gezentsvey, and M. K. Yukhat. Statistical characteristics of concentration fields of a passive impurity introduced artificially into the ocean, Izv. Akad. Nauk SSSR, Fizika atm. i okeana, **7**, No. 9, 963–973 (1972).
27. Monin, A. S. Interaction between vertical and horizontal diffusions of impurities in the ocean, Okeanologiya, **9**, No. 1, 76–81 (1969).

28. Batchelor, G. K. Computation of the energy spectrum in homogeneous two-dimensional turbulence, Phys. Fluids, Suppl. II, 233-239 (1969).
29. Kraichnan, R. Inertial ranges in two-dimensional turbulence, Phys. Fluids, **10**, No. 1, 14-23 (1967).

3.4 THE VERTICAL MICROSTRUCTURE OF THE OCEAN

The improvement of instruments for continuous sounding of vertical temperature and salinity variations (the bathythermograph in 1941; the first temperature and salinity sonde in 1948) made it possible to observe something that had for a long time remained inaccessible to observation and study by cruder means (for example, a series of sampling bottles), namely, the fine thermohaline structure of the waters of oceans and seas. Oceanologists had long neglected the fine details of the stratification and rejected all observations that showed inversions in the temperature or density distribution. However, laboratory and theoretical studies of the last 10 to 20 years tended to indicate that there actually exists a microstructure of the sea water. This gave impetus to the development of special sondes which, in many cases, made it possible to observe [1-24] a stepped microstructure—something that eventually proved to be an almost universal phenomenon.

Back in 1956, Stommel et al. [25] drew attention to the possibility of development of convection in a layer of salt water with stable density stratification but with vertical temperature and salinity gradients of opposite signs. Convection could develop in this case because of "double diffusion," that is, because of the difference in the diffusion coefficients for heat and salt in water (the diffusion coefficient for heat is 100 times larger than that for salt). This idea, which Stommel then developed in [26], gave rise to a series of theoretical [27-34] and laboratory [35-43] investigations of thermohaline convection.

It was established in the laboratory experiments of Turner and Stommel [35, 36] that "layering" occurs when cold and relatively fresh water lies above warm, salt (denser) water. This "layering" involves the formation of a sequence of convective and laminar layers because the relatively rapid diffusion of heat from below generates convection at certain levels, but the upward penetration of the latter is limited by the stable salinity gradient that persists in the laminar layers due to the slowness of the salt diffusion. A convective salt flux F is produced in addition to the convective vertical heat flux H. Reduced to dimensionless form via division, respectively, by $D_T(\delta T/h)$ and $D_s(\delta s/h)$ (where D_T and D_s are the diffusion coefficients of heat and salt, δT and δs are the vertical temperature and salinity differences, and h is the thickness of the layer), these fluxes are proportional to $\mathrm{Ra}^{1/2}$ ($\mathrm{Ra} = \alpha_T g h^3 \delta T/\nu\chi_T$ is the Rayleigh number, α_T is the thermal expansion coefficient of water, and ν and χ_T are the kinematic viscosity and kinematic thermal diffusivity), with proportionality coefficients that depend on the ratio $\alpha_s\delta_S/\alpha_T\delta T$ of the contributions of salinity and temperature to the vertical density difference, where $\alpha_s = 1/\rho(\partial\rho/\partial s)_{T,p}$. Experiments showed that the ratio $\alpha_s F/\alpha_T H$ of the potential energy changes due to salt and heat transport first decreases rapidly with increasing $\alpha_s\delta S/\alpha_T\delta T$, and at $\alpha_s\delta S/\alpha_T\delta T > 2$ becomes constant at approximately 0.15, that is, 15 percent of the potential energy released in convective heat transport is expended in lifting salt.

It was found in the experiments of Turner and Stern [37, 40] that when warm salt water is situated above denser cold and less saline water, convection develops in the form of long, narrow vertical cells known as "salt fingers." This is because the relatively rapid horizontal smoothing of temperature anomalies, coupled with persistence of salinity anomalies owing to the slowness of salt diffusion, generates density anomalies. The experiment indicated that in this case the ratio $\alpha_T H/\alpha_s F$ is almost independent of the stability parameter $\alpha_T \delta T/\alpha_s \delta S$ and is approximately equal to 0.56, that is, more than half of the potential energy released in salt convection is used to transport heat upward against the temperature gradient. Thus, salt fingers prove to be an efficient mechanism for vertical transport not only of salt, but also of heat. The convective layers should appear as vertical steps on temperature and salinity profiles. In Stern's approximate theory [32], the maximum step thickness is given by

$$H_m \approx \nu^2 (\chi_T \chi_s)^{-3/4} \left(g\alpha_s \frac{\partial s}{\partial z} \right)^{-1/4}, \tag{3-4-1}$$

from which we obtain $H_m \sim 20$ m for $\nu = 1.5 \cdot 10^{-2}$ cm^2/sec, $\chi_T = 1.3 \cdot 10^{-3}$ cm^2/sec, $\chi_s = 1.3 \cdot 10^{-5}$ cm^2/sec, and $\alpha_s(\partial S/\partial z) \sim 10^{-8}$ cm^{-1}.

Ocean and lake observations summarized by Fedorov in [14, 23] (the warm saline Lake Vanda, the hot brine-filled sea floor depressions in deep parts of the Red Sea, Mediterranean waters in the Atlantic, Red Sea waters in the Gulf of Aden, etc.) indicate that stepped structures are in fact encountered quite frequently, both in layers with inversion-type temperature distribution and in layers in which temperature and salinity decrease with depth. Figure 3-4-1 shows two characteristic profiles with a stepped thermohaline structure.

Woods [1, 4–6] observed microstructure of very thin layers in the seasonal

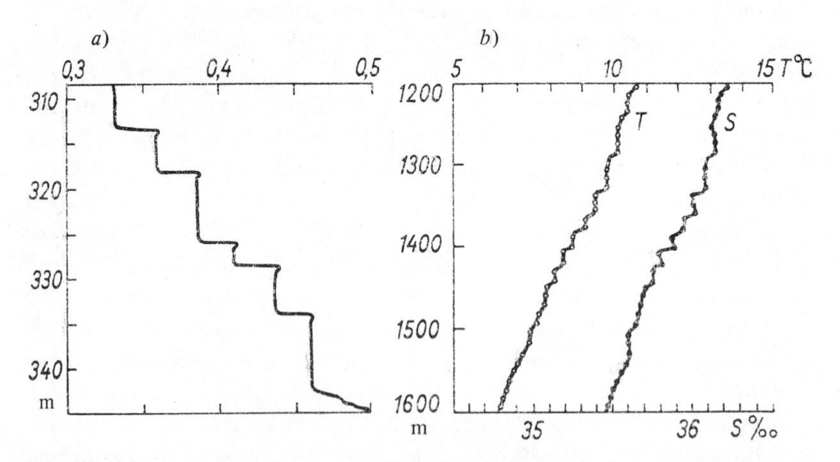

FIG. 3-4-1 Examples of stepped thermohaline structure in a temperature inversion layer of Atlantic origin in the Arctic basin under the T-3 ice island (*a*) (after Denner, Neal, and Neshyba [44]) and in a layer favoring the development of "salt fingers," where both temperature and salinity decrease with increasing depth (*b*) (after Zenk [13]).

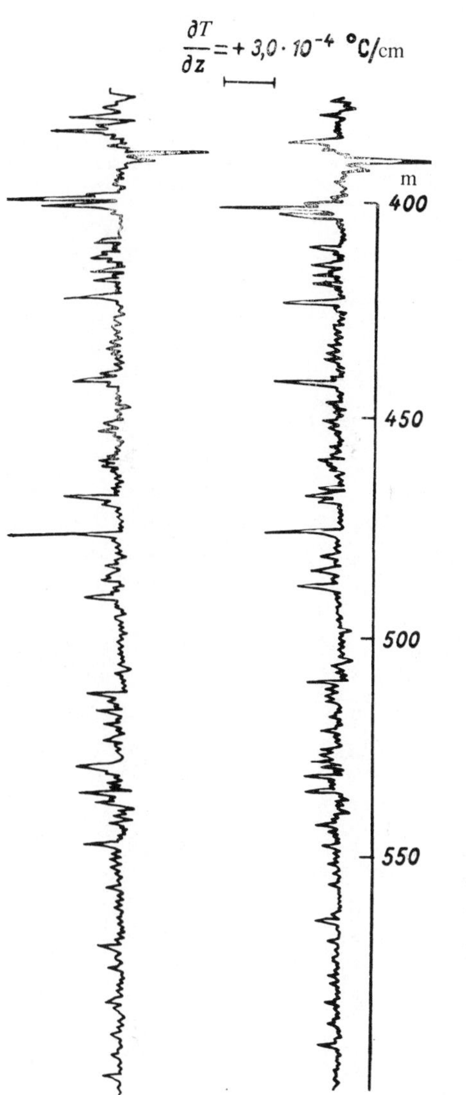

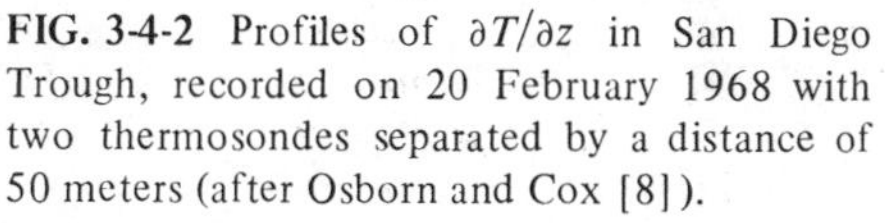
FIG. 3-4-2 Profiles of $\partial T/\partial z$ in San Diego Trough, recorded on 20 February 1968 with two thermosondes separated by a distance of 50 meters (after Osborn and Cox [8]).

thermocline of the Mediterranean, by taking underwater motion pictures of dyed layers. Nasmyth [12] recorded the microstructure in the upper ocean with transducers mounted on a periodically diving pod towed behind a ship. Since certain distortions are introduced into the microstructure record as a result of pitching of the ship when an STD is lowered from it on a cable, Woods [11, 15-16] and Cox, Nagata, and Osborn [7, 8] used *free-falling* low-lag thermosondes to observe the microstructure. Figure 3-4-2 shows profiles of the vertical temperature gradient $\partial T/\partial z$ at depths of 400 to 600 m 40 km west of San Diego, obtained from records of sondes in free fall at approximately 30 to 50 cm/sec with a time constant of about 0.1 sec. Figure 3-4-3

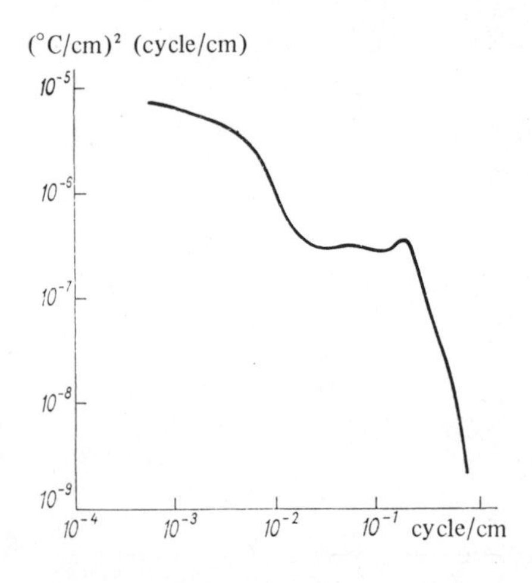

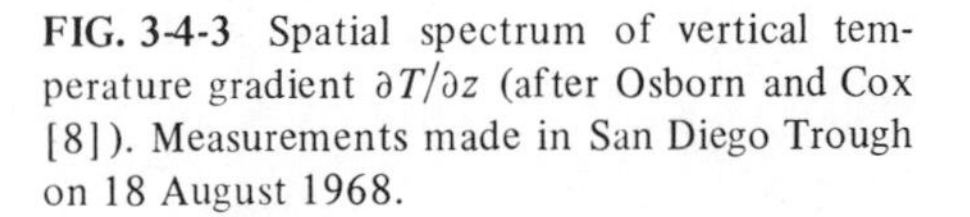
FIG. 3-4-3 Spatial spectrum of vertical temperature gradient $\partial T/\partial z$ (after Osborn and Cox [8]). Measurements made in San Diego Trough on 18 August 1968.

shows a spatial spectrum of $\partial T/\partial z$, derived from the records of a sonde [8] with a fall velocity of 3 to 20 cm/sec and a time constant of about 0.02 sec; the dropoff of the spectrum at $l > 10^{-1}$ cycles/cm that is clearly visible here indicates that this sonde registers practically the entire fine structure of the temperature profile.

The available data indicate that double diffusion of heat and salt may not always be responsible for the formation of the stepped microstructure (and is perhaps important only in some cases). Thus, Simpson and Woods [15] found a stepped microstructure on temperature profiles in the *fresh-water* lake Loch Ness (Fig. 3-4-4), where the double-diffusion mechanism cannot be operative. On the fifth voyage of the *Dmitriy Mendeleyev* in the Pacific Ocean, a stepped microstructure was recorded in several cases of stable stratification of both temperature and salinity, in which this mechanism is again inoperative. One possibility is that in layers with vertical gradients of temperature and *current velocity*, the microstructure is produced by the difference between the molecular diffusion coefficients for heat and momentum (the coefficient

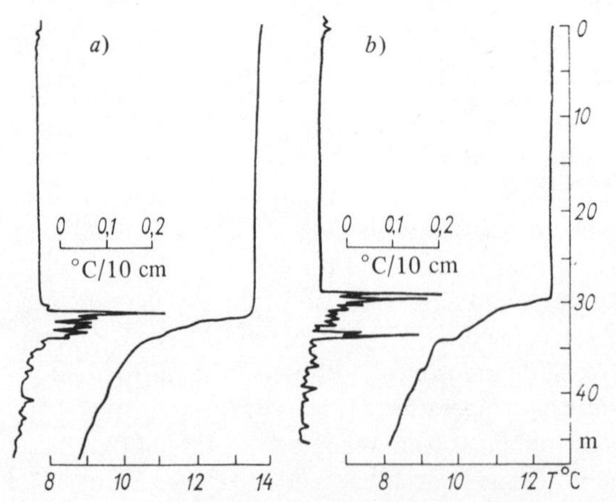

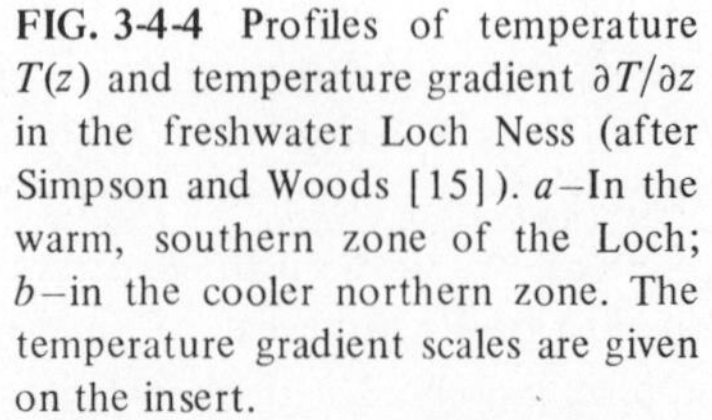
FIG. 3-4-4 Profiles of temperature $T(z)$ and temperature gradient $\partial T/\partial z$ in the freshwater Loch Ness (after Simpson and Woods [15]). *a*–In the warm, southern zone of the Loch; *b*–in the cooler northern zone. The temperature gradient scales are given on the insert.

of momentum diffusion in water, i.e., the viscosity, is 7 times larger than the diffusion coefficient for heat), but this theory has not yet been worked out.

The third and perhaps simplest hypothesis of the origin of the stepped microstructure is "lateral convection," that is, a process in which horizontal differences between neighboring differently stratified water columns are equalized by quasi-horizontal displacements of individual layers or lenses of water. Such displacements may be due to independent meandering of the currents in the different layers, for example, as a result of baroclinic instability of currents or disturbances to their geostrophic equilibrium in the upper ocean by moving wind stresses at the surface. The displacements could also be produced as a result of sliding of layers that are heavier than their horizontal neighbors along inclined isopycnic or, more probably, isentropic surfaces, leading to the formation of a stratification more stable than that in the original water columns.

The lateral-convection theory was suggested by Stommel and Fedorov [2], who explained the existence of a 10 meter-thick temperature inversion under a 120-meter upper mixing layer in the Timor Sea in terms of downsliding of warm waters from the Australian shelf, these waters being displaced over horizontal distances in the hundreds of miles after becoming more saline because of evaporation. A similar explanation for the existence of layers of Arctic water in the North Atlantic was given by Cooper [3]. Fedorov [18] used the lateral-convection idea to explain the temperature inversions at depths of 40 to 50 m at *Crawford* station No. 308 and *Atlantis* station No. 5806 in the Atlantic, and also to explain the origin of thin partings of low-salinity water at a depth of 75 m at the *Academik Kurchatov* stations Nos. 561 and 567 in the North Equatorial current in the Atlantic [19] (Fig. 3-4-5). Woods and Wiley [16] used a lateral convection of gravitational or dynamic nature to explain the quasihomogeneous layers several meters thick, separated by "sheets" 10 to 20 cm thick exhibiting sharp vertical temperature, salinity, and velocity gradients.

The "lateral convection" hypothesis received its first direct confirmation on the

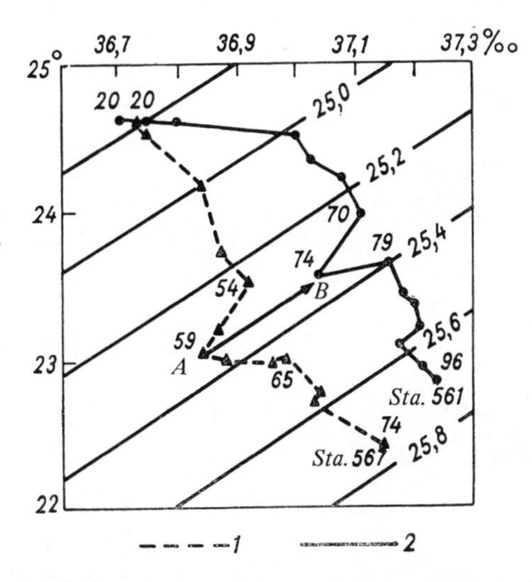

FIG. 3-4-5 *T, S* curves, plotted from the data of the *Academik Kurchatov* stations Nos. 567 (1) and 561 (2), showing low-salinity layers *A* and *B* on the same isopycnic line (after Fedorov [19]).

seventh voyage of the *Dmitriy Mendeleyev* (January–March 1972) in the Indian Ocean. Measurements of the vertical mesostructure of the current-velocity field with an acoustic Doppler sonde showed vertical inhomogeneities in this field with scales down to 5 to 10 meters and less. These correlated fairly well with inhomogeneities in the temperature and electrical conductivity of the water, so that different quasi-homogeneous layers do indeed appear to move with horizontal velocities that differ both in magnitude and in direction. Development of a theory for this independent motion of different layers is still awaited.

Lateral convection, generated by double diffusion of heat and salt, was detected in small-scale laboratory experiments by Thorpe, Hutt, and Soulsby [45] and by Gubin and Khaziev [46]. However, it is unclear whether double diffusion of heat and salt can be a factor of similar importance in lateral exchange on much larger scales.

A fourth theory on the nature of quasihomogeneous layers (vertical steps on temperature and salinity profiles) is offered by Bretherton [47] and by Orlansky and Bryan [48]. Their mechanism envisages breaking of finite-amplitude internal waves, which results in mechanical mixing in a given layer of water and in subsequent density gradient-driven convection in that layer. Orlansky and Bryan (who, however, deal only with the thermal stratification) suggest that an internal wave breaks when the orbital velocity at its crest exceeds its phase velocity. This occurs when $\mathrm{Ri} \ll 1 + k_x^2/l^2$.

A similar idea, though applied to the fine structure of the "sheets" separating quasihomogeneous layers rather than to the layers themselves (which are treated in this case as specified and laminar), is developed in a series of papers by Woods et al. ([1, 4–6, 9–11, and especially 16]; see also the discussion of this series in Fedorov's paper [24]). This theory postulates that as an internal wave propagates along a "sheet," the velocity gradients may become unstable in the neighborhoods of its crests and troughs (at $\mathrm{Ri} < 1/4$), and the "sheet" is thereby turbulized. Because of entrainment of water into the now turbulent layer, it becomes thicker (by a factor of 4 to 5). When Ri is increased to about unity, the turbulence in the layer degenerates,

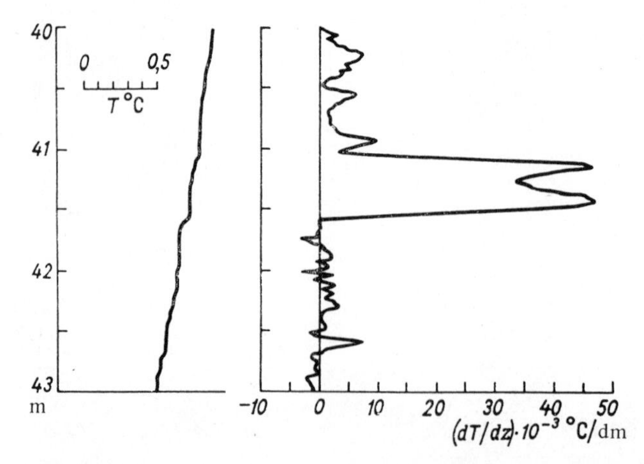

FIG. 3-4-6 Example of a split "sheet" (after Woods and Wiley [16]).

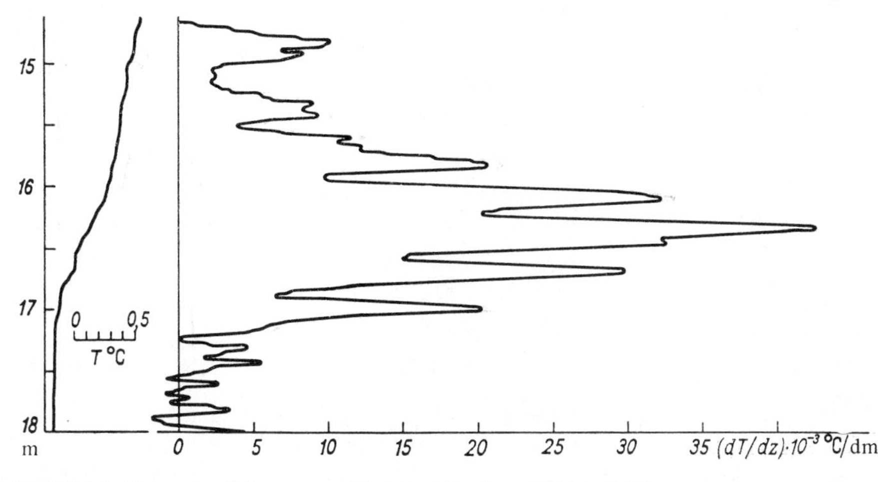

FIG. 3-4-7 Example of sheet ensemble (after Woods and Wiley [16]).

and two new sheets form on its boundaries. Repetition of this process creates whole ensembles of sheets. Woods [4] illustrates this mechanism with underwater films of loss of stability in packets of steep internal waves with lengths of about 5 meters, periods of a few minutes, and phase velocities of several centimeters per second on artificially colored sheets a few centimeters thick in the thermocline of the Mediterranean Sea. Confirmation of sheet splitting was obtained by Woods and Wiley in [16] with the aid of a free-falling (at a rate of 5 cm/sec) thermosonde with a time constant of 0.06 sec. Figure 3-4-6 shows a split sheet, and Fig. 3-4-7 a case of a sheet ensemble.

Meteorologists had long suspected that turbulence can develop in a stably stratified medium only in the form of isolated layers or "blini" that create steps on temperature profiles (see [49]). A theory similar to Woods' was developed for clear air turbulence by Ludlam [50] and has been confirmed by a number of measurements (most of them by radar), which were summarized in [16] and in Phillips' paper [51].

Whereas in the absence of internal waves the vertical temperature profile (or the profile of any other hydrodynamic characteristic) at a given station in the sea and a given layer around a fixed depth z_1 is described by $T(z) = \Gamma \cdot (z - z_1) + \vartheta(z)$ where Γ is the mean vertical temperature gradient and $\vartheta(z)$ describes the microstructure (in the presence of turbulence, ϑ and T may also depend on the time t), the appearance of internal waves, which exhibit in this layer vertical displacements of $\zeta(t)$, will cause a sensor at depth z_1 to record fluctuations $\theta(t) = T[z_1 - \zeta(t)] = \Gamma\zeta(t) + \vartheta[z_1 - \zeta(t)]$ that depend both on the internal waves and on the microstructure. Approximating the fluctuations $\vartheta(t)$ by a sequence of random uncorrelated jumps $\delta\theta$, Phillips [51] derived their spectrum

$$E(\omega) = \frac{\nu\,\overline{(\delta\theta)^2}}{2\pi}\,\omega^{-2} \tag{3-4-2}$$

(where ν is the mean frequency of the jumps) over the frequency range $\tau_l^{-1} \ll \omega \ll \tau_s^{-1}$ (τ_l is the mean interval between the jumps and τ_s is the mean width of the jumps). Roughly speaking, the spectrum of the fine-scale turbulence is simply added to (3-4-2). Reid [52] confirms Eq. (3-4-2) in the particular case of a two-layer model with Gaussian displacement $\zeta(t)$. A more detailed overall analysis was done by Garrett and Munk [53], who treated $\zeta(t)$ and $\vartheta(z)$ as stationary random processes [$\zeta(t)$ being Gaussian] with the correlation functions $B_{\zeta\zeta}(\tau)$ and $B_{\vartheta\vartheta}(\zeta)$. They assumed $\zeta(t)$ and $\vartheta[z - \zeta(t)]$ to be uncorrelated, so that the correlation function of the fluctuations $\theta(t)$ had the form $\Gamma^2 B_{\zeta\zeta}(\tau) + \langle B_{\vartheta\vartheta}(\zeta_1 - \zeta_2)\rangle$, where the angle brackets indicate averaging over the probability distribution for $\zeta_1 = \zeta(t)$ and $\zeta_2 = \zeta(t+\tau)$. The analysis assumed that the typical scale of the microstructure is small compared to the typical internal-wave height z. Garrett and Munk obtained the following expression for the contribution of the microstructure to the spectrum of the fluctuations of $\theta(t)$:

$$E(\omega) = \sqrt{\frac{2}{\pi}} \frac{sz}{\omega^2} \int_0^\infty e^{-x} F\left(\frac{\omega}{sz\sqrt{2x}}\right) dx, \qquad (3\text{-}4\text{-}3)$$

where $F(k)$ is the spectrum of the microstructure gradient $\partial\vartheta/\partial z$, and s is some typical internal-wave frequency (s^{-1} is their Taylor time scale). From this, in the case with the spectrum $F(k)$ constant in the interval $k_l \leqslant k \leqslant k_s$ and equal to zero outside it, we obtain the Phillips expression, whereby the microstructure contribution to the spectrum $\theta(t)$ proves to be larger than the internal-wave contribution at high frequencies. Even in the case of full coherence of the internal waves, the coherence between the fluctuations $\theta(t)$ at two levels degenerates with distance and with frequency, being constant over a distance inversely proportional to the frequency.

REFERENCES

1. Woods, J. D. and G. G. Fosberry. Structure of the thermocline, Rep. Underwater Assn., **2**, 5-18 (1967).
2. Stommel, H. and K. N. Fedorov. Small-scale structure in temperature and salinity near Timor and Mindanao, Tellus, **19**, No. 2, 306–325 (1967).
3. Cooper, L. H. N. The Physical Oceanography of the Celtic Sea. In: Oceanogr. Mar. Biol. Ann. Rev., London, **5**, 99–110 (1967).
4. Woods, J. D. Wave-induced instability in the summer thermocline, J. Fluid Mech., **32**, Part 4, 791–800 (1968).
5. Woods, J. D. An investigation of some physical processes associated with the vertical flow of heat through the upper ocean, Met. Mag., **97**, No. 1, 65–72 (1968).
6. Woods, J. D. Diurnal behavior of the summer thermocline off Malta, Deutsche Hydr. Zeit., **22**, No. 3, 106–108 (1969).
7. Cox, C., Y. Nagata, and T. Osborn. Oceanic Fine Structure and Internal Waves. In: Bull. Japan Soc. Fish. Oceanogr., Prof. Uda's Commemor. Pap., 67–72 (1969).
8. Osborn, T. and C. Cox. Oceanic fine structure, Geophys. Fluid Dynam., **3**, No. 4, 321–345 (1972).
9. Woods, J. D. On Richardson's number as a criterion for laminar-turbulent-laminar transition in the ocean and atmosphere, Radio Sci., **4**, 1289–1298 (1969).
10. Woods, J. D. Fossil turbulence, Radio. Sci., **4**, 1365–1367 (1969).
11. Woods, J. D. On designing a probe to measure ocean microstructure, Underwater Sci. and Technology, **1**, 6–12 (1969).
12. Nasmyth, P. W. Some Observations on Turbulence in the Upper Layers of the Ocean. In: Rapports et process-verbaux des reunions, **162**, 19–24 (1972).
13. Zenk, W. On the temperature and salinity structure of the Mediterranean water in the North-East Atlantic, Deep-Sea Res., **17**, No. 3, 627–632 (1970).
14. Fedorov, Stepped structure of temperature inversions in the ocean, Izv. Akad. Nauk SSSR, Fizika atm. i okeana, **6**, No. 11, 1178–1188 (1970).
15. Simpson, J. H. and J. D. Woods. Temperature microstructure in freshwater thermocline, Nature, **226**, No. 5248, 832–833 (1970).
16. Woods, J. D. and R. L. Wiley. Billow turbulence and ocean microstructure, Deep-Sea Res., **19**, No. 2, 87-122 (1972).

17. Fedorov, K. N. A case of convection with formation of a temperature inversion because of local instability in the oceanic thermocline, Doklady Akad. Nauk SSSR, **198**, No. 4, 822–826 (1971).
18. Fedorov, K. N. On the origin of thin temperature inversions under the topmost homogeneous layer of the ocean, Okeanologiya, **11**, No. 1, 16–22 (1971).
19. Fedorov, K. N. New evidence of the existence of lateral convection in the ocean, Okeanologiya, **11**, No. 6, 994–998 (1971).
20. Brekhovskikh, L. M. and K. N. Fedorov. Poligon-70: An experiment in the ocean, Zemlya i Vselennaya, No. 3, 6–17 (1971).
21. Brekhovskikh, L. M., G. N. Ivanov-Frantskevich, M. H. Koshlyakov, K. N. Fedorov, L. M. Fomin, and A. D. Yampol'skiy. Certain results of a hydrophysical experiment on a test range in the tropical Atlantic, Izv. Akad. Nauk SSSR, Fizika atm. i okeana, **7**, No. 5, 511–528 (1971).
22. Brekhovskikh, L. M., M. N. Koshlyakov, K. N. Fedorov, L. M. Fomin, and A. D. Yampol'skiy. A hydrophysical test-range experiment in the tropical zone of the Atlantic, Doklady Akad. Nauk SSSR, **198**, No. 6, 1434–1439 (1971).
23. Fedorov, K. N. Thermohaline convection in the form of salt fingers, and its possible manifestation in the ocean, Izv. Akad. Nauk SSSR, Fizika atm. i okeana, **8**, No. 2, 214–230 (1972).
24. Fedorov, K. N. Internal Waves and the Vertical Thermohaline Microstructure of the Ocean. In: *Abstracts of Soviet-French Symposium on Internal Waves in the Ocean*, Novosibirsk, 90–118 (1971).
25. Stommel, H., A. B. Arons, and D. Blanchard. An oceanographic curiosity: The perpetual salt fountain, Deep-Sea Res., **3**, No. 2, 152–153 (1956).
26. Stommel, H. Examples of mixing and self-excited convection on the T-S diagram, Okeanologiya, **2**, No. 2, 206–209 (1962).
27. Groves, G. M. Flow estimate for the perpetual salt fountain, Deep-Sea Res., **5**, No. 3, 209–214 (1959).
28. Stern, M. E. The "salt fountain" and thermohaline convection, Tellus, **12**, No. 2, 172–175 (1960).
29. Walin, G. Note on stability of water stratified by both salt and heat, Tellus, **16**, No. 3, 389–393 (1964).
30. Veronis, G. On finite amplitude instability in thermohaline convection, J. Marine Res., **25**, No. 1, 1–17 (1965).
31. Stern, M. E. T-S gradients in the micro-scale, Deep-Sea Res., **15**, No. 3, 245–250 (1968).
32. Stern, M. E. Collective instability of salt fingers, J. Fluid Mech., **35**, Part 2, 209–218 (1969).
33. Stern, M. E. Salt fingers convection and the energetics of the general circulation, Deep-Sea Res., Suppl. Vol. 16, 263–268 (1969).
34. Phillips, O. M. On flows induced by diffusion in stably stratified fluid, Deep-Sea Res., **17**, No. 3, 324–444 (1970).
35. Turner, J. S. and H. Stommel. A new case of convection in the presence of combined salinity and temperature gradients, Proc. U.S. Nat. Acad. Sci., **52**, 49–53 (1964).
36. Turner, J. S. The coupled turbulent transport of salt and heat across a sharp density interface, Int. J. Heat Mass Transfer, **8**, No. 5, 759–767 (1969).
37. Turner, J. S. Salt fingers across a density interface, Deep-Sea Res., **14**, No. 5, 599–612 (1967).
38. Turner, J. S. and E. B. Kraus. A one-dimensional model of the seasonal thermocline, I, A laboratory experiment and its interpretation, Tellus, **19**, No. 1, 88–97 (1967).
39. Turner, J. S. The behavior of a stable salinity gradient heated from below, J. Fluid Mech., **33**, Part 1, 183–200 (1968).
40. Stern, M. E. and J. S. Turner, Salt fingers and convecting layers, Deep-Sea Res., **16**, No. 5, 497–512 (1969).
41. Baines, P. G. and J. S. Turner. Turbulent buoyant convection from a source in a confined region, J. Fluid Mech., **37**, Part 1, 51–80 (1969).
42. Stern, M. E. Optical measurements of salt fingers, Tellus, **22**, No. 1, 76–81 (1970).

43. Shiftcliffe, T. G. and J. S. Turner. Observations of the cell structure of salt fingers, J. Fluid Mech., **41**, Part 4, 707–720 (1970).
44. Denner, W. W., V. T. Neal, and S. T. Neshyba. Modification of the expendable BT for thermal microstructure studies, Deep-Sea Res., **18**, No. 3, 375–378 (1971).
45. Thorpe, S. A., P. K. Hutt, and R. Soulsby. The effect of horizontal gradients on thermohaline convection, J. Fluid Mech., **38**, Part 2, 375–400 (1969).
46. Gubin, V. Ye. and N. N. Khaziev. Thermal-concentration convection, Izv. Akad. Nauk SSSR, Mekhanika zhidkosti i gaza, No. 3, 166–169 (1970).
47. Bretherton, F. R. Momentum transport by gravity waves, Q.J. Roy. Met. Soc., **95**, No. 404, 213–243 (1969).
48. Orlansky, I. and K. Bryan. Formation of thermohaline step structure by large amplitude gravity waves, J. Geophys. Res., **74**, No. 28, 6975–6983 (1969).
49. Monin, A. S. Influence of the temperature stratification of the medium on turbulence. In Atmosfernaya turbulentnost' i rasprostraneniye radiovoln (*Atmospheric Turbulence and Radio Propagation*), Nauka Press, Moscow, 113–120 (1967).
50. Ludlam, F. H. Characteristics of billow clouds and their relation to clear-air turbulence, Q.J. Roy. Met. Soc., **93**, No. 398, 419–435 (1967).
51. Phillips, O. M. On spectra measured in an undulating layered medium, J. Phys. Oceanogr., **1**, No. 1, 1–6 (1971).
52. Reid, R. O. A special case of Phillips general theory of sampling statistics for a layered medium, J. Phys. Oceanogr., **1**, No. 1, 61–62 (1971).
53. Garrett, C. and W. Munk. Internal wave spectra in the presence of fine-structure, J. Phys. Oceanogr., **1**, No. 3, 196–202 (1971).

4 MESOSCALE PHENOMENA

4-1 CLASSIFICATION OF FREE SMALL OSCILLATIONS IN THE SEA

Some of the free small oscillations in the sea, namely short and short-period surface and internal waves, have already been analyzed in the preceding chapter. We shall now derive a general classification of free small oscillations in a sea of constant depth H without imposing any limitations on the lengths and periods of the waves. For a correct description of very long and long-period waves we will have to take into account the global spherical curvature of the ocean's surface and the earth's diurnal rotation. In spherical coordinates on a rotating earth, the equations of motion of the water, linearized with respect to the state of rest and neglecting viscous forces, are

$$\frac{\partial u}{\partial t} - 2\Omega v \sin\varphi - 2\Omega w \cos\varphi = -\frac{1}{a\rho_0 \cos\varphi}\frac{\partial p'}{\partial\lambda};$$

$$\frac{\partial v}{\partial t} + 2\Omega u \sin\varphi = -\frac{1}{a\rho_0}\frac{\partial p'}{\partial\varphi};$$

$$\frac{\partial w}{\partial t} + 2\Omega u \cos\varphi = -\frac{1}{\rho_0}\frac{\partial p'}{\partial z} + g\frac{\rho'}{\rho_0}, \tag{4-1-1}$$

where λ and φ are the longitude and latitude $(0 \leqslant \lambda < 2\pi,\ -\pi/2 \leqslant \varphi \leqslant \pi/2)$, the z axis points downward (left-hand coordinate system), a is the earth's radius, and the remaining symbols are the same as in Sec. 3-2. If we neglect the earth's rotation, $\Omega \to 0$, and the earth's spherical curvature, $a\cos\varphi\, d\lambda \to dx$, $a d\varphi \to dy$, these equations degenerate into the first equation of (3-2-5).

The linearized continuity equation in spherical coordinates is

$$\frac{\partial \rho'}{\partial t} + w\frac{d\rho_0}{dz} = -\rho_0\left(\frac{1}{a\cos\varphi}\frac{\partial u}{\partial\lambda} + \frac{1}{a\cos\varphi}\frac{\partial v\cos\varphi}{\partial\varphi} + \frac{\partial w}{\partial z}\right). \tag{4-1-2}$$

The expression in parentheses on the right-hand side is div **u** so that if we neglect the left-hand side, which filters acoustic waves out of the solutions of the dynamic equations, the continuity equation reduces to the second equation of (3-2-5). In the present discussion, however, we shall retain the acoustic waves in order to preserve completeness of the picture.

The linearized entropy (or salinity) conservation equation has the same form as in (3-2-5), that is,

$$\frac{\partial p'}{\partial t}+g\rho_0 w=c_0^2\left(\frac{\partial \rho'}{\partial t}+w\,\frac{d\rho_0}{dz}\right). \tag{4-1-3}$$

Equations (4-1-1) to (4-1-3) are to be solved for the functions u, v, w, p' and ρ' under the same boundary conditions as for (3-2-5), that is,

$$\frac{\partial p'}{\partial t}+g\rho_0 w=0 \quad \text{at} \quad z=0; \quad w=0 \quad \text{at} \quad z=H. \tag{4-1-4}$$

We introduce the energy integral that generalizes (3-2-6):

$$E=\frac{1}{2}\,\rho_s g\zeta^2+\int_0^H\left[\frac{u^2+v^2+w^2}{2}+\frac{1}{2}\left(\frac{p'}{\rho_0 c_0}\right)^2+\frac{g^2}{2}\left(\frac{p'-c_0^2\rho'}{\rho_0 c_0^2 N}\right)^2\right]\rho_0 dz, \tag{4-1-5}$$

where, as above, $\zeta=-p'/g\rho_0$ are the vertical displacements of the sea surface. The chief difference between this expression and (3-2-6) is the new term $\dfrac{\rho_0}{2}\left(\dfrac{p'}{\rho_0 c_0}\right)^2$ under the integral sign, which describes the so-called elastic energy associated with the pressure fluctuations in acoustic waves (see [1]). As we see from Eqs. (4-1-1) to (4-1-4), the quantity $\bar{E}$ integrated over the surface of the sphere does not vary with time, so that the frequencies ω of all elementary waves described by these equations will be real (it will be recalled that the sea is assumed to be stably stratified, and $N^2>0$).

Equations (4-1-1) to (4-1-5) have elementary wave solutions that depend on longitude and time as $e^{i(m\lambda-\omega t)}$ (where the m are arbitrary integers and the ω are real frequencies to be determined later). However, these equations do not have solutions with separable independent variables φ and z, that is, solutions in which the complex amplitudes of the elementary waves for the fields u, v, w, p' and ρ' can be represented in the form of products of functions of φ and functions of z. This gives rise to serious difficulties in analysis of the wave solutions. It has therefore become traditional to neglect the horizontal projection of the angular velocity vector of the earth's rotation, that is, to neglect in the equations of motion (4-1-1) the term with the factor $\Omega \cos \varphi$ [this does not affect the energy integral (4-1-5)]. This approximation is justified in the case of all short-period oscillations, in which the effects of the earth's rotation are totally immaterial, and also for large-scale motions, which prove to be quasistatic and quasihorizontal. However, this "traditional approximation" seems of doubtful validity in the case of internal waves with $\omega \sim 2\Omega \sin \varphi$, since Munk and Phillips [2] showed that its error increases with increasing mode number of the internal waves. Still, we shall use here this "traditional approximation," because in it the variables φ and z can be separated. Thus, we shall seek the solution in the form

$$(u,\ v)=\frac{1}{\rho_0(z)}P(z)\,[U(\varphi),\ V(\varphi)]\exp i(m\lambda-\omega t);$$

$$w=i\omega W(z)\,\Pi(\varphi)\exp i(m\lambda-\omega t);$$
$$p'=P(z)\Pi(\varphi)\exp i(m\lambda-\omega t);$$
$$\rho'=\left[\frac{1}{c^2(z)}P(z)+\rho_0(z)N^2(z)\,W(z)\right]\Pi(\varphi)\exp i(m\lambda-\omega t). \qquad (4\text{-}1\text{-}6)$$

For simplicity, the sign Re (denoting the real part of the complex number) in front of the right-hand sides of relations (4-1-6) is omitted.

Substituting (4-1-6) into Eqs. (4-1-1) to (4-1-3), we derive the so-called *Laplace tidal equations*[1] for U, V, and Π:

$$-i\omega U-2\Omega V\sin\varphi=-\frac{im\Pi}{a\cos\varphi};\qquad -i\omega V+2\Omega U\sin\varphi=-\frac{1}{a}\frac{\partial\Pi}{\partial\varphi};$$

$$-i\omega\varepsilon\Pi+\left(\frac{1}{a\cos\varphi}\right)\left[imU+\frac{\partial}{\partial\varphi}(V\cos\varphi)\right]=0. \qquad (4\text{-}1\text{-}7)$$

Similarly, for P and W we have

$$\frac{dW}{dz}+\frac{g}{c_0^2}W+\left(\varepsilon-\frac{1}{c_0^2}\right)\frac{1}{\rho_0}P=0,$$

$$\frac{dP}{dz}-\frac{g}{c_0^2}P+(\omega^2-N^2)\rho_0 W=0, \qquad (4\text{-}1\text{-}8)$$

where ε is the constant of the separation of variables and has the dimensions of the reciprocal of the square of velocity.

Since we are concerned with waves in an ocean without coasts, the boundedness conditions for the functions U, V, and Π will serve as boundary conditions for (4-1-7). The boundary conditions for (4-1-8) are obtained directly from (4-1-4).

Equations (4-1-7), which contain no variables describing the vertical stratification of the ocean, define the horizontal structure of the free oscillations. They have nonzero solutions only at certain special values of ω that form a discrete set and are functions of the parameter ε (which has real values, see [1]). In the $(\omega,\ \varepsilon)$ plane they are represented by certain corresponding families of curves that are known as characteristic curves of Eq. (4-1-7). Given appropriate boundary conditions, Eqs.

[1] This term will be explained in Sec. 4-3.

(4-1-8) that do not depend on the curvature or rotation of the earth and contain the functions $\rho_0(z)$, $c_0(z)$ and $N(z)$ in their coefficients, describe the vertical structure of free oscillations in a stratified ocean. They also have nonzero solutions only at certain discrete values of ω that depend on ε, and there are families of corresponding curves in the (ω, ε) plane that are called the characteristic curves of these equations. The points of intersection of each characteristic curve of Eqs. (4-1-7) with each characteristic curve of Eqs. (4-1-8) (under appropriate boundary conditions) give the frequencies ω of the possible free oscillations.

The Laplace tidal equations (4-1-7) were analyzed by a number of authors. The most complete recent results are those of Longuet-Higgins [3] and Dikiy [1]. For a nonrotating planet, we obtain $\omega = \pm[n(n+1)/a^2\varepsilon]^{1/2}$, $n = m, m+1, \ldots$, and the eigenfunctions $\Pi(\varphi) = P_n^m(\sin\varphi)$ are associated Legendre functions of the first kind. The number $n - m$ of their zeroes on the interval $-\pi/2 < \varphi < \pi/2$ acts as the latitudinal wave number. For a rotating flat earth, the eigenfunctions depend on y as $e^{ik_y y}$ and the frequencies have the form $\omega = \pm(4\Omega^2 + k_x^2 + k_y^2/\varepsilon)^{1/2}$. The case $\omega = \pm 2\Omega$ (oscillations with the inertial period) is degenerate–$w = p' = \rho' = 0$, and $u = \pm iv$ depends on z in an arbitrary manner.

For a rotating spherical planet, we have three families of characteristic curves, which we shall enumerate for the case $m > 0$ (these curves are shown in Fig. 4-1-1 for $M = 1$). The curves of the first type are defined at $\varepsilon > 0$; as $\varepsilon_* = 4a^2\Omega^2\varepsilon \to 0$ they behave as $\omega/2\Omega \sim \pm[n(n+1)/\varepsilon_*]^{1/2}$, $n = m, m+1, \ldots$, and as $\varepsilon_* \to \infty$ they behave as $\omega/2\Omega \sim m\varepsilon_*^{-1/2}$, $\nu'' = 0$ and $\omega/2\Omega \sim \pm(2\nu+1)^{1/2}\varepsilon_*^{-1/4} + m(4\nu+2)^{-1}\varepsilon_*^{-1/2}$, $\nu = 0, 1, 2, \ldots$ at $\omega > 0$ and $\nu = 1, 2, \ldots$ at $\omega < 0$. These curves also exist in the absence of rotation and in the plane model (for which $|\omega| > 2\Omega$ always). Curves of the second type are defined at all $-\infty < \varepsilon < \infty$ and behave as $\omega/2\Omega \sim -m/n'(n'+1)$; $n' = m, m+1, \ldots$ as $\varepsilon_* \to 0$; $\omega/2\Omega \sim -m(2\nu'+1)^{-1}\varepsilon_*^{-1/2}$, $\nu' = 1, 2, \ldots$ and $\omega/2\Omega \sim -\varepsilon_*^{-1/4} + (m/2)\varepsilon_*^{-1/2}$, $\nu = 0$ as $\varepsilon_* \to \infty$; $\omega/2\Omega \sim -1 + (m+2\nu) \times (-\varepsilon_*)^{-1/2}$, $\nu = 0, 1, 2, \ldots$ as $\varepsilon_* \to -\infty$. These curves vanish in the absence of rotation and in the case of the plane model. Finally, curves of the third type are defined at $\varepsilon < 0$; as $|\varepsilon_*| \to \infty$ they behave as $\omega/2\Omega \sim 1 - (m+2\nu+2)(-\varepsilon_*)^{-1/2}$, $\nu = 0, 1, 2, \ldots$ and $\omega/2\Omega \sim -m\varepsilon_*^{-1} + 2m(m+2\nu+1)(-\varepsilon_*)^{-3/2}$, $\nu = 0, 1, 2, \ldots$. These curves also exist in the case of the plane model (in which case they are symmetric about the straight line $\omega = 0$), but vanish in the absence of rotation. In the case $m = 0$ (purely zonal motion), all characteristic curves are symmetric with respect to the line $\omega = 0$. Curves of the second type vanish. With $\varepsilon_* \to \infty$, curves of the first type behave as $\omega/2\Omega \sim \pm(2\nu+1)^{1/2}\varepsilon_*^{-1/4}$, $\nu = 0, 1, 2, \ldots$, whereas curves of the third type behave as $\omega/2\Omega \sim \pm 1 \mp (2\nu+2)(-\varepsilon_*)^{-1/2}$, $\nu = 0, 1, 2, \ldots$.

To save space, we shall not derive the corresponding asymptotic relations for U, V and Π (for these, see [3]).

Equations (4-1-8) have also been analyzed by various authors. Thus, Monin and Obukhov [4] analyzed the case of an isothermally stratified atmosphere. Dikiy [1] also analyzed these equations and computed solutions for a realistic (the so-called standard) atmosphere. The behavior of the characteristic curves of these equations was considered by Eckart [5] and by Kamenkovich and Odulo [6] in the case of an

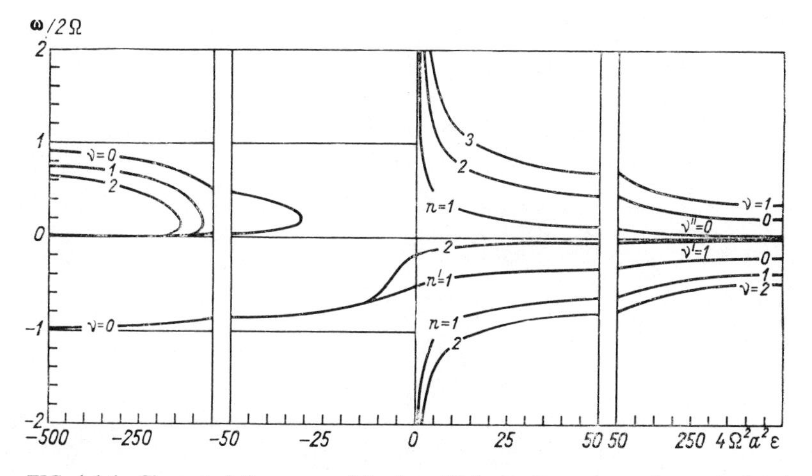

FIG. 4-1-1 Characteristic curves of Laplace tidal equations at $m = 1$, as calculated by Longuet-Higgins [3]. The curve captions at small and large $|\epsilon|$ indicate the range of validity of the corresponding asymptotic expressions.

arbitrarily stratified ocean (both with a free surface and with a solid cover). The characteristic curves of Eqs. (4-1-8) for our problem of a free surface and the typical stratification parameters that are indicated in Fig. 4-1-2 are plotted on Fig. 4.1-3. In this case, the characteristic curves can be classified into three families. The curve for $n = 0$ crosses the axis $\omega = 0$ to the right of the point $\varepsilon = 1/gH$ and at large ε behaves as $\omega \sim g\varepsilon^{1/2}$. Its behavior is controlled by the free-surface effect of the sea and is almost independent of its stratification or compressibility. Characteristic

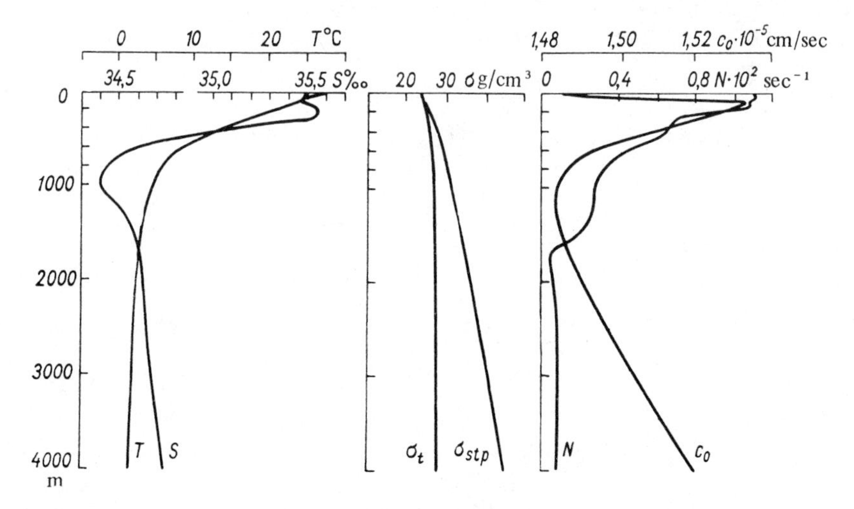

FIG. 4-1-2 Sea stratification at *Vityaz'* station No. 3823 in the Pacific Ocean (23° 10′S, 174°51′W) in the winter of 1957 [temperature T, salinity s, density $\sigma_{stp} = 10^3 \cdot (\rho - 1)$, density reduced to atmospheric pressure σ_T, Brunt-Väisälä frequency N, and sonic velocity c_0 as functions of depth z].

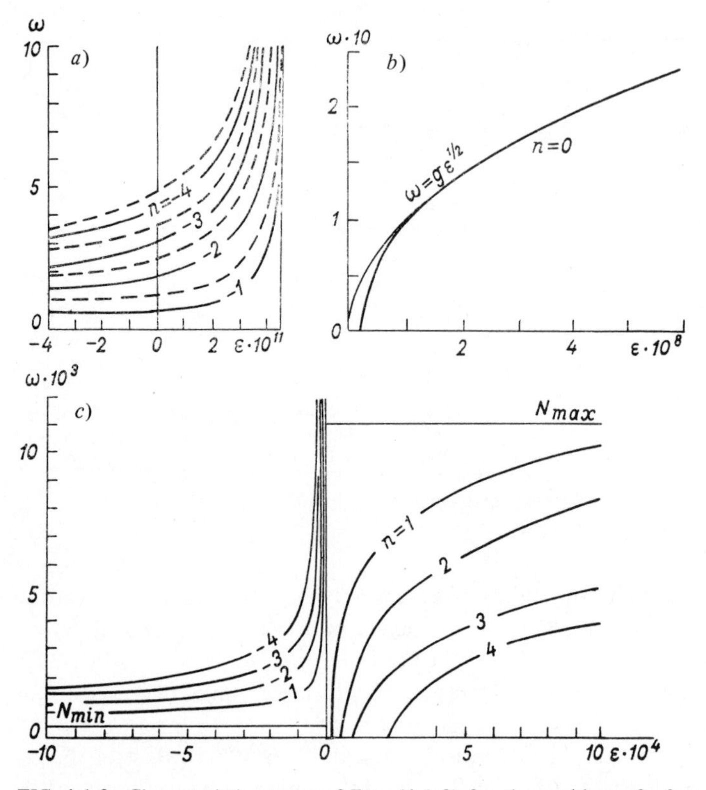

FIG. 4-1-3 Characteristic curves of Eqs. (4-1-8) for the problem of a free surface and the sea stratification indicated in Fig. 4-1-2. *a*–Curves for $n = -1, -2, \ldots$ at small ε, equation of vertical asymptote $\varepsilon = c_{0\min}^{-2}$; the dashed curves indicate the corresponding characteristic curves of Eqs. (4-1-8), obtained on replacement of the first boundary condition (4-1-4) by the condition $w = 0$ at $z = 0$; *b*–curve for $n = 0$ (its asymptote $\omega = g\varepsilon^{1/2}$ is indicated); *c*–curves for $n = 1, 2, 3, \ldots$ and $n = -1, -2, -3, \ldots$ [on this scale, the replacement of the first boundary condition in (4-1-4) by the condition $w = 0$ at $z = 0$ is not reflected in the behavior of the curves]; ω is in rad · sec^{-1}, and ε in sec^2 · cm^{-2} (A. V. Kulakov and A. B. Odulo participated in the numerical calculations).

curves for $n = 1, 2, 3, \ldots$ lie to the right of the line $\varepsilon = c_{0\max}^{-2} + \dfrac{\pi^2}{H^2 N_{\max}^2} \dfrac{\hat{\rho}_{\min}}{\hat{\rho}_{\max}}$, where $\hat{\rho} = \rho_0 e^{2\int_z^H \frac{g\,dz}{c_0^2}}$ and, as $\varepsilon \to \infty$, exhibit the horizontal asymptote $\omega = N_{\max}$. Their behavior is governed by the stratification of the sea and depends little on the compressibility. Finally, the characteristic curves for $n = -1, -2, -3, \ldots$ exhibit the vertical asymptote $\varepsilon = c_{0\min}^{-2}$ as $\varepsilon > 0$ and the horizontal asymptote $\omega = N_{\min}$ as $\varepsilon \to -\infty$; at $|\varepsilon| \sim c_0^{-2}$ they are strongly influenced by the compressibility of the medium (in an incompressible medium they are displaced into the region of $\varepsilon < 0$).

A diagram of the intersections of the characteristic curves of Eqs. (4-1-7) (dashed) and (4-1-8) (solid lines) for the free-surface problem is shown in Fig. 4-1-4. The intersections of the characteristic first-type curves of Eqs. (4-1-7) with the solid lines $n = -1, -2, \ldots$ gives points $1^{\pm}$, which correspond to *acoustic waves* generated because of the compressibility of the medium (as $c_0 \rightarrow \infty$, these points vanish). Kamenkovich and Odulo [6] state that for these $\omega^2 > \omega_a^2 = N^2_{\text{min}} + \frac{\pi^2 c_{0\,\text{min}}^2}{4H^2} \frac{\hat{\rho}_{\text{min}}}{} \sim 1$ rad/sec. The intersections of the same dashed line with the solid line $n = 0$ give the points $2^{\pm}$, which corresponds to *gravity waves on the surface of the ocean.* Neither stratification nor compressibility affect them significantly, but the rotation and sphericity of the earth are significant, especially at large ε (in the flat model, there is the constraint $|\omega| > 2\Omega$). The intersections of the same dashed lines with the solid lines $n = 1, 2, \ldots$ gives the points $3^{\pm}$, which correspond to *internal gravity waves* generated because

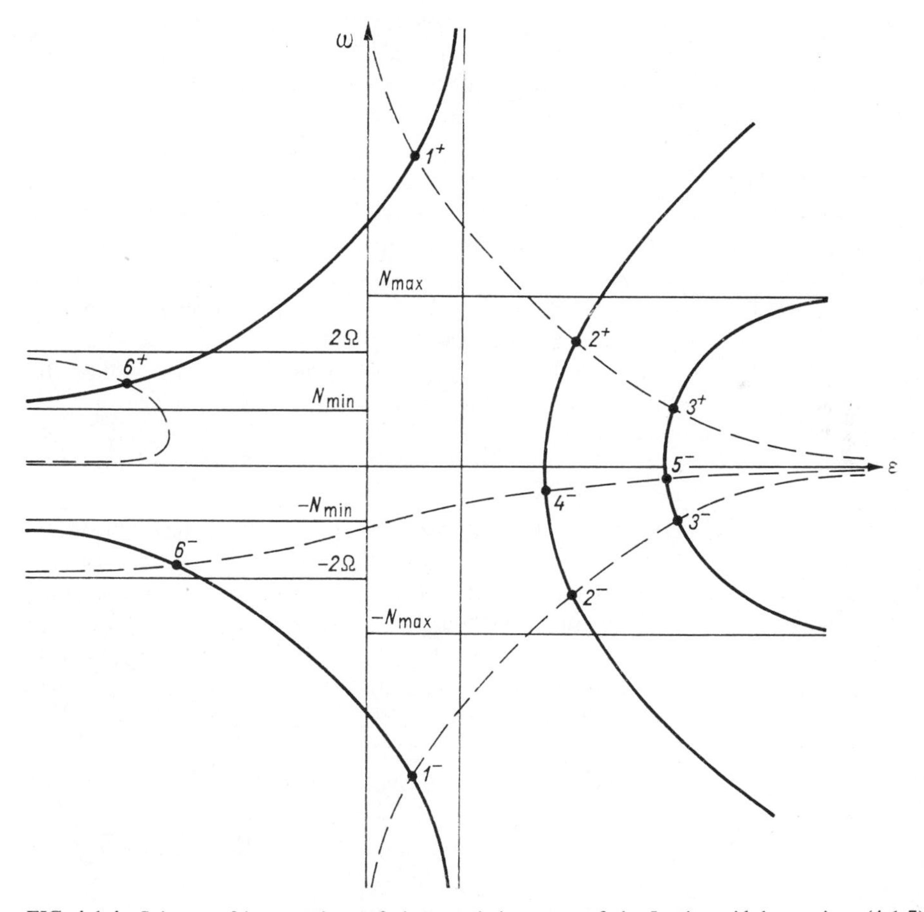

FIG. 4-1-4 Scheme of intersections of characteristic curves of the Laplace tidal equations (4-1-7) (dashed curves) with the characteristic curves of Eqs. (4-1-8) for a problem with a free surface and typical sea stratification (solid curves).

of the stratification of the ocean (they vanish at $N \equiv 0$). Their frequencies lie in the range $0 < |\omega| < N_{max}$. The compressibility has little effect in this case, but the rotation and sphericity of the earth are significant, especially at large ε (in the flat model $|\omega| > 2\Omega$).

The intersections of second-type characteristic curves of Eqs. (4-1-7) with the solid line $n = 0$ give the points 4^-, which correspond to *barotropic (or surface) Rossby waves* with frequencies in the range $0 \leqslant |\omega| \leqslant \Omega$, generated by the rotation of the spherical earth. These are virtually independent of compressibility or stratification. The intersections of the same dashed lines with the solid lines $n = 1, 2, \ldots$ yield the points 5^-, which correspond to *baroclinic (or internal) Rossby waves* with frequencies in the range $0 < |\omega| < \min(\Omega, N_{max})$ which vanish in a homogeneous ocean. Finally, the intersections of the dashed lines of the second and third types with the solid lines $n = -1, -2, \ldots$ give the points $6^{\pm}$, which correspond to *gyroscopic (or inertial) waves* with frequencies in the range $N_{min} < |\omega| < 2\Omega$, which may be produced as a result of the earth's rotation when $2\Omega > N_{min}$, among others in a homogeneous incompressible ocean and in the absence of gravity.[2] We note that surface, internal, and gyroscopic waves can propagate either westward or eastward, but Rossby waves can propagate only westward.

We now briefly analyze certain approximations that can be made in the original equations (4-1-1) to (4-1-3) (see [6]). First, we can replace Eq. (4-1-2) by div $\mathbf{u} = 0$, leaving the other equations unchanged. It is easily shown that Eqs. (4-1-7) and the second equation of (4-1-8) will not thereby be changed, while the first equation of (4-1-8) will assume the form $dW/dz + (\varepsilon/\rho_0)P = 0$. As a result (see Fig. 4-1-3), the vertical asymptotes of curves for $n = -1, -2, \ldots$ will change (we shall have $\varepsilon = 0$ instead of $\varepsilon = c_{0\,min}^{-2}$), but at large $|\varepsilon|$ there will be practically no change in these curves; curves for $n = 0, 1, 2, \ldots$ will also remain practically unchanged. Thus, the approximation filters out acoustic waves without really distorting waves of other types. We used this approximation in Chap. 3 (Secs. 3-1 and 3-2). Eliminating P from the second equation of (4-1-8) by means of $dW/dz + (\varepsilon/\rho_0)P = 0$, we obtain the equation of internal-wave theory (3-2-9) with $k^2 = \varepsilon\omega^2$.

Second, we can replace the third equation of (4-1-1) with the hydrostatic equation $\partial p'/\partial z = g\rho'$, leaving the other equations unchanged. This is the so-called *quasistatic approximation.* This approximation corresponds to deletion of ω^2 in Eqs. (4-1-8), but causes no corresponding change in (4-1-7). Obviously, this will leave only the eigencurves that intersect the ε axis in Fig. 4-1-3. They become straight lines parallel to the ω axis. This means that the acoustic and gyroscopic waves vanish, while the low-frequency ($\omega \ll N$) gravity and Rossby waves remain virtually undistorted (see [4, 8]).

Finally, let us consider the so-called *solid-cover approximation* [substitution of the condition $w = 0$ at $z = 0$ for the first boundary condition of (4-1-4), the equations and other boundary conditions remaining the same]. Then (see Fig. 4-1-3), curves for

---

[2] To avoid confusion, we shall follow Tolstoy [7] in calling these waves gyroscopic rather than inertial (the latter term is widely used in the literature), reserving the term "inertial" for oscillations with frequencies equal to or near $2\Omega \sin \varphi$ (see Sec. 4-2).

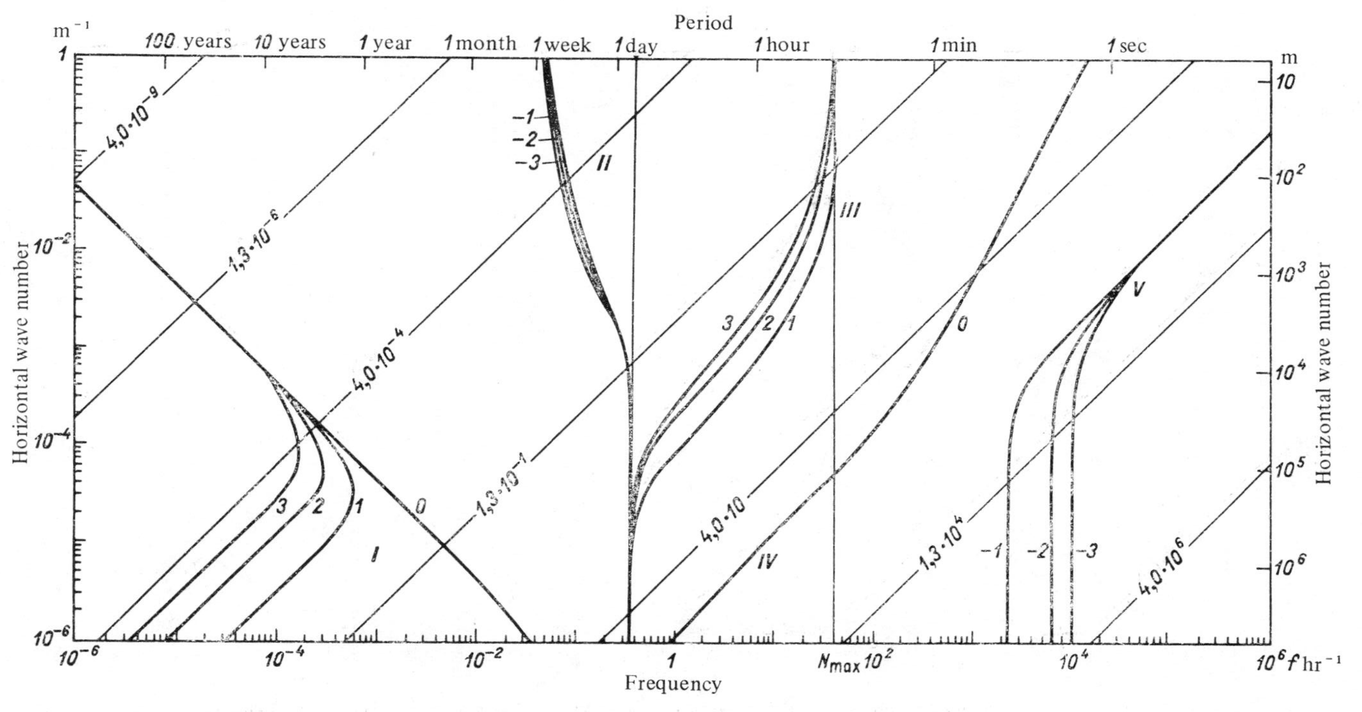

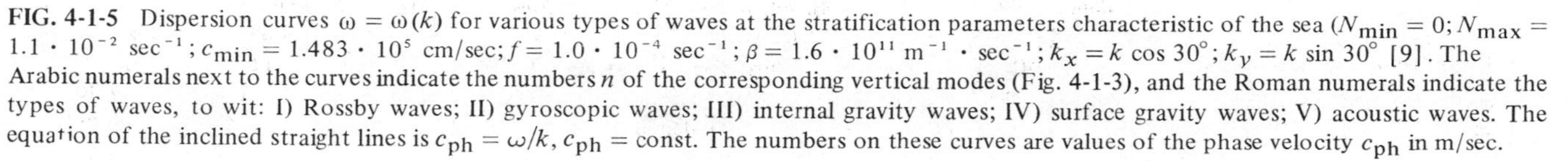

FIG. 4-1-5 Dispersion curves $\omega = \omega(k)$ for various types of waves at the stratification parameters characteristic of the sea ($N_{\min} = 0; N_{\max} = 1.1 \cdot 10^{-2}$ sec^{-1}; $c_{\min} = 1.483 \cdot 10^5$ cm/sec; $f = 1.0 \cdot 10^{-4}$ sec^{-1}; $\beta = 1.6 \cdot 10^{11}$ m^{-1} · sec^{-1}; $k_x = k \cos 30°$; $k_y = k \sin 30°$ [9]. The Arabic numerals next to the curves indicate the numbers n of the corresponding vertical modes (Fig. 4-1-3), and the Roman numerals indicate the types of waves, to wit: I) Rossby waves; II) gyroscopic waves; III) internal gravity waves; IV) surface gravity waves; V) acoustic waves. The equation of the inclined straight lines is $c_{ph} = \omega/k$, $c_{ph} =$ const. The numbers on these curves are values of the phase velocity c_{ph} in m/sec.

$n = -1, -2, \ldots$ are deformed only at $|\varepsilon| \sim c_0^{-2}$, but will, as before, have the vertical asymptote $\varepsilon = c_{0\,\min}^{-2}$. The curves for $n = 0$ will be replaced by a curve with the vertical asymptotes $\varepsilon = c_{0\,\min}^{-2}$ and $\varepsilon = c_{0\,\max}^{-2}$ and will degenerate at $c_0(z) = \text{const}$ into the straight line $\varepsilon = c_0^{-2}$ with $W \equiv 0$. Curves for $n = 1, 2, \ldots$ will be changed very little. This means that: 1) internal gravity, gyroscopic, and baroclinic Rossby waves are distorted very little, while acoustic waves are slightly modified; 2) so-called Lamb waves (or *two-dimensional waves*, see [4, 5]) are produced instead of the surface gravity waves; these waves disappear as $c_0 \to \infty$; 3) barotropic Rossby waves are replaced by *two-dimensional Rossby waves*, which, as $c_0 \to \infty$, become *nondivergent Rossby waves* (see Chap. 9 for a discussion of the effect of the solid-cover approximation on the computed behavior of barotropic Rossby waves).

Note that when these approximations are used in combination, their effects are additive.

Among the important characteristics of free waves are their dispersion relations (these, of course, are the relationships between the frequencies of the waves and their wave numbers). For the not very long ($1/k < 1000$ km) waves that are of most interest to oceanology, the characteristic curves of the Laplace tidal equations on a sphere can be found using the WKB approximation or the β-plane approximation (see [9, Chap. III]). For the problem with the free surface, the characteristic curves of Eqs. (4-1-8) are usually found numerically with no particular difficulty (Fig. 4-1-3). After this, it is not difficult to derive the dispersion relations themselves for all of the principal forms of free waves in the ocean (Fig. 4-1-5).[3]

REFERENCES

1. Dikiy, L. A. Teoriya kolebaniy zemnoy atmosfery (*The Theory of Oscillations of the Earth's Atmosphere*), Gidrometeoizdat Press, Leningrad (1969).
2. Munk, W. and N. Phillips. Coherence and band structure of internal motion in the sea, Rev. Geophys., **6**, No. 4, 447–472 (1968).
3. Longuet-Higgins, M. S. The eigenfunctions of Laplace's tidal equations over a sphere, Phil. Trans. Roy. Soc. London, **A262**, No. 1132, 511–607 (1968).
4. Monin, A. S. and A. M. Obukhov. Small oscillations of the atmosphere and adaptation of the meteorological fields, Izv. Akad. Nauk SSSR, Ser. geofiz., No. 11, 1360–1373 (1958).
5. Eckart, C. *Hydrodynamics of Oceans and Atmospheres*, Pergamon, Russian edition (1960).
6. Kamenkovich, V. M. and A. B. Odulo. Contribution to theory of free oscillations in a stratified compressible ocean of constant depth, Izv. Akad. Nauk SSSR, Fizika atm. i okeana, **8**, No. 11, 1188–1201 (1972).
7. Tolstoy, I. The theory of waves in stratified fluids, including effects of gravity and rotation, Rev. Mod. Phys., **35**, No. 1, 207–230 (1963).
8. Hendershott, M. and W. Munk. Tides. In: Ann. Rev. Fluid Mech., Palo Alto, Calif., **2**, 205–224 (1970).
9. Kamenkovich, V. M. Osnovy dinamiki okeana (*Fundamentals of the Dynamics of the Ocean*), Gidrometeoizdat Press, Leningrad (1973).

---

[3] The authors are indebted to N. P. Fofonoff (USA) for a helpful discussion of certain problems of derivation of dispersion relations.

4-2 INERTIAL OSCILLATIONS

Inertial current-velocity oscillations which have, at a fixed latitude φ, periods approaching $2\pi/f$ ($f = 2\Omega|\sin\varphi|$ is the Coriolis parameter at this latitude) occupy a special position among mesoscale oscillations with periods of many hours. These oscillations can occur by virtue of the fact that the inertial force in the motion of the water relative to the rotating earth may be offset by the inertial force in the transport motion (the rotation of the planet), that is, by the Coriolis force; thus, these oscillations are due entirely to the earth's rotation.

It appears that oscillations with the inertial period were first observed in the ocean by Ekman in 1930. These observations were made at a latitude of about $\varphi = 30°$, where the inertial period is about a day, so that the recorded oscillations were initially taken for diurnal tides. However, the random variations of the amplitude and especially of the phase of these oscillations, which were far from typical of tides, indicated that they were of inertial origin. Later, inertial oscillations were observed repeatedly at all latitudes and at various depths in the sea. Webster's summary [1] lists 23 measurements of this kind as of 1966. The amplitudes V of the observed velocity fluctuations ranged into the tens of centimeters per second, the oscillating water particles exhibiting a tendency to circular horizontal orbits (with a radius on the order of V/f, which is derived by equating the centrifugal and Coriolis forces) and clockwise motion in the Northern Hemisphere.

Special measurements, made by Webster and Fofonoff [2, 3] in the Sargasso Sea near latitude $\varphi = 30°$, to test Hendershott's theory [4] of the resonant excitation of inertial oscillations by tidal processes, demonstrated clear-cut intermittency of the inertial oscillations with typical excitation and decay times of several days. In the cases calculated by Webster [1], the coherence between inertial oscillations at different depths decreased rapidly with increasing vertical distance δz and was only 0.3 at $\delta z =$ 80 m; with increasing horizontal distance δx, the coherence decreased much more slowly and was still 0.7 at $\delta x = 3$ km.

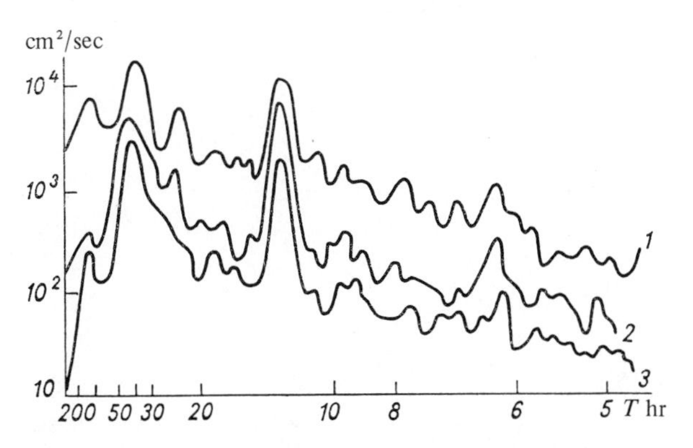

FIG. 4-2-1 Spectral energy density of horizontal current fluctuations at depths of 50 (1), 300 (2), and 1500 m (3), measured at the central buoy of the 1970 Atlantic test range (after Brekhovskikh et al. [12]).

The USSR Academy of Sciences Institute of Oceanology has been engaged in study of inertial oscillations for several years. Among such studies were the early measurements in the Black Sea and the Atlantic Ocean [5-8], the measurements on the seventh through ninth voyages of the *Akademik S. Vavilov* in the Mediterranean Sea, described by Titov [9, 10], who drew attention to small departures of the observed periods from $2\pi/f$ and to the presence of harmonics of the fundamental period (with amplitudes $A_n \approx A_1 n^{-5/3}$), and, finally, recent measurements on the long-term test ranges in the Indian Ocean [11] and in the Atlantic [12].

To cite an example from [12], Fig. 4-2-1 presents a plot of the spectral energy density of the horizontal-current fluctuations at depths of 50, 300, and 1500 m at one of the Atlantic test range buoys ($\varphi = 16°30'$N, $\lambda = 33°30'$W), as calculated from a five-month measurement series (February–September 1970); the plot shows a well-defined peak corresponding to the inertial period (about 42 hrs), and its harmonics up to the seventh order (the semidiurnal and four-day tidal periods are also distinct). Figure 4-2-2 shows the vertical energy variations of fluctuations with the inertial and semidiurnal tidal periods at another buoy on the Atlantic test range, calculated from a series of measurements extending over a month and a half. In this case the energy of the inertial oscillations exhibits a peak at the depth of the density discontinuity layer $z = 100$ m and can still be detected at least down to one-and-one-half kilometers. Figure 4-2-3 demonstrates the variability of the current velocity vector at a depth of 300 m over 12 days, with the tidal and still higher-frequency oscillations filtered out. This plot is fully consistent with the idea of circular liquid-particle orbits in inertial oscillations with a period of about 40 hours and with clockwise motion. We note the intermittency of the inertial oscillations, namely, the absence of these oscillations between the second and fourth days on this record. The coherence between the velocity components u and v in the inertial oscillations, which varied from 0.59 to 0.97 at various depths in the 100- to 1500-meter layer, that is, was quite high (averaging 0.86), and the phase difference, which varied from 249 to 292°, averaging 267° over the layer, that is, was close to the theoretical 270°, were calculated by Brekhovskikh et al. [12] from the same data as plotted on Fig. 4-2-2.

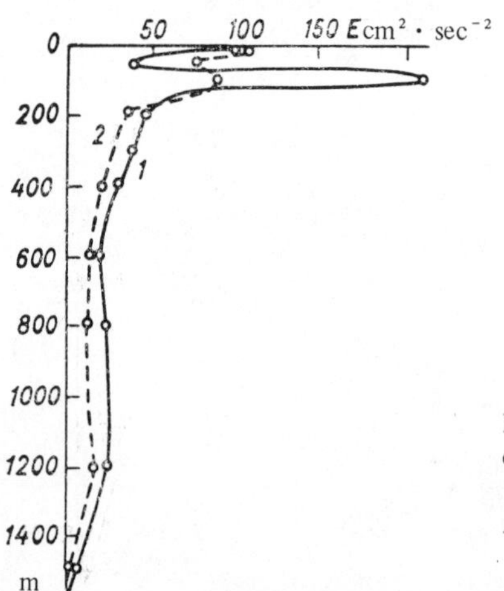

FIG. 4-2-2 Vertical distribution of the energy of oscillations with the inertial period (1) and with the 12.4-hr tidal period (2) at one of the buoys in the Atlantic test range (after Brekhovskikh et al. [12]).

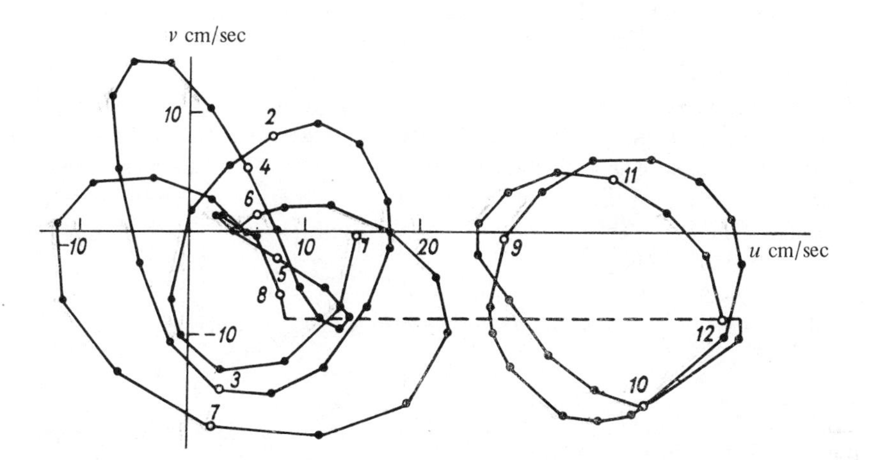

FIG. 4-2-3 Variability of current-velocity vectors (with tidal and high-frequency oscillations filtered out) over 12 days at a depth of 300 m, measured at the central buoy of the Atlantic test range (after Brekhovskikh et al. [12]). The dark circles are the tips of the 3-hr vectors; the numbers at the open circles are days of the month.

The rapid horizontal decay of the coherence of the inertial oscillations, which indicates that they are local, means that they can be described in approximation in terms of a flat-earth model having a constant Coriolis parameter. In this model, it is unnecessary to resort to the "traditional approximation" in the expression for the Coriolis force, since it is possible to use Eqs. (3-2-5) of internal-wave theory, with acoustic waves filtered out by the condition div $\mathbf{u} = 0$ and with the Coriolis acceleration $2\mathbf{\Omega} \times \mathbf{u}$ with $\mathbf{\Omega} =$ const added to the left-hand side of the first equation. The corresponding generalization of Eq. (3-2-7) for w takes the form

$$\frac{\partial^2 \Delta w}{\partial t^2} + [4(\mathbf{\Omega} \cdot \nabla)^2 + N^2 \Delta_h]\, w = -\frac{N^2}{g} \frac{\partial}{\partial t} \frac{\Delta_h p'}{\rho_0}. \tag{4-2-1}$$

An elementary model of the inertial oscillations is obtained from this in the case of a homogeneous ocean $N \equiv 0$ and an unbounded medium (gyroscopic waves in "pure" form, see [13, 14]). In this case, Eq. (4-2-1) has elementary wave solutions of the form $e^{i\chi}$, $\chi = \mathbf{k} \cdot \mathbf{x} + lz - \omega t$ with frequencies $\omega = \pm 2\Omega \cos \alpha$ (α is the angle between the vectors $\mathbf{\Omega}$ and $\boldsymbol{\varkappa}$). Here the velocity field has the form $\mathbf{u} = \mathbf{U} \cos \chi \pm \boldsymbol{\varkappa} \times \mathbf{U}/\varkappa \sin \chi$ ($\mathbf{U}$ is an arbitrary constant vector orthogonal to $\boldsymbol{\varkappa}$). This equation describes transverse waves with circular polarization. The phase velocity of these waves is $\mathbf{c} = 2(\mathbf{\Omega} \cdot \boldsymbol{\varkappa}/\varkappa^3)\boldsymbol{\varkappa}$, and their group velocity $\mathbf{c}_g = (2/\varkappa^3)\boldsymbol{\varkappa} \times (\mathbf{\Omega} \times \boldsymbol{\varkappa})$. Their maximum frequency is reached at $\boldsymbol{\varkappa} \| \mathbf{\Omega}$ (the velocities $\mathbf{u}$ are then perpendicular to $\mathbf{\Omega}$) and equals $2\mathbf{\Omega}$, which corresponds to the semidiurnal period (in this case $p' \equiv 0$, $w \equiv 0$ and the motion indeed is due to inertia). As $\omega/\Omega \to 0$, the wave vector $\boldsymbol{\varkappa}$ becomes orthogonal to $\mathbf{\Omega}$, and the velocity field becomes constant in the direction of $\mathbf{\Omega}$. Gyroscopic waves may in this case undergo resonant three-wave interactions. These were calculated by Hughes [15].

Sea stratification may strongly affect the inertial waves. To describe this effect, one must use the complete equation (4-2-1), whose right-hand side can be expressed in terms of w with the aid of the horizontal projections of the equations of motion. However, the sum total of this right-hand side is relatively small, and, following Phillips [14], we shall neglect it here to simplify the arguments. But even with this simplification, Eq. (4-2-1) can be analytically solved only for rather simple models of function $N(z)$. However, given the low coherence of the inertial oscillations at even short vertical distances, we shall, again following Phillips [14], deal here only with their local description, in which the Brunt-Väisälä frequency will be taken as quasiconstant. With $N \approx$ const, the simplified equation (4-2-1) will again have wave solutions of the form $e^{i\chi}$, but the frequencies ω will now be defined by

$$\omega^2 = (2\Omega)^2 \cos^2 \alpha + N^2 \cos^2 \theta, \tag{4-2-2}$$

where θ is the slope of the wave vector $\boldsymbol{\varkappa}$ with the horizontal plane.

Note that by virtue of the condition N = const, the dispersion relation (4-2-2) describes either gyroscopic waves or internal gravity waves, depending on the relation between N and Ω. These types of waves cannot exist simultaneously in a medium with N = const (see the analysis in Sec. 4-1).

It is easily shown that the extrema ω_m of the function $\omega(\boldsymbol{\varkappa})$ are roots of the equation $\omega_m^2(\omega_m^2 - 4\Omega^2) = N^2(\omega_m^2 - 4\Omega^2 \sin^2 \varphi)$ where φ is the local geographic latitude; at $N \gg \Omega$, we have $\omega_{min}^2 \approx 4\Omega^2 \sin^2 \varphi$ and $\omega_{max}^2 \approx N^2$.

A fluid parcel in a wave with frequency ω and wave vector $\boldsymbol{\varkappa}$ lying, for example, in the meridional plane, moves in the plane orthogonal to $\boldsymbol{\varkappa}$ and is described in that plane by the coordinates $x(t)$, $y(t)$ of the parcel, where the x coordinate increases eastward and the y axis lies in the meridional plane. The linearized ($\varphi \approx$ const) Lagrangian equations of motion of fluid parcels under the action of Coriolis and buoyant forces in the plane orthogonal to $\boldsymbol{\varkappa}$ have the form [14]

$$\frac{d^2x}{dt^2} - 2\Omega \frac{dy}{dt} \cos(\varphi - \theta) = 0; \quad \frac{d^2y}{dt^2} + 2\Omega \frac{dx}{dt} \cos(\varphi - \theta) + yN^2 \cos^2\theta = 0. \tag{4-2-3}$$

The solutions of the Lagrangian equations are the functions $x = X \cos \omega t$ $y = -Y \sin \omega t$, where the frequency is given by Eq. (4-2-2) and the amplitudes stand in the ratio $X/Y = (2\Omega/\omega) \cos(\varphi - \theta)$, that is, the path of the fluid parcel is an ellipse with its minor axis on the x axis. As $\theta \to \pi/2$ the effect of buoyant forces (i.e., stratification) vanishes, leaving the purely inertial motion along a circle in the horizontal plane with frequency $\omega_{min} = f$. It is clear from this that a gravity wave generated in equatorial waters and having a fixed frequency ω can propagate along the meridian only as far as its critical latitude φ_c, which can be found from the condition 2 sin $\varphi_c = \omega$. At that latitude the above wave degenerates into a purely inertial oscillation with a vertical wave vector $\boldsymbol{\varkappa}$.

More detailed description of the propagation of internal gravity waves along meridians (if it is still necessary, in spite of the observed rapid decay of their coherence along the horizontal) requires allowance for the spherical curvature of the earth and application of the general theory derived in the preceding section. Such calculations were made (for N = const) by Hughes [15]. Munk and Phillips [16] derived an analytic model of these waves in a spherical geometry, using the equation for $V^* = V \cos \varphi$ that follows from the Laplace tidal equations (4-1-7) and has the form

$$\cos \varphi \frac{\partial}{\partial \varphi} \cos \frac{\partial V^*}{\partial \varphi} + \Big[\varepsilon a^2 \Omega^2 \cos^2 \varphi \left(\frac{\omega^2}{\Omega^2} - 4 \sin^2 \varphi \right) - \\ - \frac{2m\Omega}{\omega} \cos^2 \varphi - m^2 \Big] V^* = \frac{4 \varepsilon a^2 \Omega \omega \sin \varphi \cos^2 \varphi}{m^2 - \varepsilon a^2 \omega^2 \cos^2 \varphi} \left(\frac{\omega \cos \varphi}{2\Omega} \frac{\partial V^*}{\partial \varphi} - m V^* \sin \varphi \right). \tag{4-2-4}$$

The expression in the square brackets on the left-hand side of this equation acts as the squared local meridional wave number β_*^2, while $\alpha_* = m/\cos \varphi$ is the local latitudinal wave number. Furthermore, it can be verified with the aid of Eqs. (4-1-7), after simplifying them in accordance with the condition div $\mathbf{u} = 0$, that $l = N\varepsilon^{1/2}$ acts as the local vertical wave number and that for the discrete values of l permitted by the dispersion relation, we have large dimensionless quantities $\gamma = 2a\Omega\varepsilon^{1/2}$. One would expect the peaks of the frequency spectra of the waves to correspond to the extreme values of the function $\omega(\alpha_*, \beta_*, \gamma)$ in three-dimensional wave space [17]. Physically meaningful extremes are obtained for $\omega = f$ and for $\beta_* = 0$. The second of these conditions gives a more accurate definition than the first one for the critical latitude φ_c above which the waves cannot propagate, but as $\gamma \to \infty$ this critical latitude tends to the value defined by the condition $\omega = f$. Therefore, the peaks in the spectra of these waves would be expected to occur at the inertial frequency.

A solution of (4-2-4) for the neighborhood of the critical latitude φ_c was derived by Munk and Phillips [16]; it is accurate to terms of the order of $\gamma^{-1/2}$ (and was used to derive solutions of the dynamic equations for u, v, w, p' and ρ'). It is expressed in terms of Airy functions of the argument $\eta = \varphi - \varphi_c/L + \alpha_*^2 L^2 - \omega - 2\Omega \sin \varphi_c / 2L\Omega \cos \varphi_c$, where $L = (\gamma^2 \sin 2\varphi_c)^{-1/2}$ is the Airy scale for the latitude φ_c. The Airy function chosen by Munk and Phillips [16] for local description does not increase as $\eta \to \infty$ (this eliminates the probably short-lived situations in which the dominant peak of the wave spectrum is not near the inertial frequency). Munk and Phillips made an attempt to use the successive extremes of this Airy function in order to explain the band structure of the spectra of inertial oscillations that were presented in [3, 4]. These sometimes have secondary peaks, in addition to the fundamental one. Munk and Phillips [16] suggested that the observed small spatial scales of coherence of inertial oscillations are due to the existence of many oscillation modes with independent phases. In that case the coherence scale is inversely proportional to the span of the range of wave numbers of the modes, and if this range is on the order of an octave, a value of 200 meters is obtained for the vertical coherence scale, while the horizontal scale is of the order of the Airy scale L of about 10 miles.

An exact nonlinear theory was derived by Monin [18] for purely inertial oscillations (with a vertical wave vector) on the basis of the equations

$$\frac{du}{dt} - 2\Omega v \sin \varphi - \frac{uv \tan \varphi}{a} = 0;$$

$$\frac{dv}{dt} + 2\Omega u \sin \varphi + \frac{u^2 \tan \varphi}{a} = 0. \tag{4-2-5}$$

It is convenient to treat these equations as Lagrangian, setting $u = a(d\lambda/dt) \cos \varphi$ and $v = a(d\varphi/dt)$, where $\varphi(t)$ and $\lambda(t)$ are the spherical coordinates of the moving fluid parcel. These equations give a nonlinear spherical generalization of Eqs. (4-2-3) at

$\theta = \pi/2$. They have the following energy and angular-momentum integrals:

$$u^2+v^2=u_0^2+v_0^2;$$
$$(u+\Omega a \cos \varphi)\, a \cos \varphi=(u_0+\Omega a \cos \varphi_0)\, a \cos \varphi_0, \qquad (4\text{-}2\text{-}6)$$

where the subscripts "0" indicate the initial values of the time functions. Eliminating u and setting $u_0 = U \cos \alpha$, $v_0 = U \sin \alpha$, we obtain the following equation for $\mu = \sin \varphi$:

$$\left(\frac{d\mu}{dt}\right)^2=\omega^2 (\mu^2-\mu_1^2)(\mu_2^2-\mu^2);$$
$$\mu_{1,2}^2=1-(\xi \pm \delta)^2; \quad \xi=\frac{U}{2a\Omega};$$
$$\delta=[(\cos \varphi_0+\xi \cos \alpha)^2+\xi^2 \sin^2 \alpha]^{1/2}, \qquad (4\text{-}2\text{-}7)$$

which can be integrated in elliptic functions. The resulting paths of the fluid parcels lie in the range of latitudes defined by the inequality $|\xi - \delta| \leqslant \cos \varphi \leqslant \xi + \delta$. They do not cross the equator if $\xi + \delta \leqslant 1$, but otherwise do cross it. In the former case, the frequency of the oscillations is given by

$$\omega=\pi\Omega\mu_2 \left[K\left(\frac{\sqrt{\mu_2^2-\mu_1^2}}{\mu_2}\right)\right]^{-1} \approx 2\Omega\,|\sin \varphi_0|\left(1-\xi\frac{\cos \varphi_0 \cos \alpha}{\sin^2 \varphi_0}\right), \qquad (4\text{-}2\text{-}8)$$

where $K(s)$ is a complete elliptic integral of the first kind with modulus s, and where the fact that ξ is small in real currents has been used in defining the right-hand side of the equation.

In conclusion, a few words on the mechanism of generation of internal oscillations are in order. The previously mentioned theory of Hendershott [4], which postulates that they are produced by tidal processes via resonance excitation, is valid only for latitudes around 30°, where the inertial period is close to the period of the semidiurnal tide, but even in this particular case it does not explain the pronounced intermittency of the inertial oscillations. Webster [1] attributes their excitation to atmospheric disturbances, namely, wind friction during storms (the development of inertial oscillations after storms has been confirmed by the observations of Gonella [19], and the relation of their amplitude to the wind by the observations of Hunkins [20]. This wind friction creates an effect that should decay rapidly with increasing depth (the corresponding theoretical model was derived by Pollard [21], see also the paper by Belyayev and Kolesnikov [22]). Webster also assumes that variations of the atmospheric pressure on the sea surface of the ocean, whose effect can manifest itself at all depths, are a contributing factor. Hasselmann [23] thinks that generation of inertial oscillations involves nonlinear interaction between high-frequency gravity waves. Finally, it is pointed out by Munks and Phillips [16] that randomly distributed sources of excitation with an inertial-frequency bandwidth of the order of the Airy scale L would produce inertial oscillations with too long a lifetime (L^{-1} of about 100 days), and that local coherent generation would be inconsistent with the rapid horizontal decay of coherence. Thus the properties of real excitation should be intermediate between these two extreme cases.

REFERENCES

1. Webster, F. Observations of inertial-period motions in the deep sea, Rev. Geophys., **6**, No. 4, 473–490 (1968).
2. Day, C. G. and F. Webster. Some current measurements in the Sargasso Sea, Deep-Sea Res., **12**, No. 6, 805–814 (1965).
3. Webster, and N. Fofonoff. A compilation of moored current meter observations, Woods Hole Oceanogr. Inst. Ref., **3**, Nos. 67–66, 105 (1967).
4. Hendershott, M. C. Internal Oscillations of Tidal Period, Ph. D. Thesis, Harvard Univ., Cambridge (1964).
5. Yampol'skiy, A. D. Internal waves in the Black Sea from observations at a multiday buoy station, Trudy Inst. Okean. Akad. Nauk SSSR, **39**, 111–126 (1960).
6. Yampol'skiy, A. D. Variations of the hydrological elements with the inertial period, Izv. Akad. Nauk SSSR, Ser. geofiz., No. 3, 445–452 (1961).
7. Yampol'skiy, A. D. Internal waves in the northeastern Atlantic Ocean, Trudy Inst. okean. Akad. Nauk SSSR, **56**, 229–240 (1962).
8. Yampol'skiy, A. D. A method for calculation of inertial oscillations of the hydrological elements, Trudy Inst. okean. Akad. Nauk SSSR, **66**, 142–148 (1963).
9. Titov, V. B. Simple practical procedures for estimating the spectrum of tidal and inertial currents in the sea, Okeanologiya, **8**, No. 3, 514–520 (1968).
10. Titov, V. B. Certain features of mesoscale motions in the ocean, Okeanologiya, **13**, No. 6 (1973).
11. Shtokman, V. B., M. H. Koshlyakov, R. V. Ozmidov, L. M. Fomin, and A. D. Yampol'skiy. Extended-term measurements of the variability of the physical fields on oceanic test ranges as a new phase in study of the ocean, Dokl. Akad. Nauk SSSR, **186**, No. 5, 1070–1074 (1969).
12. Brekhovskikh, L. M., G. N. Ivanov-Frantskevich, M. N. Koshlyakov, K. N. Fedorov, L. M. Fomin, and A. D. Yampol'skiy. Certain results of a hydrophysical experiment on a test range in the tropical Atlantic, Izv. Akad. Nauk SSSR, Fizika atm. i okeana, 7, No. 5, 511–528 (1971).
13. Phillips, O. M. Energy transfer in rotating fluids by refraction of inertial waves, Phys. Fluids, **6**, No. 4, 513–520 (1963).
14. Phillips, O. *The Dynamics of the Upper Ocean*, Cambridge University Press (1966).
15. Hughes, B. A. On Inertial Waves in a Rotating Medium, Dissertation, Cambridge Univ. (1964).
16. Munk, W. and N. Phillips. Coherence and band structure of inertial motion in the sea, Rev. Geophys., **6**, No. 4, 447–472 (1968).
17. Blandford, R. Mixed gravity-Rossby waves in the ocean, Deep-Sea Res., **13**, No. 5, 941–962 (1966).
18. Monin, A. S. Inertial motions on a rotating sphere, Izv. Akad. Nauk SSSR, Fizika atm. i okeana, **8**, No. 10, 1035–1041 (1972).
19. Gonella, J. A local study of inertial oscillations in the upper layers of the ocean, Deep-Sea Res., **18**, No. 8, 775–788 (1971).
20. Hunkins, K. Inertial oscillations of Fletcher's Ice Island (T-3), J. Geophys. Res., **72**, No. 4, 1165–1183 (1967).
21. Pollard, R. T. On the generation by winds of inertial waves in the ocean, Deep-Sea Res., **17**, No. 4, 795–812 (1970).
22. Belyayev, V. S. and A. G. Kolesnikov. Cause of inertial oscillations in pure drift currents, Izv. Akad. Nauk SSSR, Fizika atm. i okeana, **2**, No. 10, 1104–1107 (1966).
23. Hasselmann, K. Wave-driven inertial oscillations, Geophys. Fluid Dyn., **1**, No. 4, 463–502 (1970).

4-3 TIDAL OSCILLATIONS

Thus far, we have considered only free oscillations in the ocean. Now, however, we shall turn to tidal oscillations, that is, forced oscillations driven by the gravitational attraction of the moon and sun, and a factor of great importance in the behavior of the sea. Their theory dates back to Newton, Laplace, and George Darwin.

At a point on the earth with geocentric radius vector $\mathbf{r}$ the potential U of the tide-generating force produced, let us say, by the moon, is the difference between the Newtonian potentials $-Gm_☽/|\mathbf{r}_☽ - \mathbf{r}|$ and $-Gm_☽/r_☽$ of the moon's attraction for a unit mass at point $\mathbf{r}$ and at the center of the earth, less the potential $-(Gm_☽/r_☽^2)\, r \cos\vartheta$ of the moon's approximately uniform gravitational field, which imparts to the entire mass of the earth an acceleration $Gm_☽/r_☽^2$ in the direction of the moon's geocentric radius vector $\mathbf{r}_☽$, that is,

$$U_☽ = Gm_☽\left(\frac{1}{r_☽} - \frac{1}{|\mathbf{r}_☽ - \mathbf{r}|} + \frac{r\cos\vartheta}{r_☽^2}\right) \approx \frac{3}{2}\frac{Gm_☽ r^2}{r_☽^3}\left(\frac{1}{3} - \cos^2\vartheta\right). \quad (4\text{-}3\text{-}1)$$

Here G is the gravitational constant, $m_☽$ is the mass of the moon, ϑ is the angle between the vectors $\mathbf{r}$ and $\mathbf{r}_☽$, that is, the zenith angle of the moon at point $\mathbf{r}$, which is given by $\cos\vartheta = \sin\varphi \sin\delta_☽ + \cos\varphi \cos\delta_☽ \cos(\tau_☽ - \pi)$, where φ is the geographic latitude of point $\mathbf{r}$, $\delta_☽$ is the declination of the moon (the angle between the line to the moon and the plane of the equator), and $\tau_☽$ is the moon's hour angle (the angle between the plane of the meridian passing through point $\mathbf{r}$ and the plane passing through the moon and the celestial poles, reckoned eastward from the moon's lower culmination). The right-hand side of Eq. (4-3-1) is the principal term in the series expansion of the exact expression for U in powers of $r/r_☽ \approx 1/60$; it is proportional to the second-order Legendre polynomial $P_2(\cos\vartheta)$. The potential of the sun's tide-generating force has a similar form. The ratio of the solar to the lunar potentials is of the order of $\frac{m_\odot}{m_☽}\left(\frac{r_☽}{r_\odot}\right)^3 \approx 0.46$, so that the solar tides are, roughly speaking, half as large as the lunar tides.

Substituting the above expression for $\cos\vartheta$ in the right-hand side of Eq. (4-3-1), we obtain

$$U_☽ \approx -\frac{3}{2}\frac{Gm_☽ r^2}{r_☽^3}\left[\frac{3}{2}\left(\sin^2\varphi - \frac{1}{3}\right)\left(\sin^2\delta_☽ - \frac{1}{3}\right) + \right.$$

$$\left. + \frac{1}{2}\sin 2\varphi \sin 2\delta_☽ \cos(\tau_☽ - \pi) + \frac{1}{2}\cos^2\varphi \cos^2\delta_☽ \cos 2\tau_☽\right]. \quad (4\text{-}3\text{-}2)$$

The first term in the square brackets is symmetric about the earth's axis and depends on latitude as $P_2(\sin\varphi)$ (vanishing at latitudes $\varphi = \pm 35°16'$); it depends on time primarily via the factor $\sin^2\delta_☽ - 1/3$ and, consequently, has a period equal to half the period of the moon's declination, that is, half of the *tropical* month or 13.660791 mean solar days (or, for the solar tide, to half of the tropical year, 182.62110 mean solar days). The corresponding tides are known as the *long-period* tides (Laplace tidal oscillations of the first kind). Note that since the time-averaged value of $\sin^2 \delta_{☽}$ is not zero, the inclination of the moon's orbit (or of the ecliptic) to the equator introduces a constant term into U.

The dependence of the second term in Eq. (4-3-2) on latitude is defined by the associated Legendre function $P_2^1(\sin \varphi)$. This term vanishes on the equator and on the meridian 90° away from the perturbing celestial body (i.e., moon or sun), and is largest on the meridian of that celestial body at latitudes $\varphi = \pm 45°$. This term depends on time primarily via the hour angle of the given celestial body so that its period is equal to the lunar or solar day. However, a long-period time dependence, primarily due to factor $\sin 2\delta_{☽}$, is also superimposed on it. The corresponding tides are called *diurnal* tides (Laplace oscillations of the second kind).

The dependence of the third term depends on latitude defined by the associated Legendre function $P_2^2(\sin \varphi)$. It vanishes on meridians 45° to the east and west of the perturbing celestial body. The time dependence of this term via the hour angle of the given celestial body has a period of half the lunar or solar day and an amplitude proportional to $\cos^2 \delta_{☽}$. The amplitude is therefore largest when the celestial body is over the equator. The corresponding tides are known as *semidiurnal* (oscillations of the third kind).

The tidal potentials of the moon and sun and (in the linear approximation) the corresponding tides are algebraically additive. Since the lunar day is 50.47 minutes longer than the solar day, the phase difference between the lunar and solar tides continues to accumulate, and the amplitude of the resultant tide varies with a period equal to half the *synodic* month (14.7653 solar days), that is, half of the interval between times at which the positions of the sun and moon coincide (i.e., between new moons). The highest amplitude is reached at *syzygy*, that is, 1 to 2 days (*the age of the tide*) after new moon and full moon, and the lowest amplitude at the *quadratures*, again about 1 to 2 days after the moon's first and last quarters.

This inequality of the amplitudes of successive semidiurnal tides is called the *semimonthly* (*phase*) inequality. The *parallactic* (*monthly*) inequality is produced by the variations of $r_{☽}$ (owing to the deviation of the actual moon's orbit from a circular orbit) with a period equal to the *anomalistic* month (29.5546 solar days); the highest tide usually occurs in this case 2 to 3 days after the moon's closest approach to the earth (the lunar perigee). The *declinational* inequality, which has a period of half the *tropical* month, is produced by variations of the moon's declination. The highest tide in this cycle occurs soon after the moon reaches its highest northern and southern declinations, and the lowest tide soon after its passage across the equator. The *diurnal* inequality in the height and time of occurrence of successive high or low tides has a period of one tropical month; it is largest soon after the moon reaches its highest northern or southern declination. There are also long-period inequalities, most important among which is the Saros cycle of about 18 years, 11 days, which corresponds to return of the sun, moon, and the nodes of the lunar orbit to the same positions on the celestial sphere.

All of the possible tidal inequalities can be precalculated by using astronomical methods to describe the time dependence of the geocentric coordinates of the moon $(r_{☽}, \delta_{☽}, \tau_{☽})$ and sun $(r_{\odot}, \delta_{\odot}, \tau_{\odot})$ in the exact expression for the total potential $U = U_{☽} + U_{\odot}$ of the tide-producing force. Motion in the sun-earth-moon system is characterized by six independent periods: $\tau_1 = 24^h 50.47^m$ (the lunar day), $\tau_2 = 27.321582$ days (the tropical month), $\tau_3 = 365.242199$ days (the year), $\tau_4 = 8.847$ years (the period of revolution of the *line of apsides* of the moon's orbit, a line which joins the orbit's perigee and apogee), τ_5 = 18.613 years (the period of revolution of the line of the moon's nodes), and τ_6 = 20,940 years (the period of variation of the longitude of the sun's perigee that results from rotation of the line of nodes and line of apsides of the earth's orbit). If ω_i are the corresponding frequencies, $\omega_1+\omega_2-\omega_3$ is the diurnal frequency. Thus, the potential U for a given locality can be expanded in a trigonometric time series

$$U=\sum_{\mathbf{n}} c_{\mathbf{n}} \cos\left(\sum_{i=1}^{6} n_i\omega_i t+\theta_{\mathbf{n}}\right), \qquad (4\text{-}3\text{-}3)$$

where $n_i = 0, \pm 1, \pm 2, \ldots$ and $\mathbf{n} = (n_1, n_2, \ldots, n_6)$.

Doodson [1] calculated the 396 largest terms of this series ("tides"), numbering them with six-digit indices n_1, n_2+5, n_3+5, n_4+5, n_5+5, n_6+5. The most important of the constant and long-period tides in this expansion are: the constant lunar tide, denoted by the symbol M_0 and having the Doodson index 055.555; the constant solar tide (S_0 = 055.555); the elliptic tide of order S_0 (Sa = 056.554); the declinational tide of S_0 (Ssa = 057.555); the evectional tide of M_0 (MSm = 063.655); the elliptic tide of order M_0 (Mm = 065.455); the variational tide of M_0 (MSf = 073.555); the declinational tide of M_0 (Mf = 075.555).

The most important among the diurnal tides are: the variational tide to O_1 (σ_1 = 127.555); the elliptic tide of the order of O_1 (Q_1 = 136.655); the evectional tide of O_1 (ρ_1 = 137.455); the diurnal principal lunar tide (O_1 = 145.555); the elliptic tide of order K_1 (NO_1 = 155.655); the diurnal principal solar tide (P_1 = 163.555); the diurnal principal declinational tide (K_1 = 165.555); the elliptic tide of order K_1 (J_1 = 175.455); and the second-order diurnal declinational tide (OO_1 = 185.555).

Finally, the most important semidiurnal tides are: the second-order elliptic tide of M_2 ($2N_2$ = 235.755); the larger variational tide of M_1 (μ_2 = 237.555); the first-order larger elliptic tide of M_2 (N_2 = 245.655); the larger evectional tide of M_2 (ν_2 = 247.455); the semidiurnal principal lunar tide (M_2 = 255.555); the smaller first-order elliptic tide of M_2 (L_2 = 265.455); the large first-order elliptic tide of S_2 (T_2 = 272.556); the semidiurnal principal solar tide (S_2 = 273.555); the semidiurnal declinational tides of M_2 and S_2 (K_2 = 275.555). Let us also mention a third, the diurnal principal lunar tide (M_3 = 355.555). All of these tides have been listed in order of increasing frequency.

The potential of the tide-producing force is not, of course, the tide itself. But in the so-called *static* theory of tides it is assumed that the free surface of the sea responds practically instantaneously to the application of the tide-producing force and assumes the equilibrium shape $z=\bar{\zeta}(\varphi, \lambda, t)$ of the equipotential surface $U+g\bar{\zeta}=$ const (here, $g\bar{\zeta}$ is the potential of the gravitational force and const does not depend on φ or λ). Hence

$$\bar{\zeta}=-\frac{U}{g}+c(t). \qquad (4\text{-}3\text{-}4)$$

"Darwin's correction" $c(t)$ is found from the condition that the total volume of the tide in the sea, $\int_S \bar{\zeta}\, d\Sigma = 0$, where the integration is carried over the entire area S of the ocean. Therefore $c = \frac{1}{gS} \int_S U d\Sigma$. For the sea covering the entire earth, $c = 0$ and $\bar{\zeta}$ differs from U only by a constant factor. To within the right-hand side of Eq. (4-3-1), this "equilibrium tide" is highest (and equal to $2H/3$ where, say, for the moon, $H_{☽} = \frac{3}{2} \frac{Gm_{☽} r^2}{g r_{☽}^3} \approx 53.7$ cm) where the perturbing body is at the zenith or nadir and lowest (equal to $-H/3$) where the perturbing body is on the horizon. The correction $c(t)$ was calculated by Thomson and Tait in [2] for the real sea. Because of this correction, the time of high tide at a given location may not coincide with that of the maximum of the perturbing force. Assuming a sea covering the entire earth, the static theory makes it also easy to correct for the gravity of the tide bulge to within the right-hand side of Eq. (4-3-1). This correction involves decreasing U by $\frac{3}{5} \frac{\rho}{\rho_*} g\bar{\zeta}$ and, consequently, means an increase in $\bar{\zeta}$ by a factor $\left(1 - \frac{3}{5} \frac{\rho}{\rho_*}\right)^{-1}$, where ρ_* is the average density of the earth and the water (see, for example, [3, Chap. VIII]); at $\rho/\rho_* = 0.18$, the correction amounts to a 12% increase in the static tide.

Most measurements of the tidal oscillations of sea level are made at coastal stations (usually at harbor entrances) with the aid of tide staffs, which are accurate to about 1 cm in single measurements. There are about 10,000 of these stations worldwide, and dozens of them have accumulated continuous series of hourly measurements for more than 50 years (totaling about 10^7 readings). However, these data apply only to coastal areas, where shorelines and sea-bottom relief produce local anomalies in the tidal amplitudes and phases and where "noise" (i.e., sea-level oscillations of nontidal origin, especially storm-produced) is strongest. Sea-bottom *marigraphs* with sensitive pressure sensors have been used in recent years to measure tides in the open sea. Thus, the instrument of Munk and Snodgrass [4, 5] produces a monthly record with noise on the order of only one thousandth of the signal. The measurements are subjected to harmonic analysis, that is, they are statistically interpreted to determine the amplitudes C_n and phases θ_n of at least the principal terms of the series Eq. (4-3-3) for the given geographic point. A series of this kind with empirically determined parameters can be used to forecast the tides at a given point. This *harmonic method* was developed long ago by Kelvin and Darwin (note that its accuracy is affected primarily by the presence of noise).

The statistical theory is correct only as far as the frequencies of the tidal oscillations are concerned. However, the amplitudes and phases of real tides have little in common with those of the equilibrium tide Eq. (4-3-4). The phases of the real tides do not coincide with those of the tide-producing forces. This phase difference may vary strongly from place to place, sometimes even varying by many hours over short distances (for example, the tides on coastal shoals usually lag far behind those in nearby open sea because of the braking action of bottom friction on the tidal currents). In some regions of the sea, the semidiurnal and diurnal tides form *amphidromes*, that is, waves with nodal lines that rotate with the period of the tide around certain geographic points (in Proudman's model [6], such an amphidrome is formed in a rectangular basin upon interference between two perpendicular standing waves with the same period and a phase difference of a quarter-period). At the centers of amphidromes and at the nodes of the standing tidal waves the tide amplitudes drop to almost zero, but they increase to 1.5 or 2 meters at the maximum distances from these points, and may exceed 10 meters at certain coastal points (especially in bays that taper into the continent).

Naturally, the above-described failure of the static tide theory is due to the fact that it totally ignores the properties of the sea acting as a mechanical system, for example, its inertia. Because of inertia, the sea surface cannot respond instantaneously to the action of the tide-producing forces. The static theory also neglects the resonant properties of the sea, that is, the possibility of development of natural vibrations with well-defined periods, propagation velocities and ability to be reflected from the coast. It also neglects frictional forces. One would tend to assume that at low frequencies, the sea response to the tide-producing forces will approach the static response. However, measurements indicate that even the semimonthly lunar tide *Mf* differs appreciably from the equilibrium tide. Its amplitudes are about a third lower than the static amplitudes because of the effect of the tides in the solid earth (which, in general, must often be taken into account in describing ocean tides, as must the obverse effect of the ocean tides on the tidal deformation of the solid earth). Moreover, according to Wunsch [7], the *Mf* and *Mm* tides have the properties of superposed fairly short Rossby waves, which are strongest in the western parts of the ocean; this may also hold for tides with still longer periods [8].

To derive a fully adequate theory of tidal oscillations, one must start with the general equations of ocean dynamics whose linear approximation is given in Sec. 4-1. To the right-hand sides of the equations of motion (4-1-1) we need to add the potential gradient $g\nabla\zeta$ of the tide-producing forces. Then the continuity equation can be taken in the form div $\mathbf{u} = 0$, thereby filtering out acoustic waves but leaving the tidal oscillations, which have much longer periods, practically undistorted. These equations will describe both the tidal oscillations of the sea surface $z = \zeta(\varphi, \lambda, t)$, as well as internal waves with tidal periods, which are very evident in the ocean and are apparently excited by tidal currents (see Figs. 4-2-1 and 4-2-2 and the theory in Sec. 4-2, as well as [9, Sec. 5.7; 10, Sec. 163]).

Since the vertical accelerations of fluid parcels in tidal oscillations are very small compared to the acceleration of gravity (and apparently even by comparison with the acceleration of the buoyant forces $g\rho'/\rho_0$), tide theory generally employs the *quasistatic approximation*, that is, the third equation of motion in Eq. (4-1-1) is replaced by the static equation $\partial p/\partial z = \rho g$. To ensure conservation of energy, the "traditional approximation" is also used here for the Coriolis force (see the discussion of these approximations in Sec. 4-1 and also paper [8]). While this approximation does not significantly simplify the description of internal waves with tidal periods in the case of a sea of variable depth, it does so in the case of tidal oscillations in a sea of *homogeneous density*. Indeed, in the quasistatic approximation for a homogeneous ocean, $(1/\rho)\nabla_h p$ is independent of the vertical coordinate (see, for example, [11]), and can be replaced by $g\nabla_h\zeta$. In this case the horizontal velocity will also be independent of z. Integrating the continuity equation div $\mathbf{u} = 0$ with this condition over the entire depth of the sea, and eliminating the values of w on its surface and floor with the aid of appropriate kinematic conditions, we derive the third of the herein cited *Laplace tidal equations* for the homogeneous ocean:

$$\frac{\partial u}{\partial t} - 2\Omega v \sin\varphi = -\frac{g}{a\cos\varphi}\frac{\partial(\zeta-\bar{\zeta})}{\partial\lambda};$$

$$\frac{\partial v}{\partial t} + 2\Omega u \sin\varphi = -\frac{g}{a}\frac{\partial(\zeta-\bar{\zeta})}{\partial\varphi};$$

$$\frac{\partial\zeta}{\partial t} + \frac{1}{a\cos\varphi}\left(\frac{\partial uH}{\partial\lambda} + \frac{\partial vH\cos\varphi}{\partial\varphi}\right) = 0, \tag{4-3-5}$$

where H is the depth of the sea and, by contrast with previous chapters, ζ is reckoned *upward* from the undisturbed level of the sea. For free oscillations ($\bar{\zeta} = 0$) in a sea of constant depth ($H = \text{const}$) with $p' = g\zeta$ and $\varepsilon = 1/gH$, this yields Eqs. (4-1-7), which for this reason we also called Laplace tidal equations. Much of the dynamic theory of tides is concerned with the solution of Eqs. (4-3-5). The analytic solutions investigated involve simple geometric models of the ocean that yield several qualitative inferences as to the properties of tides, and numerical solutions for oceans with realistic shorelines and bottom topography are derived.

Considering oscillations with a specified frequency ω whose characteristics depend on time as $e^{-i\omega t}$, and eliminating the functions u and v from Eqs. (4-3-5), we obtain the following equation for the function ζ:

$$\Delta^*(\zeta-\bar{\zeta}) + \frac{a^2\omega^2}{gH}\zeta = 0;$$

$$\Delta^* = \frac{\omega^2}{H\cos\varphi}\left[\frac{\partial}{\partial\varphi}\frac{H\cos\varphi}{\omega^2-f^2}\left(\frac{\partial}{\partial\varphi} - \frac{if}{\omega\cos\varphi}\frac{\partial}{\partial\lambda}\right) + \right.$$
$$\left. + \frac{1}{\omega^2-f^2}\frac{\partial}{\partial\lambda}H\left(\frac{if}{\omega}\frac{\partial}{\partial\varphi} + \frac{1}{\cos\varphi}\frac{\partial}{\partial\lambda}\right)\right]. \tag{4-3-6}$$

At $H = \text{const}$, the operator Δ^* becomes the Laplace tidal operator, which in the case $\Omega = 0$ degenerates into the two-dimensional Laplace operator on the surface of a unit sphere. In the latter case, the free waves in a flat model will be described by $\Delta_h\zeta + \frac{\omega^2}{gH}\zeta = 0$ and, consequently, the elementary waves $\zeta \sim e^{i\mathbf{k}\cdot\mathbf{x}}$ will have phase velocities $\omega/k = \sqrt{gH}$ which, like the surface waves, are long compared to the depth H. Thus, the theory of the tidal equations (4-3-5) proves to be part of the theory of *long waves.* Note that the velocities of propagation of free waves $\sqrt{gH}$ in a deep ocean are of the same order of magnitude as the velocities $\Omega a \cos\varphi$ of the propagation of tidal disturbances that is created by the earth's rotation. This is the qualitative explanation for the generation of amphidromes.

To solve Eq. (4-3-6), one uses the function $\bar{\zeta}(\varphi, \lambda)$ from the static theory to solve the boundary-value problem, assuming that the velocity component normal to the shoreline is zero at the shoreline (this condition is easily written in terms of the first derivatives of the function ζ). Analytic solutions were generally derived for an ocean of constant depth (and, in a few papers, for a depth H that varies only with latitude) that covers the entire globe or is bounded by fixed parallels and meridians.

Thus, Kelvin and Darwin showed that in an ocean *whose depth is constant over the entire globe*, long-period tides are smaller than the static tides by at least a factor of two and are in phase with the perturbing force (the corresponding solutions for polar and zonal basins were derived by Goldsbrough [12, 13], using power series in $\sin\varphi$, and by Sretenskiy [14], who used asymptotic expansions). For diurnal tides ($\omega \approx \Omega$) that depend on longitude as $e^{i\lambda}$, there exist resonance depths H_C such that at $H > H_C$ the tides are in phase with the perturbing force and tend to the static tide values with increasing H, while at $H < H_C$ they are out of phase with the disturbing force (according to Goldsbrough, H_C is small in polar basins, and the semidiurnal tides are not very close to resonance at real $H > H_C$). The resonance situations were investigated in detail by Hough [15].

The real seas are much better modelled by basins bounded by two meridians. In the case of such a basin with an angle of 60° between the meridians and a depth $H = H_0 \cos^2 \varphi$ (a model of the Atlantic), Goldsbrough [16] found that semidiurnal tides will exhibit resonance at $H_0 = 7610$ m, that is, at an average depth of 5080 m, which is close to the real resonance depth. Goldsbrough calculated a chart of the semidiurnal lunar tide M_2 for the same basin of constant depth and found it to be symmetric about the equator and the central meridian with four amphidromes in each hemisphere. Outside of the polar regions, the tides proved to be substantially greater than in the static case. Colborn [17] ran a similar investigation of the diurnal tides, for which ζ proved to be small. Doodson and Proudman [18] studied the K_1 and K_2 tides of a hemispherical basin and established a strong dependence of the amphidromes on its depth. Rossiter [19] and Accad and Pekeris [20] analyzed the case of basins bounded by two meridians and two parallels. Proudman [21] presented a method of formulating boundary conditions on rectilinear *liquid* boundaries of basins, starting from known values of ζ at the shores. This method was used by Fairbairn [22] to calculate the tides on the equator in the Indian Ocean, but Stretenskiy [23] showed that errors in the shore data may upset the convergence of the Fourier series used in this method.

Numerical solutions of Eq. (4-3-6) for real oceans have been derived primarily for the M_2, S_3, K_1, and K_2 tides with the following boundary conditions: the observed tides at the shores [24, 25] or vanishing of the normal velocity components at the shores [20, 26, 27, 28], a combination of these two conditions [29], or a specified coefficient of reflection of the tidal waves from the shores [30]. Most present calculations are done on four-degree grids. The values obtained for ζ are mapped on charts (Fig. 4-3-1), on which *cotidal* lines (i.e., lines of equal phase of the particular tide) and lines of equal tidal amplitude are drawn [earlier, beginning with Harris (1904), such charts were plotted by semiempirical interpolation of coastal and island data]. The amphidromes show up clearly on these charts. Further refinement of such charts requires description of effects associated with details of the ocean's boundaries–the dissipation of tidal waves on shelves, ridges, and islands, as well as description of sea-bottom friction and the conversions of energy into internal waves. The two latter effects result in dissipation of tidal kinetic energy. The resultant dissipation ε of tidal energy in the ocean and in the solid earth must be equal to the work done per unit time by the moment **K** of the tide-producing forces **F**. In the case of the lunar tides, for example, $\mathbf{K} = m_☽ \mathbf{r}_☽ \times \mathbf{F}$ and $\varepsilon = -\overline{\mathbf{K} \cdot (\boldsymbol{\Omega} - \boldsymbol{\nu})}$, where $\boldsymbol{\Omega}$ and $\boldsymbol{\nu}$ are the vectors of the angular velocities of the earth's rotation and the moon's orbital motion and the bar over the dot product indicates averaging over time. MacDonald's calculation [31] by this method (see also [32]) gave the value $\varepsilon = 2.76 \cdot 10^{19}$ erg/sec.

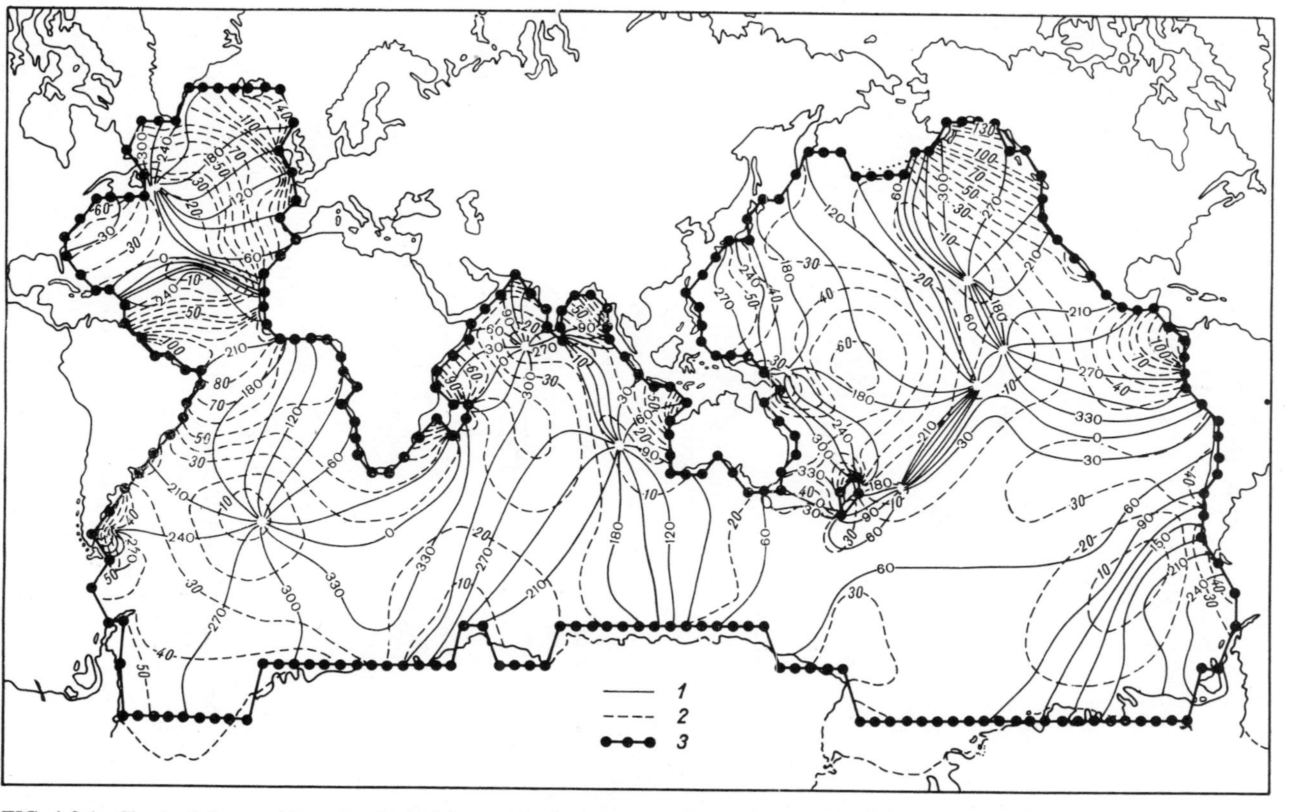

FIG. 4-3-1 Chart of the semidiurnal principal lunar tide in the ocean (after Bogdanov and Magarik [24]). 1–Cotidal lines (lines of equal tide phase, in degrees); 2–lines of equal amplitude, in cm; 3–boundary points of shores.

The dissipation of ocean tides, most of which takes place on shoals (where the tidal currents are strongly decelerated by bottom friction), has been estimated by various methods, among which calculation of the input and output energy fluxes at the outer boundaries of the shallows appears to be most accurate. Using this method, Miller [33] obtained a total tidal-energy dissipation of $1.7 \cdot 10^{19}$ erg/sec in all of the world's shallows (including $0.24 \cdot 10^{19}$ erg/sec in the Bering Sea and $0.2 \cdot 10^{19}$ erg/sec in the Sea of Okhotsk).

The solution $\zeta(\varphi, \lambda, t)$ of the Laplace tidal equations (4-3-5) (or more complete tidal equations, including nonlinear ones) can be treated as a function of the field of the tide-producing forces determined by astronomical methods, a field described by the function $\bar{\zeta}(\varphi, \lambda, t)$. This functional can be presented in the form of a functional series in powers of the function $\bar{\zeta}$, that is,

$$\zeta(M, t) = \int_0^\infty d\tau_1 \int W_1(M; M_1, \tau_1)\bar{\zeta}(M_1, t-\tau_1) \times$$
$$\times d\sum(M_1) + \int_0^\infty d\tau_1 \int_0^\infty d\tau_2 \int\int W_2(M; M_1, \tau_1; M_2, \tau_2) \times$$
$$\times \bar{\zeta}(M_1, t-\tau_1)\bar{\zeta}(M_2, t-\tau_2)\, d\sum(M_1)\, d\sum(M_2) + \ldots, \tag{4-3-7}$$

where $M, M_1, M_2, \ldots$ are points on the surface of a unit sphere, $d\Sigma(M)$ is an element of this surface at the point M, and $W_1, W_2, \ldots$ are functions of the ocean's linear, bilinear, etc., responses to deltaform disturbances at the corresponding points in space-time (within the framework of the linear theory, of course, only the first term in this series is retained). The response functions depend only on the properties of the sea, and not on those of the perturbing function $\bar{\zeta}$, which is determined by astronomical factors. Thus, representation (4-3-7) makes it possible to separate geophysical and astronomical effects. Moreover, we can now include in $\bar{\zeta}$ not only the gravitational disturbances with fixed frequencies, but also meteorological disturbances with both linear and continuous spectra, which generate "tidal noise."

The response functions W can be defined not only by solving the tidal equations, but also empirically, by the least squares method, using tide measurements (ζ), provided the perturbations $\bar{\zeta}$ are given for certain space-time regions or points. If $\bar{\zeta}_n^m(t)$ are the coefficients of the expansion of the perturbing function in spherical-function series $e^{im\lambda} P_n^m(\sin\varphi)$, the linear part of Eq. (4-3-7) assumes the form

$$\zeta(M, t) = \sum_{m,n} \int_0^\infty W_n^m(M, \tau)\bar{\zeta}_n^m(t-\tau)\, d\tau, \tag{4-3-8}$$

and the weighting coefficients $W_n^m(M, \tau)$ can be found by least squares for each fixed point M, assuming a sufficiently long record of the tides at this point is available. Then relation (4-3-8) can be used to predict the tides at this point. This method of tide prediction was suggested by Munk and Cartwright [34], who demonstrated its application to the cases of Honolulu and Newlyn. Figure 4-3-2 shows the spectrum of the semidiurnal tides at Honolulu [34], which can be described by the associated Legendre function $P_2^2(\sin\varphi)$. Here the upper diagram represents the spectrum of the equilibrium tide (with Darwin's letter symbols for the individual lines), the middle diagram is the spectrum of the observed tides (the blackened columns show the energy of oscillations coherent with the equilibrium tide), and the lower plot shows the ratio of amplitudes of the observed and equilibrium tides and their phase difference (in radians). The diagrams indicate, among others, how significant the "tidal noise" is in this case.

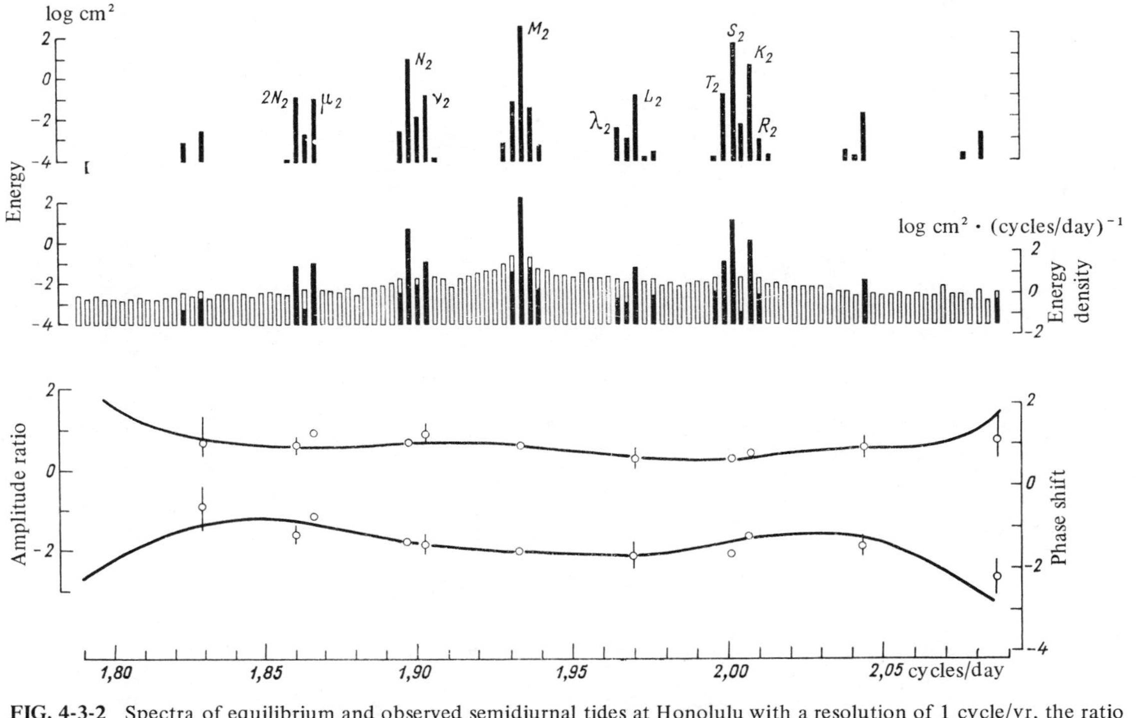

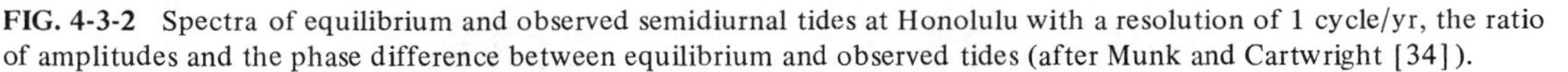

FIG. 4-3-2 Spectra of equilibrium and observed semidiurnal tides at Honolulu with a resolution of 1 cycle/yr, the ratio of amplitudes and the phase difference between equilibrium and observed tides (after Munk and Cartwright [34]).

REFERENCES

1. Doodson, A. T. The harmonic development of the tide-generating potential, Proc. Roy. Soc. London, **A100**, No. 704, 305–329 (1921).
2. Thompson, W. and P. G. Tait. Treaties on natural philosophy, Cambridge, **1**, Part 2, 2nd Edition, 386 (1883).
3. Lamb, H. *Hydrodynamics* (Translated from the English), Gostekhizdat Press, Moscow-Leningrad (1947).
4. Snodgrass, F. E. Deep-sea instrument capsule, Sciences, **162**, No. 3850, 78–87 (1968).
5. Munk, W., F. Snodgrass, and M. Wimbush. Tides off-shore: Transition from California coastal to deep-sea waters, Geophys. Fluid Dynam., **1**, No. 2, 161–235 (1970).
6. Proudman, J. *Dynamical Oceanography*, London (1953).
7. Wunsch, C. The long-period tides, Rev. Geophys., **5**, No. 4, 447–476 (1967).
8. Hendershott, M. and W. Munk. Tides. In: Ann. Rev. of Fluid Mech., Palo Alto, Calif., **2**, 205–224 (1970).
9. Phillips, O. *The Dynamics of the Upper Ocean*, Cambridge University Press (1966).
10. Krauss, W. *Internal Waves* (Translated from the German), Gidrometeoizdat Press, Leningrad (1968).
11. Monin, A. S. Variations of pressure in a barotropic atmosphere, Izv. Akad. Nauk SSSR, Ser. geofiz., No. 4, 76–85 (1952).
12. Goldsbrough, G. R. The dynamic theory of the tides in a polar basin, Proc. Lond. Math. Soc., **14**, No. 2, 31–66 (1915).
13. Goldsbrough, G. R. The dynamic theory of the tides in a zonal ocean, Proc. Lond. Math. Soc., **14**, No. 2, 207–229 (1915).
14. Sretenskiy, L. N. Motion of a free tidal wave within a polar basin; Reflection of Kelvin waves, Izv. Akad. Nauk SSSR, Ser. geofiz., No. 3, 383–402 (1937).
15. Hough, S. S. On the application of harmonic analysis to the dynamical theory of tides, Phil. Trans. Roy. Soc. London, **A189**, 201–257 (1897); **A191**, 139–185 (1898).
16. Goldsbrough, G. R. The tides in the ocean on a rotating globe, Proc. Roy. Soc. London, **A117**, No. 778, 692–718 (1928); **A122**, No. 789, 228–245 (1929); **A126**, No. 800, 1–15 (1929); **A140**, No. 841, 241–253 (1933); **A200**, No. 1061, 191–200 (1950).
17. Colborn, D. C. The diurnal tide in an ocean bounded by two meridians, Proc. Roy. Soc. London, **A131**, No. 815, 38–52 (1931).
18. Doodson, A. T. and J. Proudman. Tides in oceans bounded by meridians, Phil. Trans. Roy. Soc. London, **A235**, No. 753, 290–333 (1936); **A237**, No. 779, 311–373 (1938); **A238**, No. 797, 477–512 (1940).
19. Rossiter, J. R. On the application of relaxation methods to oceanic tides, Proc. Roy. Soc. London, **A248**, No. 1255, 482–498 (1958).
20. Accad, Y. and C. L. Pekeris. The K_2 tide bounded by meridians and parallels, Proc. Roy. Soc. London, **A278**, No. 1372, 110–128 (1964).
21. Proudman, J. A theorem in tidal dynamics, Phil. Mag., **49**, No. 6, 570–579 (1925).
22. Fairbairn, L. A. The semi-diurnal tides along the equator in the Indian Ocean, Phil. Trans. Roy. Soc. London, **A247**, No. 927, 191–212 (1954).
23. Sretenskiy, L. N. On the method proposed by Fairbairn for integration of the tidal equations, Izv. Akad. Nauk SSSR, Ser. geofiz., No. 7, 947–954 (1962).
24. Bogdanov, K. T. and V. A. Magarik. Numerical solution of the problem of propagation of semidiurnal tidal waves (M_2 and S_2) in the ocean, Dokl. Akad. Nauk SSSR, **172**, No. 6, 1315–1317 (1967).
25. Hendershott, M. C. The numerical integration of Laplace's tidal equations in idealized ocean basins. In: Proc. Symp. Math. Hydrodyn. Invest. Phys. Processes in the Sea, Moscow, 8–12 (1966).
26. Hansen, W. Gezeiten und Gezeitenströme der halbtagigen Hauptmondtide M_2 in der Nordsee (*Tides and Tidal Currents of the Semidiurnal Principal Lunar Tide M_2 in the North Sea*), Deutsche Hydrogr. Z. Erganzungsheit **I**, 1–46 (1952).

27. Hansen, W. Hydrodynamical methods applied to oceanographic problems. In: Symp. Math. Hydrodyn. Methods Phys. Oceanogr., Hamburg, 25-34 (1962).
28. Pekeris, C. L. and Y. Accad. Solution of Laplace's equation for the M_2 tide in the World Ocean, Phil. Trans. Roy. Soc. London, **A265**, No. 1165, 413-436 (1969).
29. Tiron, K. D., Yu. N. Sergeyev, and A. N. Michurin. A tidal chart for the Pacific, Atlantic, and Indian Oceans, Vestnik Mosk. gosud. Univ., **24**, No. 4, 123-135 (1967).
30. Gohin, F. P. A. Etude des marées oceaniques à l'aide de modeles mathematiques (*Study of Ocean Tides with the Aid of Mathematical Models*). In: Symp. Math. Hydrodyn. Methods Phys. Oceanogr., Hamburg, 179-194 (1962).
31. MaDonald, G. J. Tidal Friction, Rev. Geophys., **2**, No. 3, 467-541 (1964).
32. Monin, A. S. Vroshcheniye zemli i klimat (*The Rotation of the Earth and Climate*), Gidrometeoizdat Press, Leningrad (1972).
33. Miller, G. R. The flux of tidal energy out of deep oceans, J. Geophys. Res., **71**, No. 10, 2485-2490 (1966).
34. Munk, W. H. and D. E. Cartwright. Tidal spectroscopy and prediction, Phil. Trans. Roy. Soc. London, **A259**, No. 1105, 533-581 (1966).

4-4 DIURNAL FLUCTUATIONS DEPENDENT ON CHANGES IN INSOLATION

The variations of sea insolation over the day produce diurnal sea-water temperature variations. Diurnal variations of evaporation rate and water salinity are also conceivable, but very sensitive equipment would be required to measure them and very few such studies have been run.

The first specific observations of the daytime warming of the upper waters in the open sea were made by the *Meteor* expedition of 1925-1927 [1, 2]. Most later observations [3-7] date to the last ten years. It was found that when the quasi-homogeneous layer is either nonexistent or thin, the daytime warming is masked by the preexisting thermal stratification. It is also difficult to determine the extent of daytime warming or the depth to which the diurnal temperature wave propagates. At a given stratification, the amplitude of the diurnal temperature variations in the open sea also depends on cloud cover and wind speed.

Howe and Tait (see in [7]) cite the following data on the amplitudes of the diurnal temperature variations of the sea surface, derived from observations in the tropical Atlantic:

Wind and clouds	Amplitude of diurnal temperature variations, °C		
	Average	Maximum	Minimum
Moderate or fresh wind:			
Cloudy	0.39	0.6	0.0
Clear	0.71	1.1	0.3
Calm or light wind:			
Cloudy	0.93	1.4	0.6
Clear	1.59	1.9	1.2

The amplitude of the diurnal surface temperature variations varies considerably in space and time. Thus, from data from 44 stations, the average amplitude of the diurnal water temperature variations was 0.2°C in December and 0.69°C in May in the area of the British Isles [7]. From the data of weather ships in the Pacific Ocean, the annual average amplitudes of the diurnal variations were 0.28 and 0.45°C at the stations *Extra* (30°N, 153°E) and *Tango* (29°N, 135°E) [8]. The extremes of the diurnal variations occurred at 1400 and 0500 hours. The seasonal variations of the diurnal surface-temperature fluctuations were smallest in winter (0.14°C for station *Extra* and 0.3°C for *Tango*) and largest in summer (0.54 and 0.67°C, respectively). It was also observed that the daytime maximum occurs at 1500 hours in summer and 1300 hours in winter.

The table below lists the amplitudes of the diurnal *air temperature* fluctuations (in degrees), derived from North Atlantic weather ship data, summarized by Roll [9]:

Station	Mean latitude	January–February	March–April	May–June	July–August	September–October	November–December
"M," "A"	64.0	0.22	0.47	0.61	0.61	0.36	0.17
"J," "B"	57.9	0.31	0.53	0.81	0.69	0.42	0.22
"C"	52.7	0.33	0.64	0.72	0.75	0.42	0.33
"D"	44.0	0.61	0.83	1.06	1.22	0.89	0.50
"E," "H"	35.9	0.75	1.06	1.36	1.47	1.03	0.72

The range of water-surface temperature fluctuations will be smaller than that of the air. It is seen from the table that the lowest fluctuation amplitudes are observed in winter and the highest in summer, and that the amplitudes increase with decreasing latitude. There is every reason to believe that water-temperature fluctuations will behave similarly.

To calculate the depth of diurnal temperature fluctuations, one needs to conduct measurements during the period in which there exists a developed upper quasi-homogeneous layer. Such measurements should include the determination of the thermal-budget components of the surface layer, of the turbulent exchange of latent and sensible heat between the air and the sea, and of the coefficient of vertical mixing in the surface layer. An example of the diurnal water-temperature variations in the upper layer of the Atlantic Ocean (34°35′N, 21°09′ W) is Fig. 4-4-1, taken from Howe and Tait [5]. We see that the initial thickness of the homogeneous layer exceeded 32 m. The diurnal thermocline formed at a depth of 14 m, and the heat stored in the layer above this depth represented 83% of the total heat input during the period from 0800 to 2300 hours. However, the depth of the diurnal warming may depend strongly on latitude, the thickness of the homogeneous layer, and the initial turbulization in the sea. According to Defant [1], this depth is about 50 m, while Fedorov [6] gives 35 to 40 m. Others [3-5] cite values no larger than 30 m.

The development of the diurnal thermocline resembles in miniature the development of the seasonal thermocline (see Fig. 4-4-1). Therefore Fedorov [6] applied to it the theory describing the formation of the seasonal thermocline and calculated the thickness of the homogeneous layer for the diurnal discontinuity layer from the

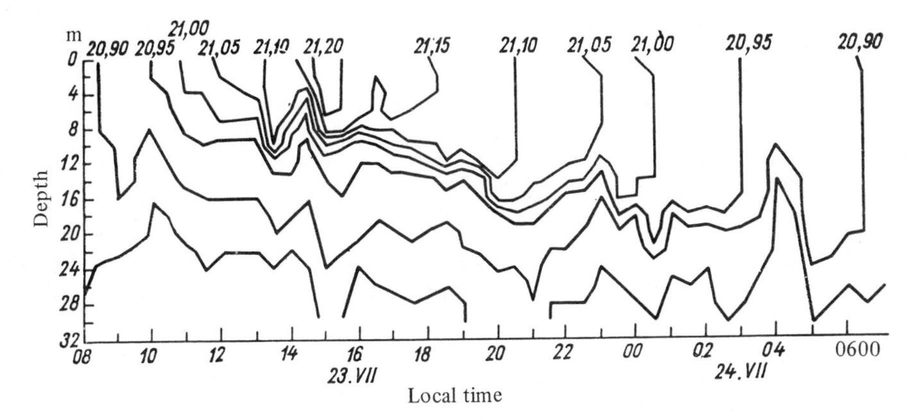

FIG. 4-4-1 Diurnal warming of the upper ocean (according to Howe and Tait [5]). The isotherms (°C) were plotted from measurements made at 30-minute intervals on 23–24 July 1966.

expressions derived by Kraus and Turner [10] and Kitaygorodskiy [11]. He obtained good agreement with observations. We do not yet have a theoretical model that fully explains the effect of all the processes that are involved in shaping the diurnal thermocline, but work cited in [12-13] allows calculation of the diurnal water temperature variations from observations in the marine atmospheric surface layer. The curve of diurnal air temperatures and total insolation, with allowance for the sea surface albedo and the thermophysical properties of the water and air, are sufficient for purposes of this calculation, in which it is assumed that there is no advective transport and that heat transport in the sea is exclusively turbulent and radiative. It is also assumed that the density discontinuity occurs below the depth of diurnal variations, and that the turbulent exchange coefficient is constant.

REFERENCES

1. Defant, A. Die Gezeiten und inneren Gezeitenwellen des Atlantischen Ozeans (*The Tides and Internal Tidal Waves of the Atlantic Ocean*), Deutsch. Atlan. Exped. "Meteor," 1925-27, Wiss. Ergeb., 7, No. 1, 318 (1932).
2. Kuklbrodt, E. and J. Reger. Die meteorologischen Beobachtungen, Deutsch. Atlant. Exped. "Meteor" (*Meteorological Observations of the German* Meteor *Expedition*), 1925-27, Wiss. Ergeb., **14**, 215-392 (1938).
3. Shonting, D. H. Some observations of short-term heat transfer through the surface layers of the ocean, Limnol. and Oceanogr., 9, 269-284 (1964).
4. Stommel, H., K. Saounders, W. Simmons, and J. Cooper. Observation of the diurnal thermocline, Deep-Sea Res., Supp. to Vol. 16, 269-284 (1969).
5. Howe, M. R. and K. J. Tait. Some observations of the diurnal heat wave in the ocean, Limnol. and Oceanogr., **14**, No. 1, 16-22 (1969).
6. Fedorov, K. N. On the summer daily heating and diurnal heat budget of the upper ocean layer. In: Studi in onore di Giuseppina Aliverti, Napoli, 27-40 (1972).
7. Sverdrup, H. U., M. W. Johnson, and R. H. Fleming. *The Oceans, Their Physics, Chemistry and General Biology*, Prentice-Hall (1955).
8. Koizumi, M. Researches on the variations of oceanographic conditions in the region of the Ocean Weather Station *Extra* in the North Pacific Ocean, Papers Meteorol. Geophys., 7, No. 7, 144-154 (1956).

9. Roll, H. *Physics of the Marine Atmosphere*, Academic Press (1965).
10. Kraus, E. B. and J. S. Turner. A one-dimensional model of the seasonal thermocline, II, Tellus, **15**, No. 1, 98–106 (1967).
11. Kitaygorodskiy, S. A. Fizika vzaimodeystviya atmosfery i okeana (*The Physics of Air-Sea Interaction*), Gidrometeoizdat Press, Leningrad (1970).
12. Dobroklonskiy, S. V. Diurnal temperature variations in the surface layer of the sea and heat fluxes at the air-sea interface, Doklady Akad. Nauk SSSR, **45**, No. 9, 391–394 (1944).
13. Kolesnikov, A. G. Calculation of the diurnal variations of temperature of the sea surface, Doklady Akad. Nauk SSSR, **57**, No. 2, 149–152 (1947).
14. Kolesnikov, A. G. and V. A. Pivovarov. Calculation of the diurnal variation of sea temperature from data on total insolation and air temperature, Doklady Akad. Nauk SSSR, **102**, No. 2, 261–264 (1955).

5 SYNOPTIC VARIABILITY

5-1 OBSERVATIONS

The synoptic variability of the sea exhibits time scales ranging from days to tens of days,[1] horizontal scales of the order of 50 to 100 km, and velocities on the order of 10 cm/sec. It would appear that synoptic variability is the principal factor in the variability of the ocean. Its systematic study did not begin until the end of the 1950's, when extended-term (multiday) current observations began to be made with recorders mounted on buoy stations, or with neutral-buoyancy floats, electromagnetic current meters (EMCM), etc. The time and space scales of the vortices that produce the synoptic variability of the ocean impose several constraints on the observations. For example, a set of current measurements at buoy stations really involves extended-period simultaneous current recording at several buoy stations properly positioned over a given sea area, called herein a test range. It is the great difficulty of arranging observations of this kind that accounts for the extreme paucity of data on "synoptic" vortices.

In the Soviet Union, the prime mover behind the organization of long-term physical measurements on test ranges in the sea was V. B. Shtokman, who as far back as the 1930's understood the need for and value of such observations. In 1935, he conducted long-term observations on the Caspian Sea. The 1956 test range in the Black Sea and the first ocean test range in the North Atlantic (1958) grew out of Shtokman's proposals. These studies made it obvious that ocean test ranges offer an extremely important means of studying the variability of the ocean. During the 1960's, a long-term program of test-range observations was developed under Shtokman's supervision. A test range in the northwestern Indian Ocean and the 1970 Atlantic test range were the results of this program. Table 5-1-1, which we have prepared for purposes of general orientation and which therefore does not pretend to be exhaustive, presents a chronological summary of extended-term instrumental

---

[1] It is natural to take the characteristic time scale as equal to one-quarter (or $\pi/2$) of the characteristic period of the fluctuations.

TABLE 5-1-1 Chronological Summary of Multiday Current Observations

Country	Year	Area of observations	Depth, m	Number of days	Nature of observations
Germany, Denmark, Sweden, Finland	1931	Southern Kattegat	30	8	Synchronous observations from five ships [1]
USSR	1935	Northwestern Caspian Sea	30–50	30	Observations from anchored ship [2]
USSR	1956	Black Sea	260	18	Observations at one buoy station [3]
USSR	1958	Northeastern part of the North Atlantic	3000	30	Test range with three buoy stations [4]
USA	1961–1962	Oceanographic section from Woods Hole to Bermuda	200–4500	From a few days to 233 days	Observations at 13 buoy stations [5, 6]
USA	1962	Bermuda area	5000	From a few days to 113 days	Observations at 12 buoy stations [6, 7]
USA	1963	Equatorial zone of Atlantic (29–32°W)	3000–5000	5–58	Observations at 26 buoy stations [7, 8]
USA	From 1965 to present	Station D (39°20′N, 70°00′W)	2600	Continuous record for about 6 months; gaps in record, about 2 months	Buoy station [7]
USSR	1967	Northwestern Indian Ocean	4500–5000	70	Observations on test range with 7 buoy stations [9]

USA	1967	Denmark Strait		35	Observations at 6 buoy stations [10]
Norway	1967	Norway Sea (66°N, 1–5°E)	750–2500	43–45	Observations at 4 buoy stations [11]
Great Britain	1967	Northeastern part of Atlantic Ocean, station B (45°N, 8°W)	4900	5–33	Observations at 5 buoy stations [12, 13]
Great Britain	1969	Northeastern part of Atlantic Ocean, station N (47°34′N, 8°20′W)	2000	72	Observations at 1 buoy station [13]
USA	1969	Gulf Stream (36–37°N, 70°W)	4000–4500	60	Observations at 4 buoy stations (as part of the overall program of study of Gulf Stream meanders) [14]
USSR	1970	Central Atlantic Ocean (16°30′N, 33°30′W)	5100–5700	195	Observations on test range with 17 buoy stations [15]
Great Britain	1970	Northeastern part of Atlantic Ocean, station J (52°30′N, 20°00′W)	3000	9–12	Observations at 3 buoy stations [13, 16]
USSR	1972	Equatorial zone of Pacific Ocean (2°30′S to 2°30′N, 166°30′E)	4000–4500	17	Observations on section with 6 buoy stations

current observations that have been made at buoy stations (unfortunately, not all of these measurements are useful for purposes of analysis of synoptic variability).

It appears that oceanologists began to speak of "synoptic" vortices after Swallow observed the drift of neutral-buoyancy floats (see [17–19]). But even now there are very few data on synoptic fluctuations with periods of the order of a month and more. This lends great interest to the paper by Thompson [20], who used a long series of observations at station D in the western Atlantic (extending over more than three years, but with large gaps in continuity) to estimate the energy of fluctuations with such periods. Thompson's procedure for estimating the spectrum from a series of observations with random gaps assumes that, in general, if there exists a stationary random process $u(t)$, its energy $\overline{u^2(t)}$ (the overbar indicates the statistical mean) can be represented as

$$\overline{u^2(t)} = \int_0^\infty E_u(\omega)\, d\omega, \tag{5-1-1}$$

where $E_u(\omega)\, d\omega$ is the fraction of energy of process $u(t)$ that is found in frequency range ω, $\omega + d\omega$ (see [21], Sec. 11).

Since the energetically significant frequency range is usually very broad, it is convenient to rewrite Eq. (5-1-1) as

$$\overline{u^2(t)} = 1/(\ln\, 10) \int_{-\infty}^{+\infty} \omega E_u(\omega)\, d \log \omega, \tag{5-1-2}$$

after which $\omega E_u(\omega)\, d \log \omega$ characterizes the energy of the process $u(t)$ in the range $(\log \omega, \log \omega + d \log \omega)$.

Thompson's calculated spectra for a depth of 500 m (similar computations were made for the depths of 100, 1000, and 2000 m) are shown in terms of a $\log \omega$ vs. $\omega E_u \omega$ plot in Fig. 5-1-1, taken from Rhines [22]. In Fig. 5-1-2 we present a temperature-fluctuation spectrum plotted by Wunsch [23] from various measurements off the Bermudas, and also given by Rhines [22]. There are good reasons to believe that both spectra are typical of the open sea (the literature does not appear to offer any other examples of spectra of this type).

It is clearly seen in both figures that most of the energy is concentrated in the ranges of periods from 15 to 100 days (Fig. 5-1-1) and from 40 to 200 days (Fig. 5-1-2). This is the synoptic range of periods (characteristic time scales from a few days to tens of days). Figure 5-1-1 also clearly shows a deep energy minimum that separates the synoptic energy peak from the energy peaks describing the semidiurnal (tidal) and inertial oscillations. It is also clear that energy very much larger than the energy of the semidiurnal and inertial oscillations is concentrated in the synoptic frequency range.

To illustrate the energy distribution over spatial scales, Fig. 5-1-3 presents a surface-temperature spectrum derived by Saunders [24] from temperature measurements made in the Ionian Sea from aircraft. Note that the energy peak is associated

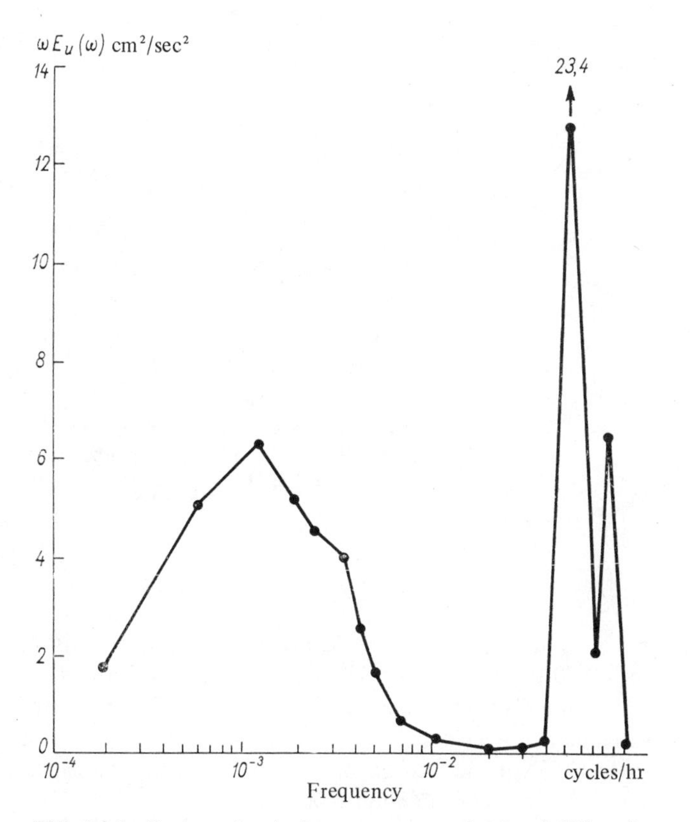

FIG. 5-1-1 Horizontal velocity spectrum at depth of 500 m for station D in the western Atlantic (after Rhines [22]). Only characteristic points are shown in the case of high frequencies.

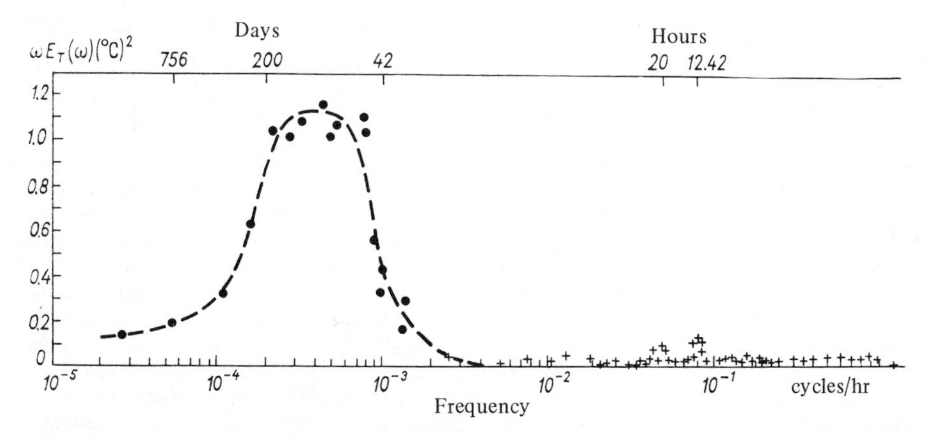

FIG. 5-1-2 Temperature fluctuations spectrum off the Bermudas (after Wunsch [23]; see also [22]). The points and crosses indicate different groups of data; the dashed curve is drawn approximately.

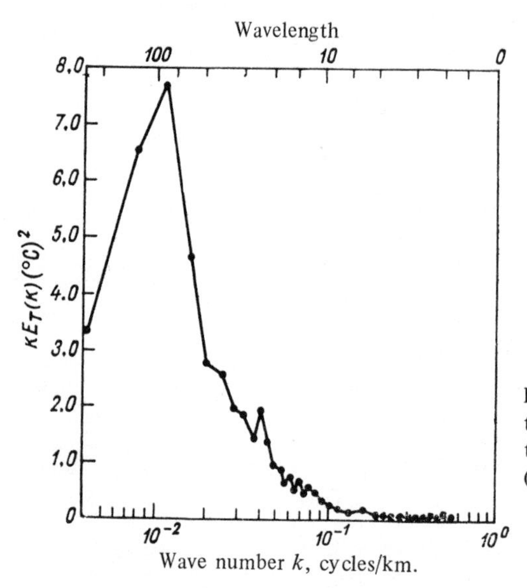

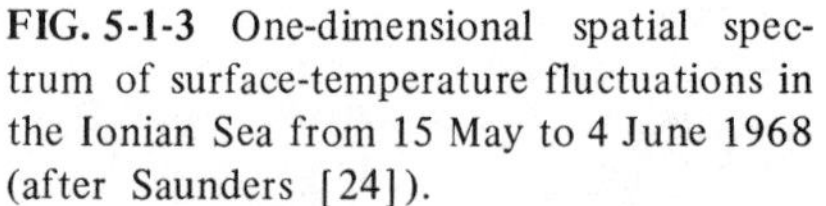
FIG. 5-1-3 One-dimensional spatial spectrum of surface-temperature fluctuations in the Ionian Sea from 15 May to 4 June 1968 (after Saunders [24]).

with perturbations having wavelengths of the order of 100 km, that is, characteristic horizontal scales of the order of 20 to 30 km. This is the magnitude of the scales of the synoptic eddies, which would be somewhat larger in the case of the ocean. Thus, the analysis of the spatial variability of the water-temperature field (on scales from 40 to 1200 km), made by Wyrtki [25] for the area of the Hawaiian Islands (10-25°N, 147-157°W) indicates an energy peak on scales of about 100 km.

Figure 5-1-4, which we have borrowed from Fofonoff and Webster's review [26] of American current observations in the western Atlantic, is quite interesting. This figure clearly demonstrates the existence of sea fluctuations with wavelengths of the order of 100 km and a period of about two weeks. This is essentially a trace of in-phase oscillations with a two-week period, recorded on two stations situated 370 km apart on different sides of the Gulf Stream, and possibly due to fluctuations of the Gulf Stream itself.

Gould [13] has summarized the current data for the eastern Atlantic, and also detected oscillations with periods of about a month. Sauskan [27, 28] and Volzhenkov and Istoshin [29] report interesting material on synoptic variability. Not too long ago, Leder [30] analyzed data from multiday (3 to 15 days) hydrological observations, primarily by Soviet expeditions to the North Atlantic over the period of 1958 through 1965. Grigorkina et al. [31] analyzed the synoptic variability with the aid of daily charts of the surface water temperature distributions, as well as depth and the strength (mean vertical temperature gradient) of the upper thermocline in the northwestern Pacific from 1 June, 1967 through 31 May, 1968 (these charts are regularly transmitted by radio by the U.S. Weather Service). Their analysis of the variability of the upper thermocline indicated that the dominant scales, with high energy levels, are 25 to 30, 12 to 15, and 4 to 8 days, and, in the case of the temperature gradient, 30 to 18, 9 to 7, and 6 to 4 days. Monthly oscillations of thermocline depth predominate north of 35° parallel. Their amplitudes vary from 10 m off the coast of Japan to 18 m in the open ocean. To the south and east of the main flow of the Kuroshio, the highest energy levels in the spectra of the depth of the thermocline occur at scales of 12 to 15 days (the highest amplitudes reach 25 m) and 4 to 8 days.

The test range that was established in the central Atlantic Ocean for six months of 1970, provided especially important data for the study of synoptic variability of the ocean. This test range was a 115 X 120-mile rectangle 450 miles west of the Cape Verde Islands. The range was centered on 16°30′N, 33°30′W. This area was chosen as typical of the open ocean and lies in the zone of the quasistationary North Equatorial current, with a rather smooth sea floor at depths of 5100 to 5700 m. A cross-shaped (with north-south and east-west arms) pattern of 17 buoy stations was emplaced on this test range. The distances between the buoy stations on each arm of the cross were the same at 5, 10, 17.5, and 24 miles, respectively. The different combinations of distances between buoy stations make it possible to study the spatial variability of the oceanological fields in the horizontal direction. Ten to twelve BPV-2 automatic current recorders and two or three thermographs were mounted at each of the buoy stations. The standard depths of the current observations were 25, 50, 100, 200, 300, 400, 500, 600, 800, 1000, 1200, and 1500 m. The measurements were taken at 30-minute intervals. Water temperature was recorded at the 50- and 200-meter depths at intervals of 10 or 15 minutes, depending on the design of the instrument. All current recorders were rearranged at an average interval of 20 to 23 days. An additional buoy station (No. 18), with more closely spaced measurement levels and deep-water (down to 4500 m) current recorders was in place at the center of the range (at a distance of 2 to 3 miles from station No. 1) during certain periods. Measurements on the range were taken from 23 February through 10 September, 1970 with minor interruptions at the individual buoy stations for technical reasons.

The test range was deliberately established in an area with the simplest possible hydrometeorological conditions. This has allowed uninterrupted long-term observations in the ocean, and has also made it possible to study the space-time structure of the oceanological fields in their purest form. Indeed, weather conditions were

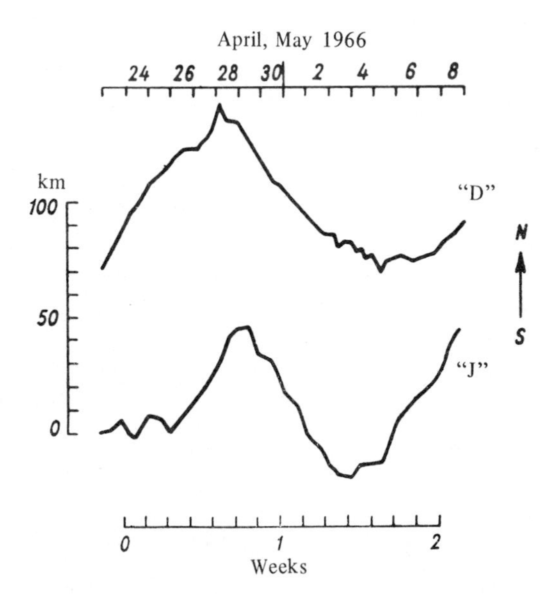

FIG. 5-1-4 Time-integrated meridional velocity component (near the surface) at stations D (39° 20′N, 70° 00′W) and J (36° N, 70° W) (after Fofonoff and Webster [26]).

relatively steady and quiet during most of the expedition's work on the range. In 63% of cases, the weather over the range was controlled by the North Subtropical high, whose southern edge extended over the range, and in 25% of cases it was controlled by the northern part of the Equatorial low. Only in 12% of cases was it disturbed by the passage of tropical storms. Therefore, a 4 to 5 point north trade wind prevailed during

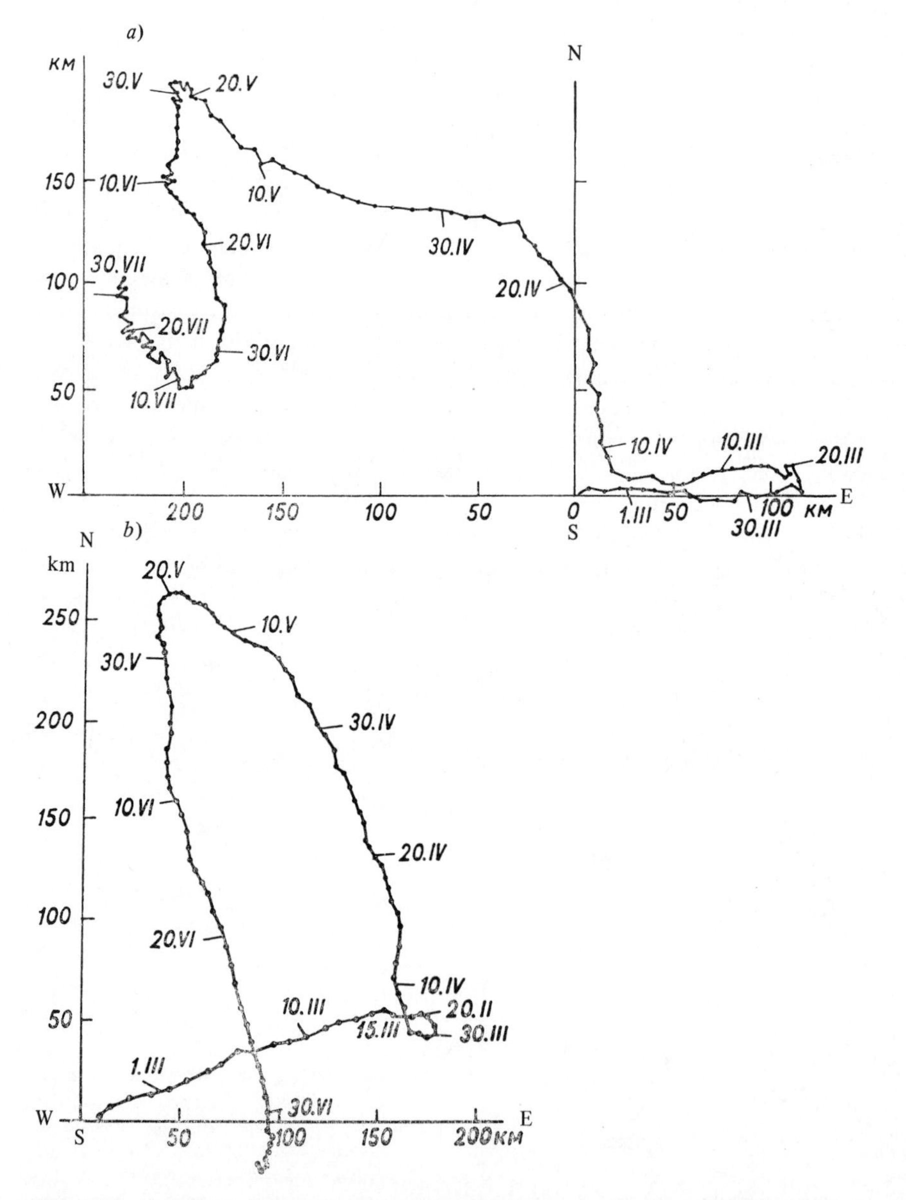

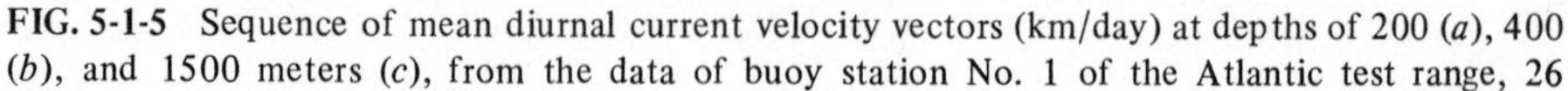
FIG. 5-1-5 Sequence of mean diurnal current velocity vectors (km/day) at depths of 200 (*a*), 400 (*b*), and 1500 meters (*c*), from the data of buoy station No. 1 of the Atlantic test range, 26

the greater part of the observations. Under such conditions, it was expected that a fairly stable east-to-west current, designated the North Equatorial current in sea atlases, would also dominate in the sea. However, analysis of the data from the central buoy station immediately indicated a more complex pattern for this current.

Figure 5-1-5*a–c*, which we have taken from Brekhovskikh et al. [32], shows a sequence of current vectors for a 155-day period (from 26 February through 1 August) at depths of 200, 400, and 1500 m.[2] Note that these plots are not paths of the current in time. The figures show that during the first measurements (from 26 February through 12 March) the current did flow against the wind at the top 1500-meter layer. In the next period (until 20 May), the current was generally west-north-westward over the entire thermocline layer. From 20 May through 30 June, the current in the top layers was again flowing against the prevailing wind direction, but it was flowing to the south at depth. Thus, several changes of current direction were observed in the center of the test range over five months. It is rather difficult to estimate the periods of such oscillations. All that can be said with certainty is that they are variations whose characteristic scale is of the order of a month.

Analysis of the time distribution of currents at all of the buoy stations (see [33]) gave synoptic patterns of currents at the various depths. For this purpose, all measurements were smoothed with an effective period of 84 hr. This filter suppressed inertial and tidal current velocity variations, which were very strongly in evidence on the Atlantic test range and had periods of 42 and 12.4 hr, respectively. By way of

[2] The corresponding diagram for the 50-meter depth has already been given in Fig. 1-1-1.

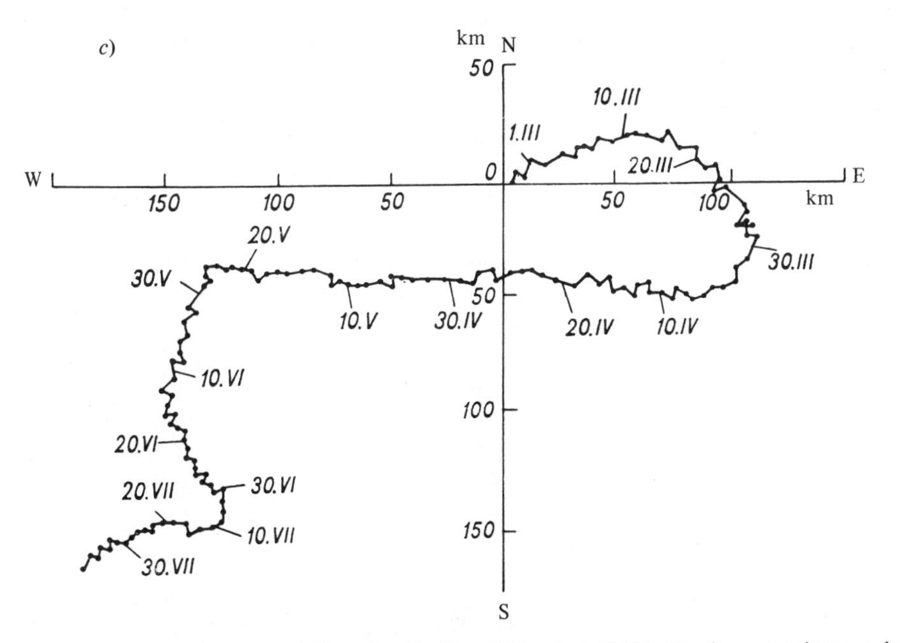

February through 1 August 1970 (after Brekhovskikh et al. [32]). Numbers at points are dates (the Roman numeral is the month).

example, Fig. 5-1-6 presents synoptic patterns of the mean current vectors at the depth of 300 m for the period from 13 March through 12 August. We see that a northeasterly (i.e., in the direction of the North Equatorial current) anticyclonic eddy passed across the range during the period from the middle of April to the first few days of July. The synoptic patterns also indicate that this eddy was asymmetrical, that is, its span in the direction perpendicular to its motion was slightly larger than its length in the direction of motion. However, even the latter dimension was obviously greater than the horizontal size of the test range (113 miles). The current velocity in the field of this high reached 25 cm/sec. Rough estimates of the direction from the synoptic charts showed the heading of the above eddy to be approximately 240° and its velocity about 4 cm/sec.

The pattern of the currents for March at the 300-meter depth probably indicates that the anticyclonic disturbance described above was preceded by a disturbance of the same sign; on the other hand, the pattern for August suggests the existence of a cyclonic disturbance moving behind the "main anticyclonic eddy." Thus there seems to have been a chain of eddies, moving from the northeast to the southwest across the test range. Approximately the same synoptic current pattern was observed at a depth of 1000 m, showing how deep the vortical velocity-field disturbances described above penetrate. The ratio of the absolute current velocities at the 1000 and 300 m depths at corresponding points and times averages about 3/5, suggesting that the baroclinic component of these disturbances is a significant factor. An estimate of the velocity and direction of motion of the main anticyclonic disturbance from the observations at 1000 m gives practically the same value as the estimate made from data collected at 300 m. However, analysis indicated that the current patterns at 1000 m appear to have a time lead over the corresponding patterns at the 300-m depth. This leads us to conclude that the axis of the main anticyclonic disturbance was tilted away from the direction of motion.

Calculations made for 200 m and all depths between 300 and 1000 m yield current patterns that are quite similar to the one discussed above. They also support the inference as to the inclination of the axis of the anticyclonic disturbance. The pattern of variations at 50 m is more complex than at greater depths, although even at 50 m one can detect "traces" of the vortical velocity-field disturbances described above (the data indicate that the current velocities near the surface are controlled to a significant extent by synoptic disturbances in the wind field).

The synoptic eddies can be traced most clearly in strong boundary currents (Gulf Stream, Kuroshio, etc.). As far back as 1937, Church found by studying thermograph records from ships crossing the Gulf Stream [34] that the axis of the Gulf Stream (beyond Cape Hatteras) is curved and describes wavy oscillations.[3] Following Iselin, these distortions of the axis of the Gulf Stream came to be known as meanders (by analogy with river meanders) (see Fig. 5-1-7, taken from [36]). The

[3]The position of the axis of the Gulf Stream is usually determined not directly, but from the position of the 15°C isotherm at a depth of 200 m, or from the position of the peak of the horizontal temperature gradient at the surface. The technique makes possible fast hydrological surveys (see [35]).

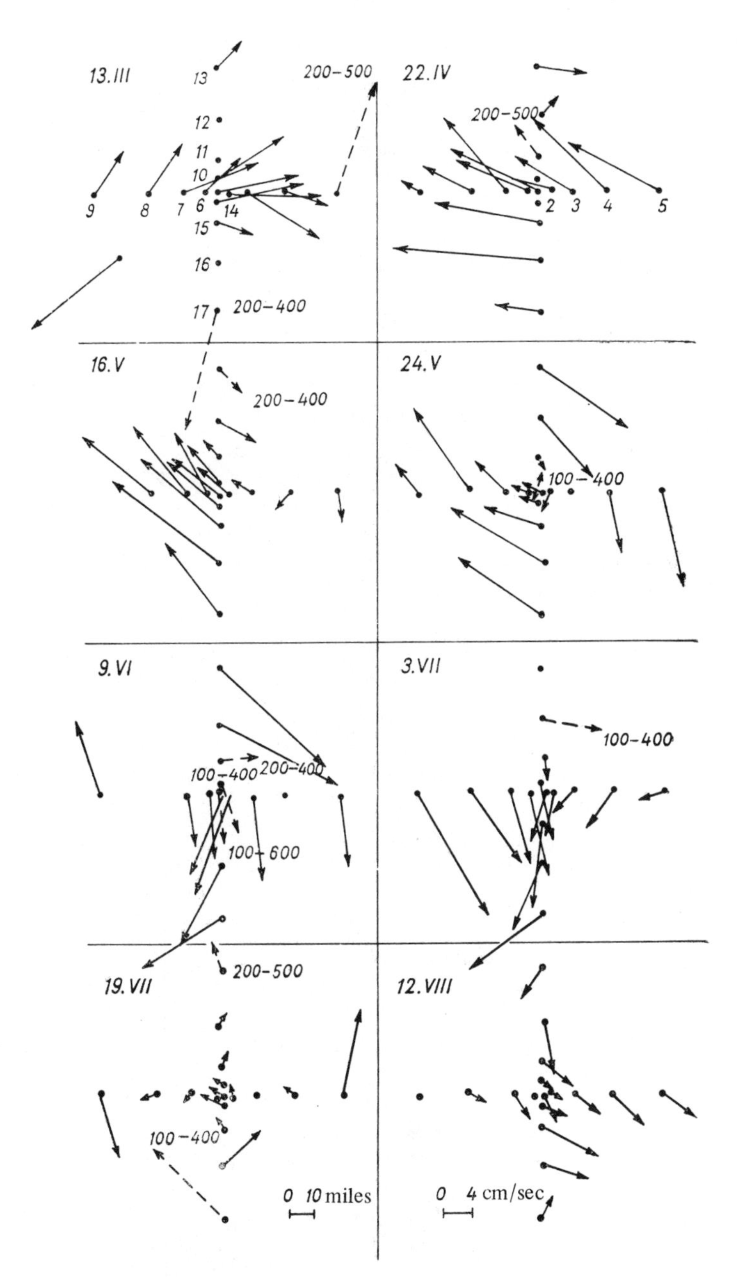

FIG. 5-1-6 Synoptic patterns of distribution of vectors of large-scale currents at depth of 300 m (after Kort et al. [33]). The velocities indicated by dashed arrows were computed by interpolation over depth (the depths are shown). The numerals next to the points indicate the positions of the buoy stations on the test range. The velocity scales and distances between stations are indicated at the bottom. Lack of arrows indicate a gap in the data.

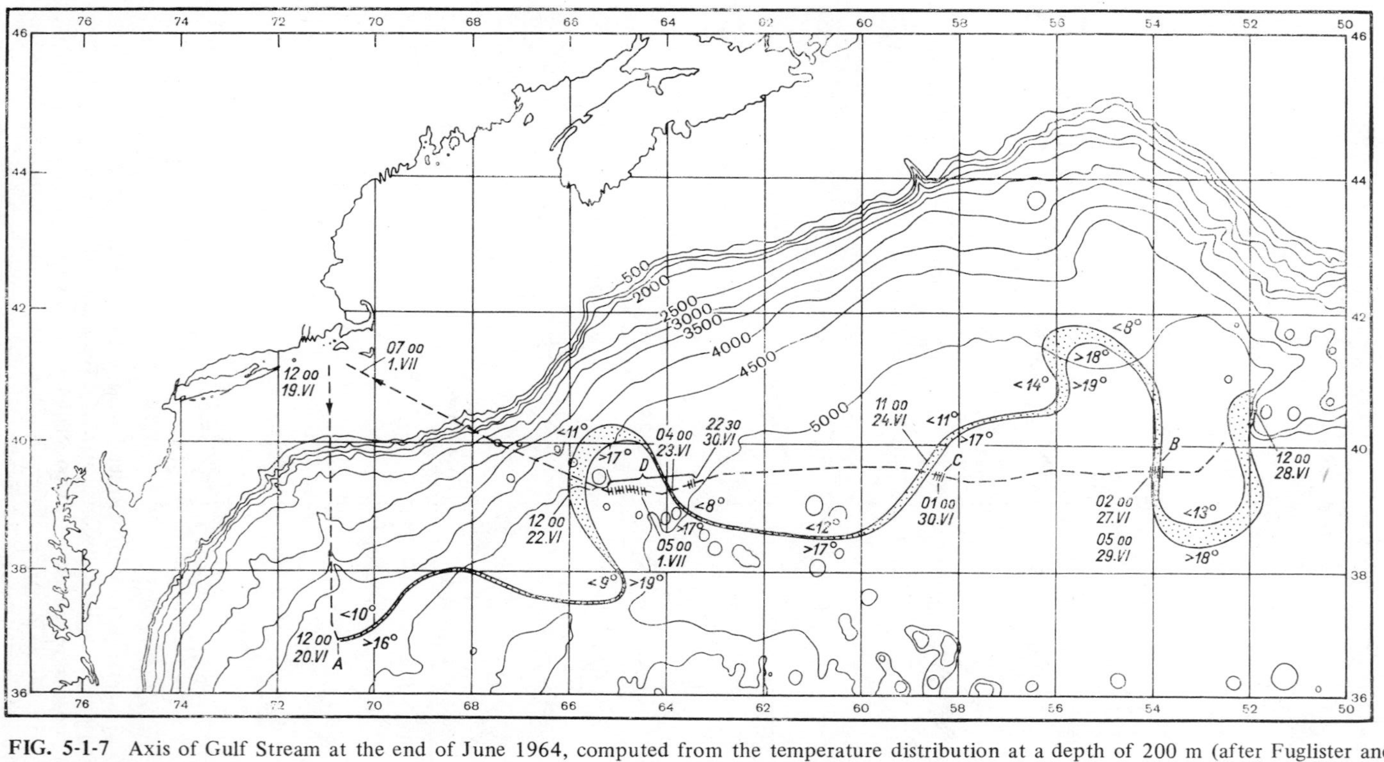

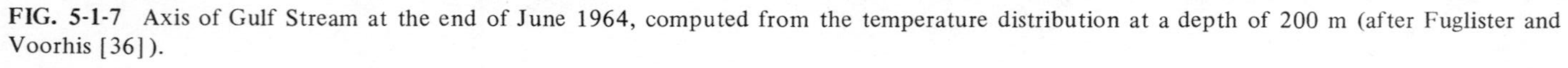

FIG. 5-1-7 Axis of Gulf Stream at the end of June 1964, computed from the temperature distribution at a depth of 200 m (after Fuglister and Voorhis [36]).

meandering of the Gulf Stream has since then been studied by numerous expeditions (see [37]).

Figure 5-1-8 (from Hansen [38]) shows the time pattern of Gulf Stream meanders on the path between 70 and 60°W during the period from September 1965 through November 1966, as reconstructed from more than 4000 bathythermograph records. The diagonal lines are supplied to permit determination of the phase velocity of the eastward movement of the meanders; on the average, this velocity was 10 km/day, or 8 cm/sec, over the 15-month period. The pattern of the equal-phase lines indicates that the shift of the meanders is nonuniform. The pattern of rises of amplitude of the meanders is also very complex.

Figure 5-1-9 (from [39]) demonstrates the curious evolution of the meanders, which results in their transformation into individual eddies. This phenomenon was also studied by Fuglister [40] and Saunders [41].

Robinson [42] reports highly interesting data from a 1969 expedition planned specifically for the study of the Gulf Stream meanders. The first instantaneous "photographs" of the Gulf Stream axis were pieced together from the data of this expedition. Among the three most interesting results, Fig. 5-1-10 shows slow changes in the position and shape of the Gulf Stream's axis that took place during the two-week expedition. Figure 5-1-11 clearly shows small-scale disturbances (with wavelengths comparable to the width of the Gulf Stream) with a period of 2 to 3 days (they are completely comparable in amplitude to the slower variations indicated on Fig. 5-1-10. Finally, Fig. 5-1-12 indicates the presence of S-shaped features that pass through the region in about a day (and are apparently moving eastward).

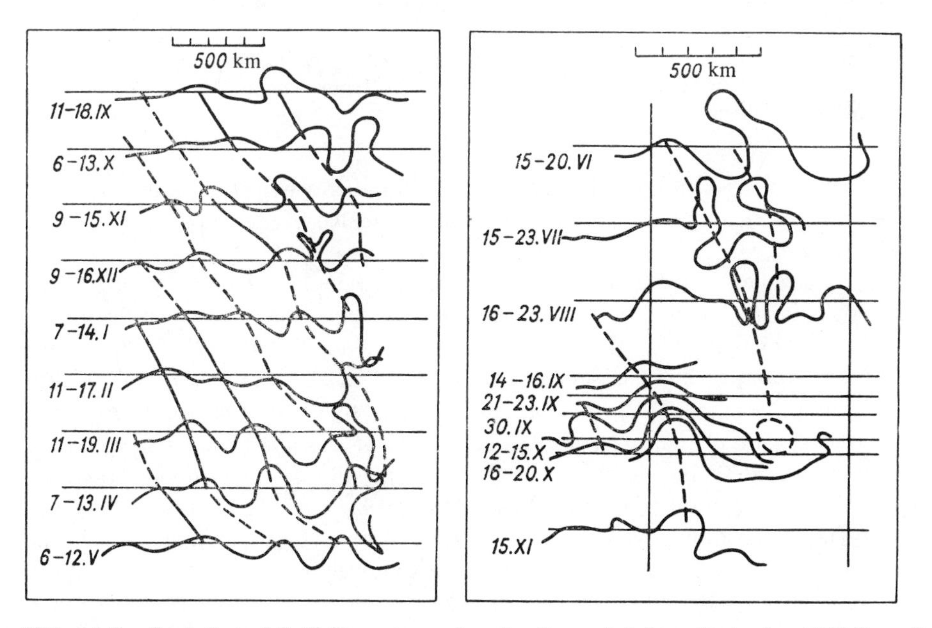

FIG. 5-1-8 Evolution of Gulf Stream meanders for the period from September 1965 through November 1966 (after Hansen [38]).

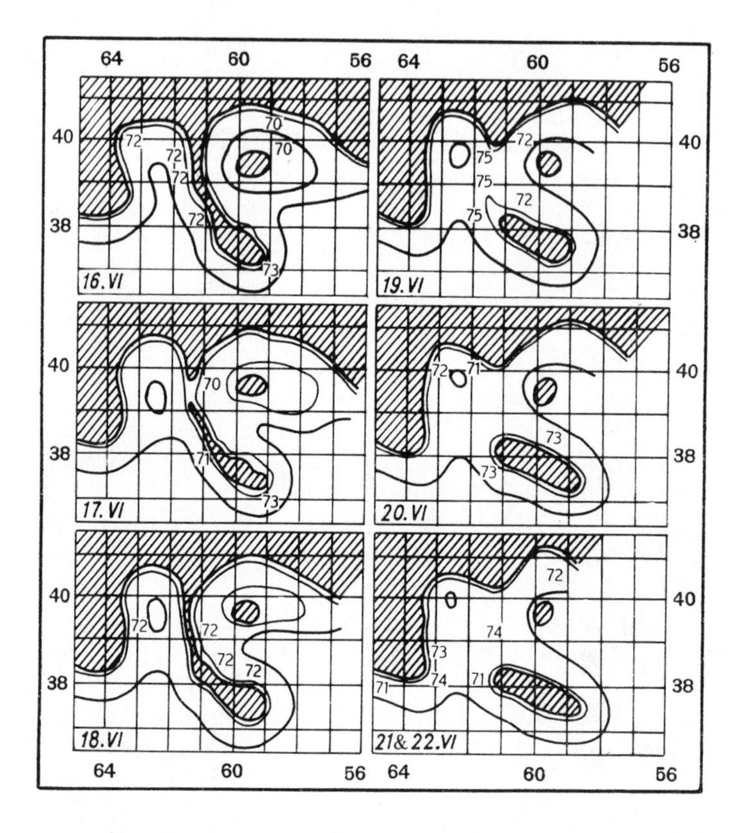

FIG. 5-1-9 Formation of an eddy from an elongated meander between 16 and 20 June 1950 (after Fuglister and Worthington [39]). Temperature is given in degrees Fahrenheit.

Finally, let us mention statistical estimates of such parameters as the degree of meandering of the Gulf Stream, etc., that were obtained by Baranov [43; 44, Chap. II].

The meanders of the Kuroshio have their own characteristic features. The synoptic variability of this current appears to be much less developed than that of the Gulf Stream [45, 46, 47, 48]. According to Robinson and Taft [49], there are two "stable" positions of the Kuroshio axis (these typical positions are indicated in Figs. 5-1-13 and 5-1-14). The Kuroshio occupies each of these positions for several years (thus, the axis of the current indicated in Fig. 5-1-15 did not move from May 1959 through May 1963), and the transition from one position to the other occurs over a period of 3 to 6 months (the Gulf Stream does not have two stable positions of this kind). These transitions, which are of special interest from the standpoint of the synoptic variability, have not yet been adequately studied. Figure 5-1-15, plotted from Yoshida's [46] data, gives some idea of the position change of the Kuroshio' axis from March through July 1959 (the rate of "advance" of the meanders from west to east reaches 4 miles/day).

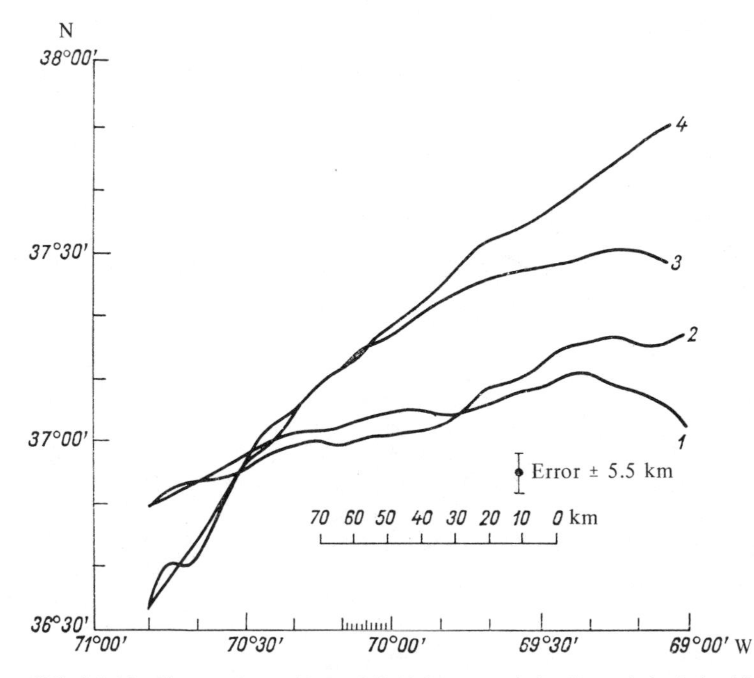

FIG. 5-1-10 Changes in position of Gulf Stream axis in the period of the 1969 experiment (from Robinson [42]). Dates and times are indicated. 1–0000 hr on 4 July; 2–0000 hr on 7 July; 3–0000 hr on 12 July; 4–1200 hr on 17 July.

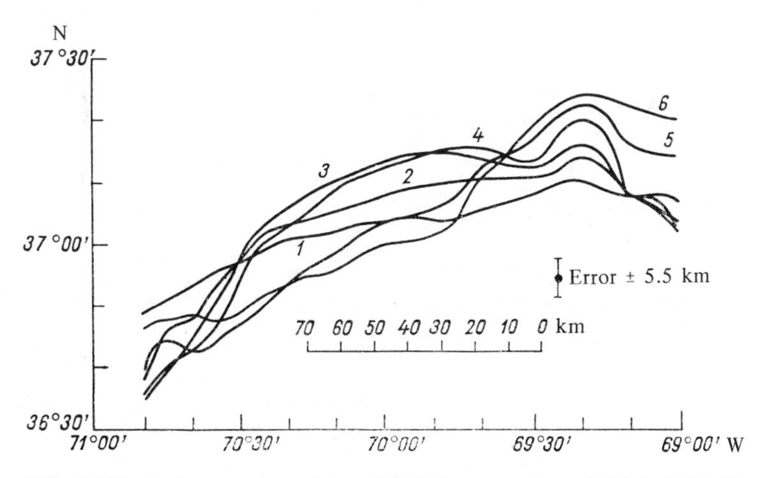

FIG. 5-1-11 Instantaneous position of Gulf Stream axis on 4-6 July 1969 (from Robinson [42]). 1–0000 hr, 4 July; 2–1200 hr, 4 July; 3–0000 hr, 5 July; 4–1200 hr, 5 July; 5–0000 hr, 6 July; 6–1200 hr, 6 July.

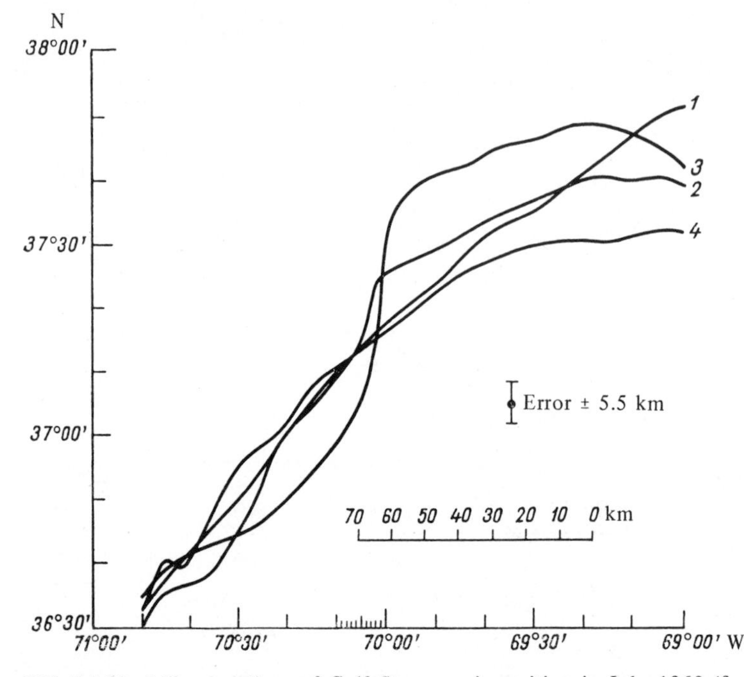

FIG. 5-1-12 S-like variations of Gulf Stream axis position in July 1969 (from Robinson [42]). 1–1200 hr, 13 July; 2–1200 hr, 15 July; 3–1200 hr, 16 July; 4–1200 hr, 17 July.

REFERENCES

1. Defant, A. and O. Schubert. Strommessungen und ozeanographische Serien-Beobachtungen der 4-Länder Untersuchung im Kattegat 10–17 August 1931 (*Current Measurements and Oceanographic Series Observations of the Quadripartite Investigation in the Kattegat from 10 to 17 August 1931*), Veroffentlichungen des Instituts fur Merreskunde in der Universitat Berlin, No. 25, 1-144 (1937).
2. Shtokman, V. B. and I. I. Ivanovskiy. Results of a stationary study of the currents near the west–central force of the Caspian Sea, Meteorologiya i gidrologiya, No. 3-5, 154-160 (1937).
3. Ozmidov, R. V. Statistical properties of the horizontal macrostructure in the Black Sea, Trudy Inst. Okeanol. Akad. Nauk SSSR, **60**, 114–129 (1962).
4. Ozmidov, R. V. and A. D. Yampol'skiy. Certain statistical properties of velocity and density variations in the ocean, Izv. Akad. Nauk SSSR, Fizika atm. i okeana, **1**, No. 6, 615–622 (1965).
5. Richardson, W. S., P. B. Stimson, and C. H. Wilkins. Current measurements from moored buoys, Deep-Sea Res., **10**, No. 4, 369–388 (1963).
6. Webster, F. A preliminary analysis of some Richardson current meter records, Deep-Sea Res., **10**, No. 4, 389–396 (1963).
7. Fofonoff, N. P. Current measurements from moored buoys, 1959–1965, Woods Hole Oceanogr. Inst., Ref. No. 68–30 [unpublished manuscript] (1968).
8. Staleup, M. C. and W. G. Metcalf. Direct measurements of the Atlantic Equatorial undercurrent, J. Mar. Res., **24**, No. 1, 44–55 (1966).
9. Shtokman, V. B., M. N. Koshlyakov, R. V. Ozmidov, L. M. Fomin, and A. D. Yampol'skiy. Extended-term measurements of the variability of the physical fields on ocean test ranges as a new step in study of the sea, Dokl. Akad. Nauk SSSR, **186**, No. 5, 1070–1073 (1969).

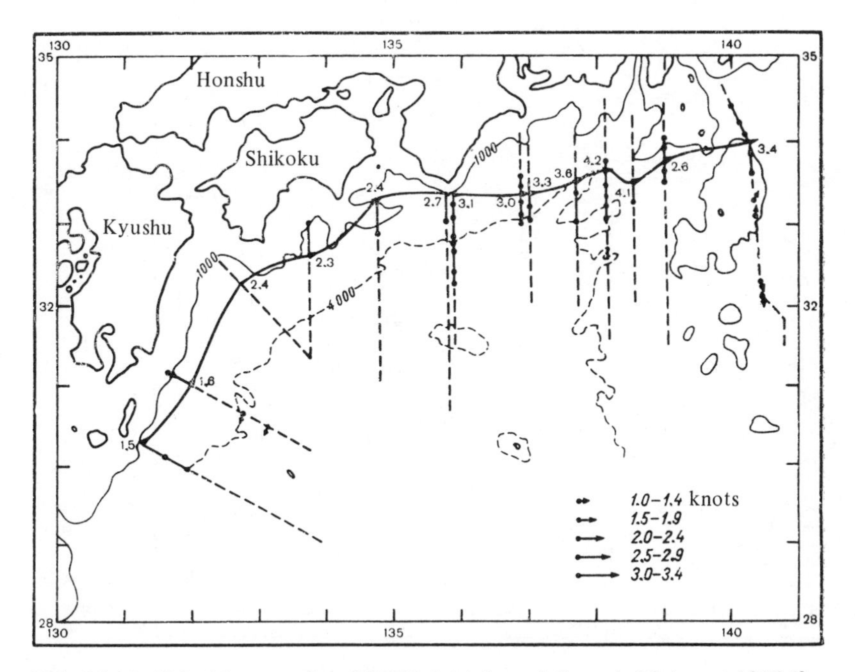

FIG. 5-1-13 Velocities at surface (EMCM data) from 4 through 26 August 1956 (from Robinson and Taft [49]). The dark triangle on each tack indicates the position of highest velocity and the direction of the current; the numbers next to the triangles indicate the velocity in knots. The dark circles on the tacks (represented as solid lines) indicate velocities higher than 0.9 knot that approximately parallel the velocities on the axis of the Kuroshio. The dashed segments of the tacks indicate areas where the velocities are less than one knot and appear to be outside of the Kuroshio; where the velocities exceeded 0.9 knot, the velocities themselves are indicated (the scale appears in the lower right-hand corner). Isobaths are in meters.

10. Worthington, L. V. An attempt to measure the volume transport of Norwegian Sea overflow water through the Denmark Strait, Deep-Sea Res., Suppl. to Vol. 16, 421–432 (1969).
11. Kvinge, T., A. Lee, and R. Satre. Report on Study of Variability in the Norwegian Sea, April/May 1967, Universitet in Bergen Geofysisk Institutt, Bergen, 1–31 (1968).
12. Gould, W. J. Observations of an event in some current measurement in the Bay of Biscay, Deep-Sea Res., **18**, No. 1, 35–50 (1971).
13. Gould, W. J. Spectral characteristics of some deep current records from the eastern North Atlantic, Phil. Trans. Roy. Soc. London, **A270**, No. 1206, 437–450 (1971).
14. Schmitz, W. J., Jr., A. R. Robinson, and F. C. Fuglister. Bottom velocity observations directly under the Gulf Stream, Science, N.Y., **170**, No. 3963, 1192–1194 (1970).
15. Kort, V. G. A hydrophysical experiment in the Atlantic, Vestnik Akad. Nauk SSSR, No. 2, 92–101 (1971).
16. *R.R.S. Discovery Cruise 34 Report*, 2–28 June 1970, N.I.O. Cruise Report, No. 34 [unpublished manuscript] (1970).
17. Swallow, J. C. A neutral-buoyancy float for measuring deep current, Deep-Sea Res., **3**, No. 1, 74–81 (1955).
18. Swallow, J. C. Some further deep current measurements using neutrally buoyant floats, Deep-Sea Res., **4**, No. 2, 93–104 (1957).

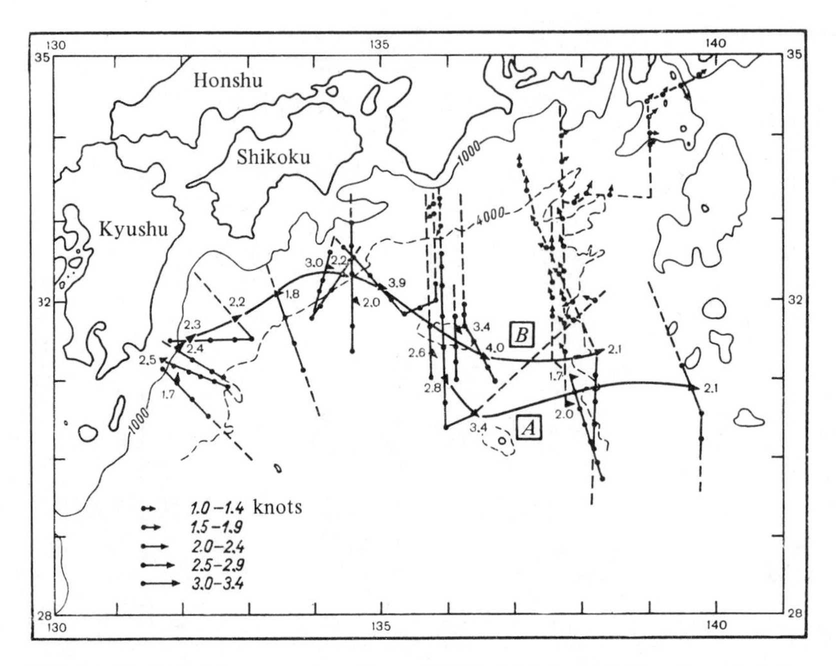

FIG. 5-1-14 Velocities at surface (from EMCM data) from 2 May through 1 June 1960 (from Robinson and Taft [49]). The two positions of the axis are indicated for the two time intervals *A* (27 May–1 June) and *B* (2–18 May). Refer to the caption of Fig. 5-1-13 for the other items of the legend.

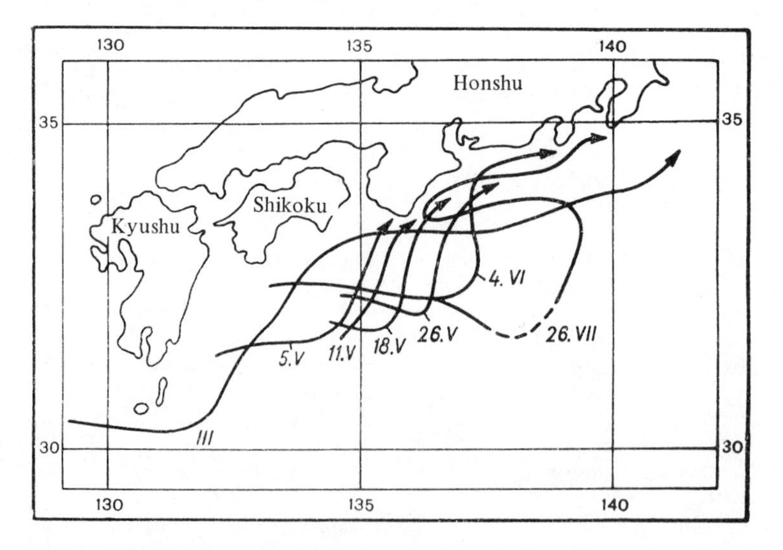

FIG. 5-1-15 Changes in position of Kuroshio axis from March through July 1959 (from Yoshida [46]).

19. Crease, J. Velocity measurements in the deep water of the western North Atlantic, J. Geophys. Res., **67**, No. 8, 3173–3176 (1962).
20. Thompson, R. Topographic Rossby waves at a site north of the Gulf Stream, Deep-Sea Res., **18**, No. 1, 1–19 (1971).
21. Monin, A. S. and A. M. Yaglom. *Statistical Fluid Mechanics*, Part 2, Nauka, Moscow [English translation, MIT Press] (1967).
22. Rhines, P. Observations of the energy-containing oceanic eddies and theoretical models of waves and turbulence. In: *IUCRM Conference on Atm. Waves*, La Jolla, Calif. USA, June 1972 (unpublished manuscript).
23. Wunsch, C. The spectrum from two years to two minutes of temperature fluctuations in the main thermocline at Bermuda, Deep-Sea Res., **19**, No. 8, 577–593 (1972).
24. Saunders, P. M. Space and time variability of temperature in the Upper Ocean, Deep-Sea Res., **19**, No. 7, 467–480 (1972).
25. Wyrtki, K. The spectrum of ocean turbulence over distances between 40 and 1000 kilometers, Deutsche Hydrogr. Zeitschrift, **20**, No. 4, 176–186 (1967).
26. Fofonoff, N. P. and F. Webster. Current measurements in the western Atlantic, Phil. Trans. Roy. Soc. London, **A270**, No. 1206, 423–436 (1971).
27. Sauskan, E. M. Study of currents in the Atlantic Ocean from multiday buoy station data, Okeanologiya, **6**, No. 1, 53–61 (1966).
28. Sauskan, E. M. Short-period variations of currents in the northwestern Pacific and their prediction, Trudy Gidromettsentra SSSR, No. 51, 3–16 (1969).
29. Volzhenkov, V. A. and Yu. V. Istoshin. Application of spectral functions to study of the variability of oceanographic characteristics, Trudy TsIPa, No. 142, 103–107 (1965).
30. Leder, I. Z. The short-period variability of water temperature and salinity in certain regions of the North Atlantic, Trudy Gosud. Okeanogr. Inst., No. 96, 106–125 (1969).
31. Grigorkina, R. G., A. N. Michurin, P. P. Provotorov, and V. R. Fuks. Short-period variability of oceanological conditions in the fishing grounds of the Kuroshio System, Part 1, Izv. TINRO, **68**, 45–67 (1969).
32. Brekhovskikh, L. M., G. N. Ivanov-Frantskevich, M. N. Koshlyakov, K. N. Fedorov, L. M. Fomin, and A. D. Yampol'skiy. Certain results of a hydrophysical experiment on a test range in the tropical Atlantic, Izv. Akad. Nauk SSSR, Fizika atm. i okeana, **7**, No. 5, 511–528 (1971).
33. Kort, V. G., M. N. Koshlyakov, G. N. Ivanov-Frantskevich, V. I. Byshev, and Yu. M. Grachev. Variability of large-scale currents on a hydrophysical test range in the tropical Atlantic, Izv. Akad. Nauk SSSR, Fizika atm. i okeana, **9**, No. 1, 105–109 (1973).
34. Church, P. E. Temperature of the western north Atlantic from thermograph records, Assoc. d'Oceanogr. Phys. UGGI, Publ. Sci., No. 4, Liverpool, 32 (1973).
35. Fuglister, F. C. Gulf Stream "60." In: *Progress in Oceanography*, Ed. Mary Sears, **1**, 263–373 (1963).
36. Fuglister, F. C. and A. D. Vorrhis. A new method of tracking the Gulf Stream, Limnol. and Oceanogr., Suppl. to **10**, 115–124 (1965).
37. Stommel, H. *The Gulf Stream, A Physical and Dynamical Description*, Univ. Calif. Press, Cambr. Univ. Press, Second Ed. (1966).
38. Hansen, D. V. Gulf-stream meanders between Cape Hatteras and the Grand Banks, Deep-Sea Res., **17**, No. 3, 495–511 (1970).
39. Fuglister, F. C. and L. V. Worthington. Some results of a multiple ship survey of the Gulf Stream, Tellus, **3**, No. 1, 1–14 (1951).
40. Fuglister, F. C. Cyclonic eddies formed from meanders of the Gulf Stream, Trans. Am. Geophys. Union, **48**, No. 1, 123 (1967).
41. Saunders, P. M. Anticyclonic eddies formed from shoreward meanders of the Gulf Stream, Deep-Sea Res., **18**, No. 12, 1207–1219 (1971).
42. Robinson, A. R. The Gulf Stream, Phil. Trans. Roy. Soc. London, **A270**, No. 1206, 351–370 (1971).

43. Baranov, E. I. Short-period oscillations of the Gulf Stream front during the Winter-Spring season of 1963, Okeanologiya, **6**, No. 2, 228-233 (1966).
44. Barnov, E. I., V. A. Bubnov, R. P. Bulatov, and I. V. Privalova. A study of the circulation and water transport of the Atlantic Ocean, Okeanologicheskiye issledovaniya, No. 22, Nauka Press, Moscow (1971).
45. Uda, M. On the Nature of the Kuroshio, Its Origin and Meanders. In: *Studies on Oceanogr.*, Hidaka Anniv. Vol., Tokyo, 89-107 (1964).
46. Yoshida, K. On the variation of Kuroshio and cold water mass off Enshunada, Hydrogr. Bull., No. 67, 11-18 (1961).
47. Moriyasu, Sh. The fluctuation of hydrographic conditions in the Sea South of Honshu, Japan (Review), Oceanogr. Mag., **15**, No. 1, 11-29 (1963).
48. Shoji, D. Description of the Kuroshio (Physical Aspect). In: *Proc. Symposium on the Kuroshio*, Oceanogr. Soc. Japan and UNESCO, 1-11 (1965).
49. Robinson, A. R. and B. A. Taft. A numerical experiment for the path of the Kuroshio, J. Mar. Res., **30**, No. 1, 65-101 (1972).

5-2 THEORETICAL ANALYSIS

Synoptic eddies in the open sea. To summarize the preceding section, we might state that synoptic eddies (i.e., motions with time scales t_0 ranging from a few days into the tens of days, horizontal scales L of the order of 100 km, vertical scales H of the order of the depth of the ocean, and characteristic velocities V of the order of 10 cm/sec) are very important in the overall dynamics of the ocean. We shall begin with derivation of equations that describe these motions.

These motions are described by a Kibel' number $\mathrm{Ki} = U/fL$ of the order of 0.01. Since the characteristic time scale t_0 can be taken as equal to L/U (L/U of about 10 days), we can say without too much error that synoptic motions are *geostrophic*. In addition, the applicability of the *quasistatic* approximation to analysis of synoptic eddies (H/L of about 0.01) is obvious.

The vorticity equation, obtained by eliminating the pressure from the equations of motion along longitude λ and latitude φ, is an important role in the study of geostrophic motions. Neglecting frictional forces and using the Boussinesq approximation, we obtain in accordance with Eqs. (3-2-1) [supplemented by the Coriolis force written in the traditional approximation (see Sec. 4-1)]

$$\frac{\partial\zeta}{\partial t}+\frac{u}{a\cos\varphi}\frac{\partial\zeta}{\partial\lambda}+\frac{v}{a}\frac{\partial\zeta}{\partial\varphi}+w\frac{\partial\zeta}{\partial z}+\beta v-$$

$$-(\zeta+f)\frac{\partial w}{\partial z}+\frac{\partial v}{\partial z}\frac{\partial w}{a\cos\varphi\partial\lambda}-\frac{\partial u}{\partial z}\frac{\partial w}{a\,\partial\varphi}=0, \qquad (5\text{-}2\text{-}1)$$

where

$$\zeta=\frac{\partial v}{a\cos\varphi\partial\lambda}-\frac{1}{\cos\varphi}\frac{\partial}{a\,\partial\varphi}(u\cos\varphi)$$

is the vertical component of the relative vorticity. The incompressibility condition $\mathrm{div}_h(u, v) + \partial w/\partial z = 0$ was assumed for sea water in deriving Eq. (5-2-1).

Estimating the orders of magnitude of the terms, we have

$$\frac{\zeta}{f} \sim \text{Ki}, \quad \frac{\dfrac{\partial v}{\partial z}\dfrac{\partial w}{a\cos\varphi\,\partial\lambda} - \dfrac{\partial u}{\partial z}\dfrac{\partial w}{a\,\partial\varphi}}{f\dfrac{\partial w}{\partial z}} \sim \text{Ki}.$$

Equation (5-2-1) can therefore be rewritten as

$$\frac{d\zeta}{dt} + \beta v - f\frac{\partial w}{\partial z} = 0, \tag{5-2-2}$$

where the meaning of the total-derivative symbol d/dt is obvious.

Following Burger [1] (see also the review by N. A. Phillips [2]), we shall classify geostrophic motions into two types. For motions of the first type, all three terms in Eq. (5-2-2) are assumed to be of the same order. Since $d\zeta/dt \sim U^2/L^2$ and $\beta \sim f/a$, we have

$$W = \frac{H}{L}\text{Ki}U, \quad \frac{L}{a} \sim \text{Ki}. \tag{5-2-3}$$

Generally, it is sufficient to assume that $L/a \sim$ Ki. Then $d\zeta/dt \sim \beta v$ and, from Eq. (5-2-2), we immediately obtain the first relation of (5-2-3). The condition $L/a \sim$ Ki is well satisfied for synoptic eddies ($L \sim 100$ km, Ki ~ 0.01), which are therefore geostrophic motions of the first type (they are still often called *mesoscale* geostrophic motions).

Geostrophic motions of the second type are described by the condition

$$\frac{L}{a} \sim 1. \tag{5-2-4}$$

These are large-scale geostrophic motions; in analyzing them, the term $d\zeta/dt$ in Eq. (5-2-2) can be dropped [$(d\zeta/dt)/\beta v \sim$ Ki] and we can assume $W = (H/L)U$. Numerical modelling of large-scale geostrophic motions is discussed in Chap. 9.

Let us return to synoptic motions. Given their scales, it is natural to assume that such motions exist against a background of a *specified* stratification of the sea. We shall therefore write the density equation as [cf. Eq. (4-1-3)]

$$\frac{\partial \rho^*}{\partial t} + \frac{u}{a\cos\varphi}\frac{\partial \rho^*}{\partial \lambda} + \frac{v}{a}\frac{\partial \rho^*}{\partial \varphi} + w\left(\frac{\rho_s}{g}N^2 + \frac{\partial \rho^*}{\partial z}\right) = 0, \tag{5-2-5}$$

where $\rho_s(z)$ is the standard density distribution in the ocean, $N^2(z)$ is the Brunt-Väisälä frequency corresponding to distribution $\rho_s(z)$, and ρ^* is the deviation of the density from $\rho_s(z)$.

By virtue of the quasistatic and geostrophic approximations, the characteristic magnitude of the density perturbation ρ^* becomes

$$\frac{\delta\rho^*}{\rho_0} = \frac{(fL)^2}{gH}\text{Ki}. \tag{5-2-6}$$

Assuming further that, in Eq. (5-2-5), the horizontal advection ρ^* and its time variation, which have the same order of magnitude (the time scale L/U), are commensurate with the term $w(\rho_s/g)N^2$, we find

$$L = L_R, \tag{5-2-7}$$

where $L_R = H(\bar{N}/f)$, $\bar{N}$ is the mean Brunt-Väisälä frequency.

The quantity L_R is known as the Rossby scale.[1] For the ocean $\bar{N} \simeq 2 \cdot 10^{-3}$ $\sec^{-1}$, and if $H = 4$ km, we have $L_R \simeq 80$ km. Thus, the characteristic horizontal scale of the synoptic eddies is equal to the Rossby scale.

Obviously, all of the above applies outside of a narrow equatorial belt and only for open-sea conditions. The quantity f that appears in the above formulas should be understood as some characteristic value of the Coriolis parameter; it will henceforth be denoted f_0.

Relations (5-2-3), (5-2-6), and (5-2-7) can be used in derivation of equations for synoptic eddies in the ocean. For this purpose, we represent the unknowns in the form (see [5, 6])

$$\begin{aligned} (u,\ v) &= U(u',\ v'); \\ w &= \frac{H}{L}\,\mathrm{Ki}\,Uw'; \\ p &= p_s(z) + \mathrm{Ki}\,\Gamma g H \rho_s(z)\,p'; \\ \rho &= \rho_s(z) + \mathrm{Ki}\,\Gamma \rho_s(z)\,\rho', \end{aligned} \tag{5-2-8}$$

where $\Gamma = (H\bar{N}^2)/g$ is the stratification parameter ($\Gamma \simeq 10^{-3}$; we shall assume $\Gamma \sim \mathrm{Ki}$); all primed quantities are dimensionless.

Since $L \sim 100$ km, we can confine ourselves to the β-plane approximation and assume, by virtue of the second relation of Eq. (5-2-3), that $f = f_0(1 + \beta^*\mathrm{Ki}y')$, where β^* and y' are dimensionless and of the order of unity.

Formal expansion of the primed variables in a Kibel'-number series (for example, $u' = u_0 + \mathrm{Ki}u_1 + \ldots$) and substitution of these series into the original equations yield [see [2, 6]):

in the zeroth approximation

$$\begin{aligned} -v_0 &= -\frac{\partial p_0}{\partial x} \\ +u_0 &= -\frac{\partial p_0}{\partial y}; \\ \frac{\partial p_0}{\partial z} = \rho_0; &\qquad \frac{\partial u_0}{\partial x} + \frac{\partial v_0}{\partial y} = 0; \\ \frac{\partial \rho_0}{\partial t} + u_0\frac{\partial \rho_0}{\partial x} &+ v_0\frac{\partial \rho_0}{\partial y} + N^2 w_0 = 0; \end{aligned} \tag{5-2-9}$$

[1] The importance of this characteristic scale was apparently first noted by Rossby [3], who called it the radius of deformation. This scale is also implicit in Prandtl's [4] work.

and in the first approximation

$$-\beta^* y v_0 - v_1 + \frac{\partial u_0}{\partial t} + u_0 \frac{\partial u_0}{\partial x} + v_0 \frac{\partial u_0}{\partial y} = -\frac{\partial p_1}{\partial x};$$

$$+\beta^* y u_0 + u_1 + \frac{\partial v_0}{\partial t} + u_0 \frac{\partial v_0}{\partial x} + v_0 \frac{\partial v_0}{\partial y} = -\frac{\partial p_1}{\partial y};$$

$$\frac{\partial u_1}{\partial x} + \frac{\partial v_1}{\partial y} + \frac{\partial w_0}{\partial z} = 0. \tag{5-2-10}$$

The analogs of the third and fifth equations of system (5-2-9) are omitted here from the first approximation because they will not be used below.

Eliminating the pressure p_1 from the first two equations of (5-2-10), we obtain

$$\left(\frac{\partial}{\partial t} + u_0 \frac{\partial}{\partial x} + v_0 \frac{\partial}{\partial y}\right)\left(\frac{\partial v_0}{\partial x} - \frac{\partial u_0}{\partial y} + \beta^* y\right) = \frac{\partial w_0}{\partial z}.$$

This is essentially the dimensionless notation for Eq. (5-2-2).

Dividing the fifth equation of Eq. (5-2-9) term by term by N^2 and differentiating it with respect to z, we find by virtue of the first three equations of (5-2-9)

$$\left(\frac{\partial}{\partial t} + u_0 \frac{\partial}{\partial x} + v_0 \frac{\partial}{\partial y}\right)\left[\frac{\partial}{\partial z}\left(\frac{\rho_0}{N^2}\right)\right] = -\frac{\partial w_0}{\partial z}.$$

Adding these equations, we have

$$\left(\frac{\partial}{\partial t} + u_0 \frac{\partial}{\partial x} + v_0 \frac{\partial}{\partial y}\right)\left[\frac{\partial v_0}{\partial x} - \frac{\partial u_0}{\partial y} + \beta^* y + \frac{\partial}{\partial z}\left(\frac{\rho_0}{N^2}\right)\right] = 0. \tag{5-2-11}$$

This is the final form of the equation for description of the synoptic eddies. By virtue of the first three equations of (5-2-9), the functions u_0, v_0, and ρ_0 are simply expressed in terms of p_0. Equation (5-2-11) is nothing but the equation of conservation of potential vorticity of a fluid parcel (for greater detail, see [2]).

The applicability of equations of the type (5-2-11) to analysis of synoptic eddies was discussed in several papers. Thus, Koshlyakov and Grachev [7] estimated certain parameters of the eddy shown in Fig. 5-1-6 and showed that this eddy could be interpreted as a baroclinic Rossby wave over an uneven sea floor [linear analysis of Eq. (5-2-11)]; Bretherton and Karweit [8] showed that an equation of type (5-2-11) satisfactorily describes the evolution of observed eddies (see Figs. 5-1-6 and 5-1-9).

Equation (5-2-11) should be solved with the following boundary conditions:

at the sea surface

$$w_0 = 0 \quad \text{at} \quad z = 0, \tag{5-2-12}$$

and at the sea floor

$$w_0 = u_0 \frac{\partial h'}{\partial x} + v_0 \frac{\partial h'}{\partial y} \quad \text{at} \quad z = 1, \tag{5-2-13}$$

where h' is the dimensionless disturbance of the ocean depth.

The condition (5-2-12) is readily derived from the usual boundary conditions at the free surface (see Sec. 4-1), since the characteristic scale for the depth of the ocean equals $H\Gamma$Ki. We write condition (5-2-13) for $z = 1$, since in a region with a characteristic horizontal scale of the order of 100 km, the sea-floor irregularities are usually small ($h \sim 50$ m). We shall therefore set $h/H \sim$ Ki. The analysis of synoptic eddies in a region with sharp depth differences will be found in the review by N. A. Phillips [2, p. 138].

The statement of the horizontal boundary conditions is very specific and will not be discussed here.

It will be helpful to describe the energy equation for the motions herein discussed. Multiplying the first two equations of (5-2-10) by u_0 and v_0, respectively, and the last equation of (5-2-9) by ρ_0 and adding the results, we find after simple algebra

$$\frac{\partial}{\partial t}\left(\frac{u_0^2 + v_0^2}{2} + \frac{\rho_0^2}{2N^2}\right) = -\operatorname{div}_h \left\{ \mathbf{v}_0 \left(\frac{u_0^2 + v_0^2}{2} + \frac{\rho_0^2}{2N^2} + p_1\right) + \mathbf{v}_1 p_0 \right\} - \\ - \frac{\partial}{\partial z}(p_0 w_0), \tag{5-2-14}$$

where the symbol div_h stands for the two-dimensional divergence of the vectors and $\mathbf{v}_0 = (u_0, v_0)$; $\mathbf{v}_1 = (u_1, v_1)$. The quantity $\rho_0^2/2N^2$ readily expresses the density of the *available potential energy* (see [9]).

If it is assumed that the liquid is enclosed in a finite volume V with vertical walls at which the normal component of the horizontal velocity vanishes, we can integrate (5-2-14) over the volume V and apply condition (5-2-12) and the fact that

$$p_0 w_0 = \operatorname{div}_h (p_0 h' \mathbf{v}_0) \quad \text{at} \quad z = 1$$

by virtue of (5-2-13) and the first, second, and fourth equations of (5-2-9). We thereby obtain

$$\frac{\partial}{\partial t} \int_V \left(\frac{u_0^2 + v_0^2}{2} + \frac{\rho_0^2}{2N^2}\right) dV = 0. \tag{5-2-15}$$

This is the integrated form of the energy conservation law.

Using the results of Sec. 4-1, let us now define the types of waves that will be filtered out under our assumptions. For this purpose, we linearize Eq. (5-2-11); by virtue of the first three equations of (5-2-9), we have

$$\frac{\partial}{\partial t}\left[\Delta_h p_0 + \frac{\partial}{\partial z}\left(\frac{1}{N^2}\frac{\partial p_0}{\partial z}\right)\right] + \beta^* \frac{\partial p_0}{\partial x} = 0.$$

In the case of a sea of constant depth, the solution of this equation is easily found by separation of variables:

$$p_0 = \sum_{1}^{\infty} \Psi_n (x, y, t) Z_n (z),$$

where the functions $\Psi_n (x, y, t)$ satisfy the equations

$$\frac{\partial}{\partial t} (\Delta_h \Psi_n - \varepsilon_n \Psi_n) + \beta^* \frac{\partial \Psi_n}{\partial x} = 0, \quad n = 1, 2, \ldots, \tag{5-2-16}$$

and ε_n and $Z_n(z)$ are the eigenvalues and corresponding eigenfunctions of the problem

$$\frac{d}{dz}\left(\frac{1}{N^2}\frac{dZ}{dz}\right) + \varepsilon Z = 0;$$

$$Z'(0) = 0, \quad Z'(1) = 0; \tag{5-2-17}$$

The problem (5-2-16) is completely equivalent to the overall problem (4-1-8) in the quasistatic approximation. Equation (5-2-16) describes the propagation of barotropic and baroclinic Rossby waves in the ocean. For not very long waves with wave numbers of the order of 1/1000 km and higher, this equation can be derived in the WKB approximation from the Laplace general tidal equations (for the corresponding analysis, see paper [10] by N. A. Phillips).

Thus, the approximations that we have used in deriving Eqs. (5-2-9) and (5-2-10) and the boundary conditions (5-2-12) and (5-2-13) filter out acoustic, gyroscopic, and gravity waves completely and leave only low-frequency Rossby waves (barotropic and baroclinic). In the linear approximation, our equations describe Rossby waves with practically no distortion (see Fig. 4-1-5 for dispersion relations for Rossby waves). It is, of course, clear that one can separate the solution into noninteracting barotropic and baroclinic Rossby waves only in the linear approximation. Let us emphasize that the basic equation (5-2-11) is nonlinear and that, in the general case, various Rossby waves will interact with one another.

It is also interesting to note that the definition of the energy density as in Eq. (5-2-14) agrees with the definition of energy density for wave motions of small amplitude [see Eq. (4-1-5)] if, of course, the compressibility of the medium is neglected and the quasistatic approximation is used.

The most important problem of ocean dynamics is that of the generation of synoptic eddies or, in other words, the problem of the source of energy producing the synoptic variations in the sea. It is logical to consider first the possibility of direct resonant generation of synoptic eddies in the sea by atmospheric disturbances. We refer therefore to Table 5-2-1 (see [11]), which gives the distribution of energy over characteristic ranges of the spectrum for certain atmospheric variables.[2]

This table shows a distinct energy peak for the wind velocity in the frequency range corresponding to the so-called natural synoptic period (5 to 10 days) in the atmosphere (range 3 in Table 5-2-1). This is the range of synoptic variability of the atmosphere. It is a well-known fact that the synoptic variability of the atmosphere is shaped by pressure lows and highs with characteristic horizontal scales of the order of 1000 km. Since the characteristic horizontal scale for synoptic eddies in the sea is of

---

[2] This table was derived from data of weather ships and island weather stations in the Atlantic for 50 to 100 years in the case of the monthly averages and 1 to 2 years in the case of standard observations.

TABLE 5-2-1 Distribution of Energy over Characteristic Spectral Ranges for Temperature T (deg²), Pressure p (mb²), and Wind Velocity V_w (m²/sec²) at Sea Level

Region	Spectral range					
	$6 \cdot 10^{-9}$ to $3 \cdot 10^{-8}$ Hz (6)	$3 \cdot 10^{-8}$ to $1.2 \cdot 10^{-7}$ Hz (5)	$1.2 \cdot 10^{-7}$ to $8 \cdot 10^{-7}$ Hz (4)	$8 \cdot 10^{-7}$ to $7 \cdot 10^{-5}$ Hz (3)	$7 \cdot 10^{-5}$ to $5 \cdot 10^{-3}$ Hz (2)	$6 \cdot 10^{-3}$ to $5 \cdot 10^{-1}$ Hz (1)
North Atlantic	0.50	7.75	1.34	2.44	–	–
(60–40°N)	4.35	72	61	82	–	–
	–	–	14.2	75		
Central Atlantic	0.081	1.57	3.74	5.25	–	–
(40–20°N)	0.58	7.46	–	50	–	–
	–	–	17.5	95		
Equatorial Atlantic	0.10	0.74	0.15	2.48	–	–
(10°N–10°S)	0.11	1.04	0.26	1.48	–	–
	–	–	1.7	15	0.149	0.428
South Atlantic	0.12	1.22	0.90	6.6	–	–
(40–60°S)	1.31	14.60	20.5	114	–	–
	–	–	6.4	65	–	–

Note: The first figure for each station denotes $\sigma_T^2(n)$, the second $\sigma_p^2(n)$, and the third $\sigma_v^2(n)$; n is the number of the spectral segment.

the order of 100 km, the possibility of direct resonant excitation of these eddies can evidently be discounted.

The following scheme now appears realistic. Atmospheric disturbances–lows and highs–generate large-scale currents in the sea, which are rendered *unstable* by various physical mechanisms. This results in the excitation of mesoscale geostrophic motions, which thus in effect derive their energy from the large-scale motions. However, the energy flow by far is not unidirectional. Thus, in some areas of sea the flow of energy is from mesoscale to large-scale motions (we shall return to this problem in Chap. 9).

We therefore have the problem of stability of large-scale currents or, more generally, the problems of the interaction between large-scale and mesoscale geostrophic motions. Let us first discuss the work on the *linear* theory of stability of simple zonal or meridional geostrophic currents. An extensive literature has been devoted to this problem (see the papers [5, 6, 12-17] and their bibliographies). Problems of *nonlinear* stability, recently under intensive study, are discussed in [18-21].

Let there be given, for example, a zonal geostrophic motion and $h' = h'(y)$. Given the simplicity of this motion, we can assume that it can be expanded in series in the Kibel' number Ki and that the individual terms of this series satisfy Eqs. (5-2-9), (5-2-10), etc.:

$$U_0 = U_0(y, z);\quad V_0 = 0;\quad W_0 = 0;\quad P_0 = P_0(y, z);\quad \rho_0 = \rho_0(y, z);$$

$$U_1 = U_1(y, z);\quad V_1 = 0;\quad W_1 = 0;\quad P_1 = P_1(y, z);\quad \rho_1 = \rho_1(y, z).$$

By perturbing this solution, we readily obtain equations that are *linear* with respect to such perturbations. We shall present only the equation for the total energy of these perturbations. Essentially repeating the derivation of Eq. (5-2-15), we find

$$\frac{\partial}{\partial t}\int_V \left(\frac{\tilde{u}_0^2 + \tilde{v}_0^2}{2} + \frac{\tilde{\rho}_0^2}{2N^2}\right) dV = \int_V \left(-\tilde{u}_0 \tilde{v}_0 \frac{\partial U_0}{\partial y}\right) dV + \int_V \left(-\frac{\tilde{v}_0 \tilde{\rho}_0}{N^2}\frac{\partial \rho_0}{\partial y}\right) dV, \tag{5-2-18}$$

where a quantity under a tilde is a perturbation. We see that there are two sources of perturbation energy. The first is derived from the fact that the large-scale motion varies along the horizontal. If this motion is in fact unstable (we shall not discuss the specific stability criteria, referring the reader to the papers just cited), instability of this kind is called *barotropic* because, generally speaking, the baroclinicity of the ocean water is immaterial in this case. The energy source for the perturbations is the kinetic energy of the large-scale motion.

The second source of perturbation energy is inherent in the existence of the horizontal density gradient $\partial\rho_0/\partial y$ (or, by virtue of the geostrophic relation, in the existence of the vertical velocity gradient $\partial U_0/\partial z$). An instability of this kind is called *baroclinic.* The primary source of energy for the perturbations is the available potential energy of the large-scale motion.

An interesting new mechanism for the generation of synoptic eddies was suggested by Bretherton and Karweit [8]. It involves the instability of the large-scale current under a weak ($h/H \sim$ Ki) perturbation of the sea floor topography (the presence of sea floor "roughness"). Numerical analysis has indicated not only the possibility of synoptic-eddy generation, but also a strong effect of these eddies on the large-scale motion.

Meandering of boundary currents. Finally, let us briefly review the theory of narrow meandering currents such as the Gulf Stream and the Kuroshio (see the review by Robinson [22]). The original theories of meandering treated this phenomenon within the framework of classical theory of baroclinic (or barotropic) instability of a given flow (see, for example, [14, 15, 23, 24]). However, as Robinson observes [22], the theory of baroclinic instability is consistent only with a very rapid buildup of perturbations. Like the linear theory, therefore, it is valid only in a very narrow range (in both time and space).

A nonlinear stationary theory that explains the formation of meanders in terms of the effect of sea floor topography was developed by Warren [25], Robinson and Niiler [26], and Niiler and Robinson [27]. This theory was successfully applied by Robinson and Taft [28] to explain the existence of two stable stationary positions of the axis of the Kuroshio (see Sec. 5-1). It was shown that the position of the axis depends significantly on the velocity at the floor (which is an input variable in the theory). Basically, the explanation for the existence of two stable positions of the Kuroshio's axis reduces to proof of the existence of a "forbidden zone" separating these positions. The axis finds itself in this zone exceedingly rarely if the input parameters of the theory are varied in accordance with the observed data. However, the nonlinear theory of topographic meanders does not allow for the important nonstationary effects (see the discussion of the Gulf Stream meanders in Sec. 5-1).

While the above two mechanisms are no doubt significant, we need (at least in the case of the Gulf Stream) a nonstationary nonlinear theory that would take account of both baroclinic instability and the formation of stationary meanders because of the effect of sea floor relief. Since very great difficulties are encountered in the derivation of such a theory (see Robinson's review [22, Secs. 6, 7], where the basic premises of this theory are briefly discussed), we shall analyze a simplified linear theory that appears to describe correctly the formation of meanders and eddies that are not too strong [29]. In this case the Kibel' number is assumed to be small, and all but the first terms of the series expansions of the velocity, pressure, and temperature of the current are dropped. After a few transformations, the problem reduces to analysis of the following set of equations, which it is convenient to write in dimensionless form:

$$\frac{\partial^2 \tilde{v}}{\partial z\, \partial t} + V(z) \frac{\partial^2 \tilde{v}}{\partial \nu\, \partial z} - \frac{dV}{\partial z} \frac{\partial \tilde{v}}{\partial \nu} = 0; \qquad (5\text{-}2\text{-}19)$$

$$\int_0^1 \left\{ \frac{\partial^2 \tilde{v}}{\partial \nu\, \partial t} + V(z) \frac{\partial^2 \tilde{v}}{\partial \nu^2} \right\} dz + h^* \tilde{v}(\nu,\ 0,\ t) = 0, \qquad (5\text{-}2\text{-}20)$$

where ν is measured across the isobaths; the latter are assumed to be straight lines (z = 0 corresponds to the bottom and z increases upward); $\tilde{v}(\nu, z, t)$ is the velocity along the isobaths, $V(z)$ is the initial velocity ($x = 0$) across the isobaths [the velocity across isobaths is everywhere equal to $V(z)$ in the first approximation], and h^* is a certain parameter that defines the effect of sea floor topography.

We shall confine ourselves to description of some of the most important properties of solutions of these equations that have a bearing on the meandering of currents. First of all, Eqs. (5-2-19) and (5-2-20) admit of wave solutions such as $\tilde{v} = A(z)\exp i(k\nu - \omega t)$. We find at once from Eq. (5-2-19) that $A(z) = V(z) \quad - \omega/k$. Substitution into (5-2-20) gives the dispersion relation

$$-k\omega^2 + 2\langle V\rangle k^2\omega - \langle V^2\rangle k^3 + h^*(-\omega + k\overline{V}) = 0, \qquad (5\text{-}2\text{-}21)$$

where the expression in the angle brackets denotes the vertically averaged value, and $\overline{V}$ is the value of V at $z = 0$.

It is readily shown that the waves are always stable in the barotropic case $\langle V\rangle = \sqrt{\langle V^2\rangle} = \overline{V} = V$. Buildup of oscillations may occur in the baroclinic case. This is best demonstrated by setting $h^* = 0$. Then

$$k = \frac{\omega}{\langle V^2\rangle}\{\langle V\rangle \pm \sqrt{\langle V\rangle^2 - \langle V^2\rangle}\}.$$

Since $\langle V\rangle^2 < \langle V^2\rangle$, the wave number k will be complex for real ω, which means an exponential buildup of the oscillations with increasing ν (along the current).

Note of the unusual character of the system (5-2-19), (5-2-20). Generally speaking, Eq. (5-2-19) defines the vertical structure of $\tilde{v}$ and Eq. (5-2-20) its horizontal structure. It can be shown that, after some transformations, the following equation is obtained for $\tilde{v}_0 = \tilde{v}(\nu, 0, t)$:

$$\frac{\partial^3\tilde{v}_0}{\partial\nu\,\partial t^2} + 2\langle V\rangle\frac{\partial^3\tilde{v}_0}{\partial\nu^2\,\partial t} + \langle V^2\rangle\frac{\partial^3\tilde{v}_0}{\partial\nu^3} + h^*\left(\frac{\partial\tilde{v}_0}{\partial t} + \overline{V}\frac{\partial\tilde{v}_0}{\partial\nu}\right) = 0. \qquad (5\text{-}2\text{-}22)$$

A stationary solution of this equation exists only if the sea floor has relief ($h^* \neq 0$); the wavelength of the topographic meanders is $h^*\overline{V}/\langle V^2\rangle$.

Equation (5-2-22) is greatly simplified in the barotropic case, wherein it assumes the form

$$\frac{\partial^2\tilde{v}_0}{\partial\nu\,\partial t} + V\frac{\partial^2\tilde{v}_0}{\partial\nu^2} + h^*\tilde{v}_0 = 0.$$

This is the familiar telegraph equation (except that ν acts as "time" and t becomes the "coordinate"). Therefore specification of $\tilde{v}_0$ and $\partial\tilde{v}_0/\partial\nu$ at $\nu = 0$ results in a properly conditioned problem.

In a certain sense, therefore, the meander model considered above combines the baroclinic-instability model and the model with stationary topographic meanders. It is shown in a paper by Robinson and Gadgil [29] (see also [22]) that this model permits qualitative description of several important features of the meandering process (variation of wavelength with time, effect of initial directions and curvature of the jet, effect of variation of V with time, tendency to formation of individual eddies, etc.). All of this is important both for planning and interpreting observations and for analysis of a more exact nonlinear model.

REFERENCES

1. Burger, A. P. Scale consideration of planetary motions of the atmosphere, Tellus, **10**, No. 2, 195–205 (1958).
2. Phillips, N. A. Geostrophic motion, Reviews of Geophysics, **1**, No. 2, 123–176 (1963).
3. Rossby, C. G. On the mutual adjustment of pressure and velocity distributions in certain simple current systems, II, J. Mar. Res., **1**, No. 3, 239–263 (1937–1938).
4. Prandtl, L. Beitrage zur Mechanik der Atmosphäre (*Contributions to Mechanics of the Atmosphere*), Berichte Assoc. Meteor. Union Geod. Geophys. Edinburgh, 1–32 (1936).
5. Charney, J. G. and M. E. Stern. On the stability of internal baroclinic jets in a rotating atmosphere, J. Atm. Sci., **19**, No. 2, 159–172 (1962).
6. Pedlosky, J. The stability of currents in the atmosphere and the ocean, Part I, J. Atm. Sci., **21**, No. 2, 201–219 (1964).
7. Koshlyakov, M. N. and Y. M. Grachev. Mesoscale currents at a hydrophysical polygon in the tropical Atlantic, Deep-Sea Res., **20**, No. 6, 507–526 (1973).
8. Bretherton, F. P. and M. Karweit. Midocean Mesoscale Modelling. In: *Numerical Models of Ocean Circulation*, Proc. of Durham Symp., USA Nat. Acad. Sci., 237–250 (1975).
9. Lorenz, E. Available potential energy and the maintenance of the general circulation, Tellus, **7**, No. 2, 157–167 (1955).
10. Phillips, N. A. Models for weather prediction. In: *Annual Review of Fluid Mechanics*, Palo Alto, Calif., **2**, 251–292 (1970).
11. Byshev, V. I. and Yu. A. Ivanov. Time spectra of certain characteristics of the atmosphere over the ocean, Izv. Akad. Nauk SSSR, Fizika atm. i okeana., **5**, No. 1, 17–28 (1969).
12. Charney, J. G. The dynamics of long waves in a baroclinic westerly current, J. Meteorology, **4**, No. 5, 135–162 (1947).
13. Pedlosky, J. The stability of currents in the atmosphere and the ocean, Part II, J. Atm. Sci., **21**, No. 4, 342–353 (1964).
14. Tareyev, B. A. Unstable Rossby waves and the nonstationarity of ocean currents, Izv. Akad. Nauk SSSR, Fizika atm. i okeana, **1**, No. 4, 426–438 (1965).
15. Orlanski, I. The influence of bottom topography on the stability of jets in a baroclinic fluid, J. Atm. Sci., **26**, No. 6, 1216–1232 (1969).
16. Abramov, A. A., B. A. Tareyev, and V. I. Ul'yanova. Baroclinic instability in Kochin's two-layer frontal model on the beta plane, Izv. Akad. Nauk SSSR, Fizika atm. i okeana, **8**, No. 2, 131–141 (1972).
17. Abramov, A. A., B. A. Tareyev, and V. I. Ul'yanova. Instability of a two-layered geostrophic current with an antisymmetric velocity profile in the upper layer, Izv. Akad. Nauk SSSR, Fizika atm. i okeana, **8**, No. 10, 1017–1028 (1972).
18. Pedlosky, J. Finite-amplitude baroclinic waves, J. Atm. Sci., **27**, No. 1, 15–30 (1970).
19. Pedlosky, J. Finite-amplitude baroclinic waves with small dissipation, J. Atm. Sci., **28**, No. 4, 587–597 (1971).
20. Pedlosky, J. Limit cycles and unstable baroclinic waves, J. Atm. Sci., **29**, No. 1, 53–63 (1972).
21. Pedlosky, J. Finite-amplitude baroclinic wave packets, J. Atm. Sci., **29**, No. 4, 680–686 (1972).
22. Robinson, A. R. The Gulf Stream, Phil. Trans. Roy. Soc. London, **A270**, No. 1206, 351–370 (1971).
23. Ichiye, T. On the variation of oceanic circulation (V), Geophys. Mag., **26**, No. 4, 283–342 (1955).

24. Ichiye, T. On the mechanism of a cold water domain on the northern boundary of the Kuroshio, Oceanogr. Mag., **8**, No. 1, 43–52 (1956).
25. Warren, B. A. Topographical influences on the path of the Gulf Stream, Tellus, **15**, No. 2, 167–183 (1963).
26. Robinson, A. R. and P. P. Niiler. The theory of free inertial currents, I, Path and structure, Tellus, **19**, No. 2, 269–291 (1967).
27. Niiler, P. P. and A. R. Robinson. The theory of free inertial jets, II. A numerical experiment for the path of the Gulf Stream, Tellus, **19**, No. 4, 601–619 (1967).
28. Robinson, A. R. and B. A. Taft. A numerical experiment for the path of the Kuroshio, J. Mar. Res., **30**, No. 1, 65–101 (1972).
29. Robinson, A. R. and S. Gadgil. Time-dependent topographic meandering, Geophys. Fluid. Dynam., **1**, No. 4, 411–438 (1970).

6 SEASONAL VARIATIONS

The seasonal variability of the oceanological fields results from the annual variations of insolation and of the state of the atmosphere (i.e., primarily the wind, air temperature, and precipitation). This variability has been studied more thoroughly than that of the other types, but given the scarcity of ocean measurements (described in Chap. 2), we still cannot derive a complete picture of the ocean's seasonal variability.

Among the data on seasonal variability of the various oceanological fields, those on surface temperature are most readily generalized to a global pattern. Panfilova [1] made an attempt at such a generalization, identifying five types of annual temperature variation: 1) an equatorial-tropical type with very weak seasonal variations; 2) a tropical type with a distinct annual trend having a total amplitude of 4 to 6°C. In the Northern Hemisphere it exhibits a temperature peak in August–September and a minimum in February-March, while in the Southern Hemisphere it has a peak in March–April and a minimum in July–August. In both cases, the spring warming of the water is slow, and the autumn cooling is fast; 3) a temperature-latitude type with the largest (up to 8°C) annual-variation amplitudes. This type exhibits nearly constant temperature throughout the winter, with a peak in August and a minimum in March–April in the Northern Hemisphere, and a maximum in February and a minimum in August in the Southern Hemisphere. The spring warming and fall cooling of the water are of equal duration; 4) an Antarctic–Subantarctic type with an annual variation amplitude of 2 to 3°C and nearly constant temperature in the winter (June-October) and summer (January-March) and transitional seasons of equal length; 5) a north Indian Ocean type with a temperature peak in April–May and a secondary peak in October–November, a minimum in January–February (stable winter monsoon) and a secondary minimum in August (stable summer monsoon) and a 2 to 4°C amplitude of the semiannual temperature variations.

Figure 6-1 presents a chart of these types of seasonal temperature variations, and Fig. 6-2 shows typical annual curves. These data reflect primarily the tendency to latitudinal zoning of the seasonal temperature-variation types produced by the latitudinal zoning of the seasonal solar heat flux variations (the increase in the amplitude of such variations from the equator toward the poles). The only difference

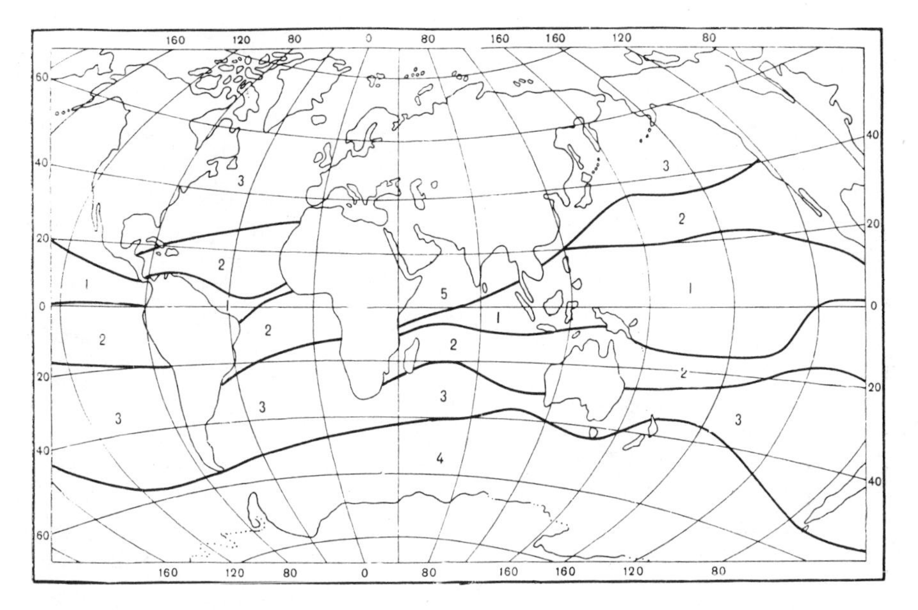

FIG. 6-1 Zoning of the ocean by types of seasonal variation of surface water temperature (after Panfilova [1]).

from the latter variations is that the amplitude of the annual water-temperature variation is not highest in the subpolar regions (where the water temperature does not become very high but cannot drop below zero either), but in temperate latitudes. The monsoon, that is, type 5 seasonal variations of surface water temperature in the northern Indian Ocean is a departure from this simple latitudinal pattern.

The above-described pattern is, of course, merely a sketch, the real variability being more complex because of both local effects and year-to-year variations. For example, the weather ship *Extra* recorded in the northwestern Pacific (in a zone with type 3 seasonal variability) annual water temperature amplitudes of 14°C (in 1948) and even 18°C (in 1950). Near the Grand Banks off Newfoundland in the Atlantic (also a type 3 zone), this amplitude reaches 10 to 15°C in some years, while the multiyear average amplitude of the annual variation in the shelf waters of the Gulf Stream is 13°C.

Because of the vertical mixing of the waters, seasonal temperature variations can usually be detected down to 200 to 300-meter depths, involving the upper mixing layer of the ocean and the upper (seasonal) thermocline. Figure 6-3, cited from Turner and Krams [2], shows two typical examples of seasonal temperature-field variations in the upper ocean, namely, near the Bermuda islands in the Atlantic and at the station of the weather ship *Papa* in the Pacific (50°N, 145°W). They indicate that the summer warming (which is greatest in July–September in these examples) spreads only through a comparatively thin upper mixing layer, bounded below by a sharply defined thermocline. The cooling of the surface waters that begins in the fall intensifies the mixing because it induces convection, the thickness of the mixing layer increases, and

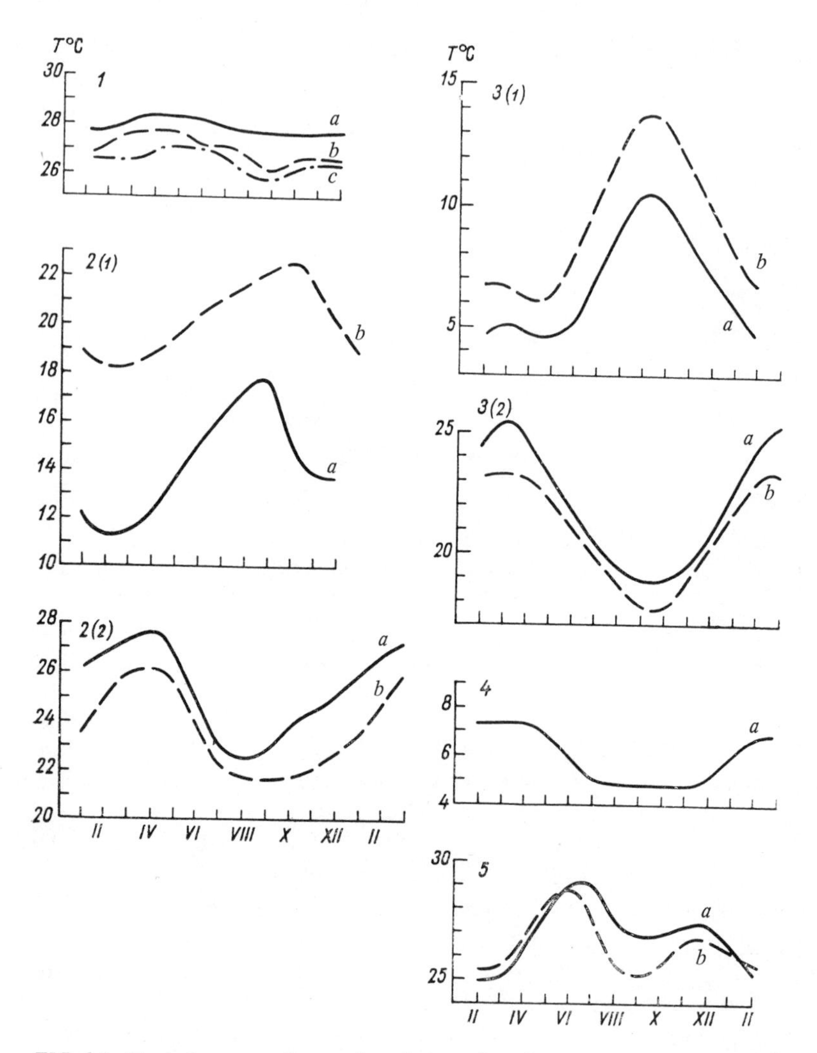

FIG. 6-2 Typical curves of annual variation of surface water temperature (after Panfilova [1]). 1) *a*–5°S, 96°E: *b*–5°S, 145°W; *c*–2°N, 30°W; 2(1)) *a*–38°N, 139°W; *b*–26°N, 17°W; 2(2)) *a*–3°S, 7°W; *b*–2°S, 106°W; 3(1)) *a*–54°N, 45°W; *b*–50°N, 136°W; 3(2)) *a*–29°S, 27°W; *b*–31°S, 177°E; 4) *a*–55°S, 77°W; 5) *a*–17°N, 66°E; *b*–12°N, 58°E.

the thermocline top boundary becomes less distinct. This process builds up throughout the winter, and by the end of winter (February-March), the entire upper thermocline becomes effaced and turns into a mixed layer. This pattern of seasonal thermocline variation is typical of high and middle latitudes, but in the tropics the upper thermocline, though not too distinct, persists year-round, and any convective mixing that may occur above the thermocline is due to changes in salinity of the surface water produced by evaporation.

Filyushkin [3] plotted the seasonal variations of the depths of the Upper (H_0) and lower (H_1) boundaries of the seasonal thermocline and the mean vertical temperature gradient $\partial T/\partial z$ in it at several points in the northern Pacific (Fig. 6-4). The curves in Fig. 6-4*a–c*, pertain to middle latitudes with well-defined seasonal thermocline variations, where the depth of the upper boundary of the thermocline varies by 60 to 120 m over the year. The plot of Fig. 6-4*d* pertains to the tropics, where the thermocline is indistinct ($\partial T/\partial z$ less than 0.1°C/m) and varies fairly irregularly over the year.

Some work has been done on the theory of formation of the seasonal thermocline (see, for example, [4, 5]). In terms of physics, this theory reduces to a description of radiative warming and vertical dynamic and convective mixing in the upper ocean. Kitaygorodskiy and Miropol'skiy [6, 7] have derived a theoretical model based on the assumption of self-similarity of the temperature

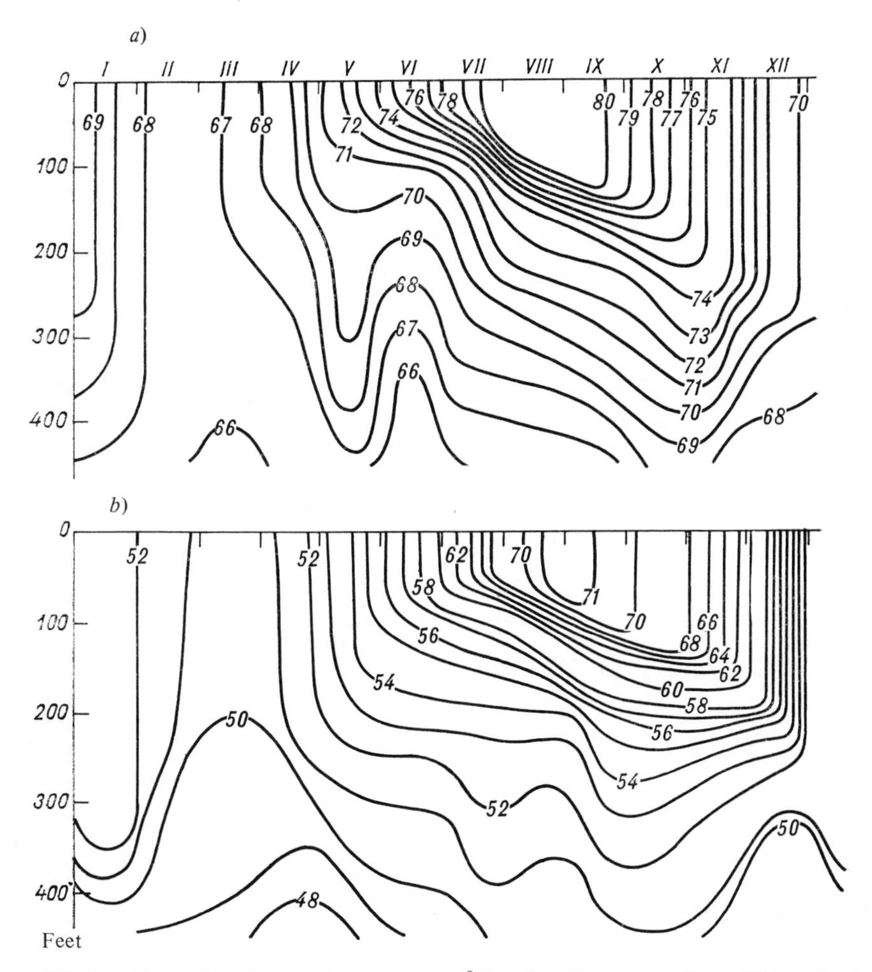

FIG. 6-3 Seasonal variation of temperature (°F) (after Turner and Kraus [2]). *a*–In the Bermuda area; *b*–in northern Pacific Ocean.

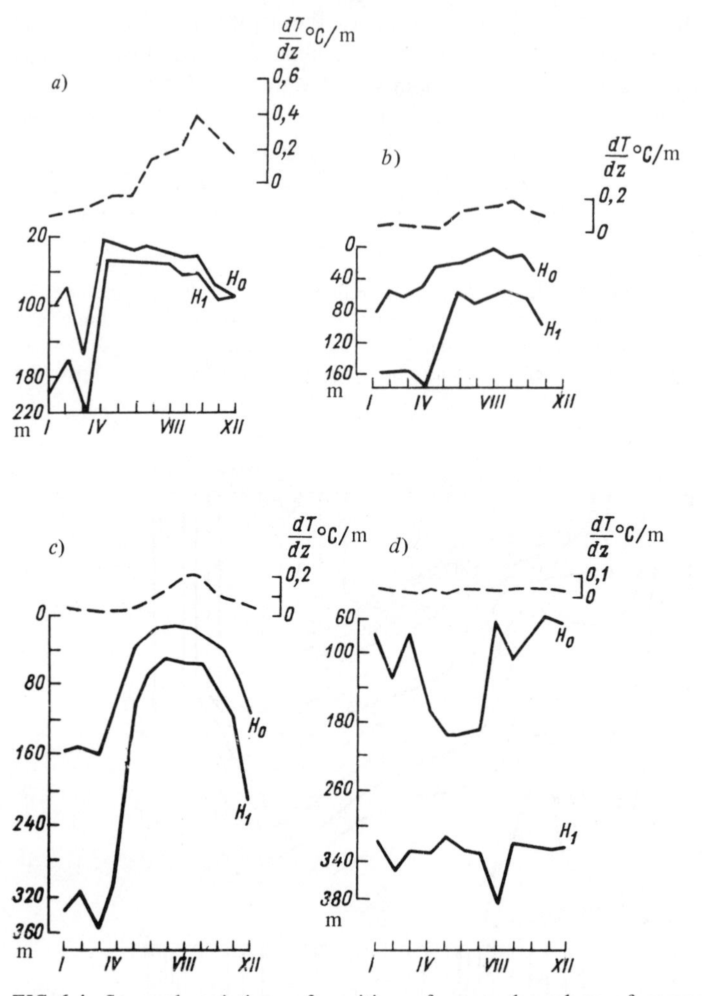

FIG. 6-4 Seasonal variations of position of upper boundary of seasonal thermocline (H_0), its lower boundary (H_1), and the average vertical gradient dT/dz (after Filyushkin [3]). *a*–At station *Papa* (50°N, 145°W); *b*–in square at 45°N, 130°W; *c*–at station *Extra* (30°N, 153°E); *d*–in square at 20°N, 155°W.

profile $T(z, t)$ in the seasonal thermocline, which is satisfactorily confirmed in a variety of observations. The model is

$$\frac{T_s(t) - T(z, t)}{T_s(t) - T_1} = f\left[\frac{z - H_0(t)}{H_1 - H_0(t)}\right], \tag{6-1}$$

where the subscript s pertains to the sea surface and 0 and 1 to the upper and lower boundaries of the seasonal thermocline, with H_1 and T_1 assumed constant in the simplest model. In this case the heat-transport equation, integrated over the upper mixing layer and over the thermocline, yields

$$c_p \rho H_0 \frac{\partial T_s}{\partial t} = q_s - q_0;$$

$$\frac{\partial H_0}{\partial t} + \frac{1-a}{a} \frac{H_1 - H_0}{T_s - T_1} \frac{\partial T_s}{\partial t} = \frac{q_0}{a c_p \rho (T_s - T_1)}, \qquad (6\text{-}2)$$

where $a = \int_0^1 f(\eta) d\eta$ is constant and about 0.73 according to observations, q_s and q_0 are the values of the vertical turbulent heat flux, and c_p is the specific heat.

Functions $H_0(t)$ and $q_0(t)$ can be determined from Eq. (6-2) by specifying $T_s(t)$ and $q_s(t)$. Such calculations were done in [6]. Miropol'skiy [7] also recommended that q_s be determined from the approximate turbulent energy balance integrated over the upper mixed layer:

$$\frac{1}{2} H_0^2 \frac{\partial T_s}{\partial t} - \frac{H_0 q_s}{c_p \rho} + (1-b) \frac{c u_*^3}{\alpha g} = 0, \qquad (6\text{-}3)$$

where b is the ratio of the turbulent energy dissipation rate to the rate $c \rho u_*^3$ of its production by surface waters and drift currents, u_* is the friction velocity on the surface of the water, α is the thermal expansion coefficient of the water (αg is the buoyancy parameter), and c is an empirical constant.

The properties of the seasonal thermocline, calculated from Eqs. (6-2) and (6-3) and T_s and u_* data from weather ship *Papa* [7] proved to agree satisfactorily with observations (except perhaps for the winter months, in which case it is best to correct for unsteadiness of the turbulence in the energy equation).

In certain parts of the sea, seasonal temperature variations can also penetrate to depths far below the seasonal therm‹ cline because of seasonal variability of warm and cold currents, as well as downwelling and upwelling. A few examples are given in Fig. 6-5. They indicate that in zones of strong currents [Kuroshio (Ia), California current (Ib)], the total amplitude of the annual temperature variations may reach 2 to 3°C even at depths of 500 to 600 m, while they can hardly be detected outside of these zones (see IIb, IIIa and b). In the East Australian current (IVa), the seasonal temperature variations can be observed to depths greater than 1000 m.

The salinity of sea water is a more conservative characteristic, and its seasonal variations (which are due chiefly to the annual evaporation and precipitation trends and, in high latitudes, to the formation and melting of ice; in coastal regions, river discharge also becomes a factor) are much less conspicuous. The total amplitude of the seasonal salinity variation in surface waters is about 0.2 to 0.3 per mille, and can be significantly higher only in certain areas (over 0.7‰ near the Grand Banks off Newfoundland, up to 1‰ in the California current, 1 to 3‰ in the Bay of Bengal and the Australasian seas, 2.7‰ in the Kuroshio current, and 5‰ in the Skagerrak).

Because of vertical mixing, the seasonal variability of currents and the vertical motions of the waters, seasonal salinity variations are observed in deep as well as surface waters. Two examples (subarctic and tropical waters of the northwestern Pacific), which we have borrowed from Panfilova [1], illustrate this in Fig. 6-6*a* and *b*. In these cases, the largest total amplitude of the seasonal salinity variations (1.0 to 1.2‰) were observed in the surface water, where a salinity maximum was reported in winter and a minimum in summer. The smallest seasonal variations were observed in

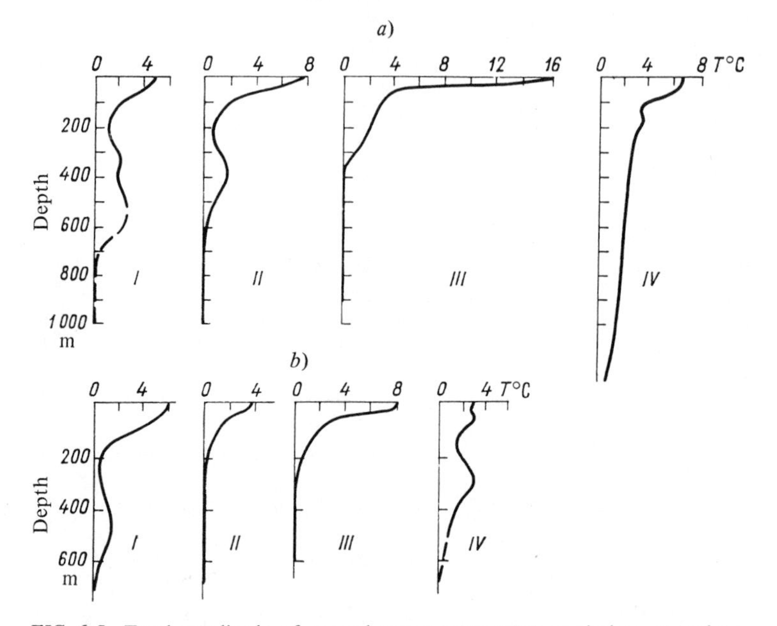

FIG. 6-5 Total amplitude of annual water temperature variations at various depths (data of Panfilova [1]). *a*–Northwestern Pacific [I) 20°N, 120°E; II) 25°N, 130°E; III) 40°N, 150°E; IV) 35°S, 155°E]; *b*–northeastern Pacific [I) 25°N, 115°W; II) 30°N, 125°W; III) 50°N, 145°W; IV) 20°N, 155°W].

subarctic waters at depths of 50 to 100 m and in tropical waters at 100 to 200 m. These variations increased with depth and reached their largest amplitude (0.3 to 0.4‰) at 400 to 500 m, where a semiannual period is conspicuous, especially in the subarctic waters. Still deeper, the variations taper out, being traceable to only 800 to 1000 m. However, there are still too few data of these phenomena to permit formulation of any general relationships governing the seasonal salinity variations in deep waters.

The seasonal variations of the currents are apparently caused primarily by the annual variation of wind velocity. In middle and high latitudes, the velocity is highest in winter and lowest in summer. In the tropics, and especially over the Indian Ocean, it contains fairly large components corresponding to the winter and summer monsoons. The seasonal variability of the surface currents was observed long ago and was used by seamen in plotting optimal sea routes. Below we shall discuss data on certain important parts of the ocean.

One of the first summaries of data on the seasonal variability of the currents in the north Atlantic was published by Fuglister [8]. Several of his curves, which he plotted from data on ship drift, are shown in Fig. 6-7. They indicate that the range of the seasonal velocity variations of north Atlantic surface currents may reach 20 to 35% of the mean velocity. It is interesting that in the Caribbean Sea and the southern part of the Gulf Stream the strongest currents were observed in the summer, in July, and the weakest currents in November.

Iselin [9] has analyzed the data on the seasonal variations of the water flow in the Gulf Stream. The data were obtained by 13 hydrological sections and by observations off Miami and Charleston in 1937–1940. Iselin estimated the range of the flow variations at 15 to $20 \cdot 10^6$ m^3/sec or 15 to 20% of the largest flow rate (later, variations with an even larger range, as much as 50% of the flow rate, were observed). Iselin suggested that the peak summer flow rate is due to a northward shift of the Azores high, the net result of which is that in summer a major fraction of water of the North Equatorial current enters the Gulf Stream directly, bypassing the Caribbean Sea and the Florida Strait.

Kutalo [10] has analyzed the seasonal variability of the currents and the atmospheric circulation in the North Atlantic and their interaction. He used data on the Gulf Stream, the variations of sea level off the American coasts, and the variability

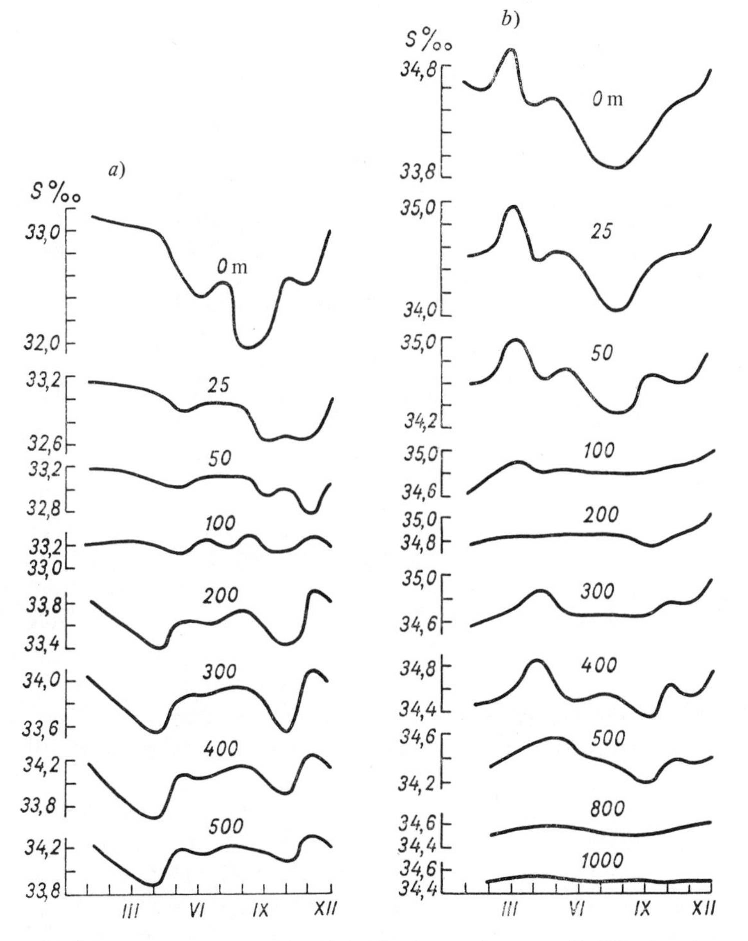

FIG. 6-6 Annual variation of salinity in the northwestern Pacific at various depths (from Panfilova [1]). *a*–Subarctic waters (50°N, 160°E); *b*–tropical waters (20°N, 120°E). The numbers above the curves are depths in meters.

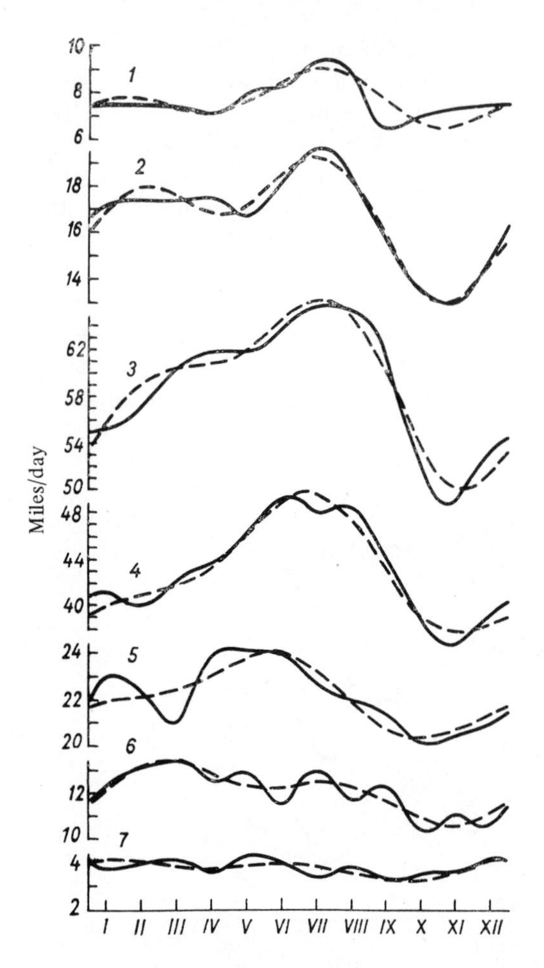

FIG. 6-7 Seasonal variations of surface currents in the north Atlantic (from Fuglister [8]). The solid curves represent observed current velocities, the dashed curves sums of the annual and semiannual current velocity components. 1) Trade winds zone; 2) Caribbean Sea; 3) Florida Strait; 4) South of Cape Hatteras; 5) northeast of Cape Hatteras; 6) southwest of Grand Banks off Newfoundland; 7) south of the Azores.

of the atmospheric circulation. He computed the annual variations of the sea level and currents from a model of nonstationary, two-layer oceanic circulation and found that the level of the North Atlantic rises in the fall off the American coast and in the spring off the coasts of Europe and Africa. The variations of the velocities of the principal currents and the position of the Gulf Stream are 180° out of phase. Among other things, he found that the Gulf Stream moves toward the American coast in the fall, out to sea in the spring, and occupies an intermediate position in the summer and winter.

The seasonal variations of the Kuroshio current in the Pacific have been studied by many authors. An extremely long series of measurements, from 1906 through 1960, was analyzed by Pavlova [11], who calculated the velocities and flow rates in this current from mean monthly average water-density profiles (to run over many years) parallel to five standard hydrologic sections through the Kuroshio between 130 and 150°E. Pavlova's results (Fig. 6-8) indicate a well-defined semiannual harmonic, with two peaks and two minima, in the seasonal variations of Kuroshio velocity and flow rate. The annual harmonic is the dominant feature east of Japan, while to the south oscillations with periods of less than 6 months also appear to exist. The multiyear mean amplitude of the seasonal variations does not exceed 10 to 15% of the annual mean, but in any given year it may reach 50 to 60% of the maximum annual velocity and flow rate.

The causes of the dominance of the semiannual harmonic in current variations

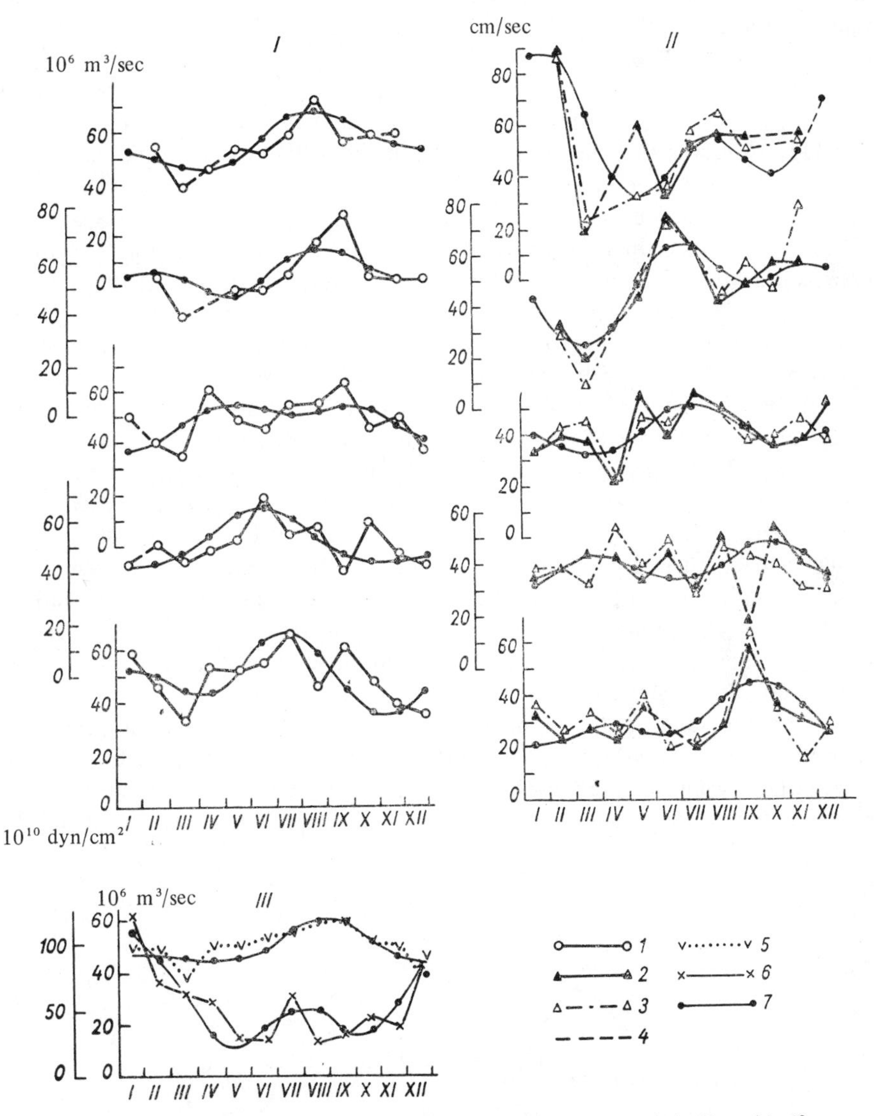

FIG. 6-8 Seasonal variations of flow rate and velocity in five zones of the Kuroshio (from Pavlova [11]). I) Flow rate on five sections through the Kuroshio (curves 1); II) mean velocity in the same sections (curves 2) and velocities at specific points (curves 3); interpolation (4); III) mean flow rate of Kuroshio (curves 5) and curl of wind stress $\text{curl}_z\tau$ (curves 6); resultant curves obtained by adding the annual and semiannual components (curves 7). See in [11] for the positions of the sections.

have been intensively discussed. Naturally, all the harmonics of the annual period are present in the seasonal variations of the solar heat input at any point on the globe, but it seems probable that these harmonics do not act directly on the currents. Instead, these harmonics would tend to affect the seasonal variations of the atmospheric circulation over the oceans, which, generally speaking, also contain all harmonics (even though the semiannual harmonic may not be the dominant one in the variations of certain properties of the circulation). This reasoning has been confirmed, among others, by Byshev [12], who plotted spectra of the fluctuations of temperature, atmospheric pressure, and its differences over the sea over a broad frequency range (from $5 \cdot 10^{-9}$ to $5 \cdot 10^{-1}$ Hz) and found both annual and semiannual harmonics. Ichiye [13] suggested that the semiannual harmonic in the seasonal variations of the Kuroshio is due to the net effect of the annual variations of the zonal and meridional wind-stress components in the northern half of the Pacific.

One would expect that differences in atmospheric circulation and in its seasonal variations over different oceans would induce differences in current variations. In the North Atlantic, for example, the semiannual harmonics of the fluctuations of the currents and of the differences in atmospheric pressure between the centers of the Azores high and the Iceland low are approximately in phase, whereas in the northern half of the Pacific the phase of the semiannual harmonic in the current variations lags considerably the variations of the pressure difference between the Honolulu high and the Aleutian low. Fedorov [14] suggested that this is due to differences in the effect of the atmospheric circulation in the zones on the Equatorial currents (the Pacific area of Equatorial currents is much larger than the Atlantic one), and showed that the product of the annual harmonics of the variations of the strength of trade winds and trade-zone width may contain a fairly significant semiannual harmonic. Some authors have suggested the semiannual solar tide as a possible cause of the semiannual harmonic in the current variations, but its amplitude is so small that it could hardly have a discernible influence on the seasonal variations of currents.

The above reasoning as to the role of the atmospheric circulation in the generation of seasonal current variations are confirmed by data on the Antarctic Circumpolar current, which were analyzed by Ivanov [15]. Figure 6-9 shows his curves of the annual variation of the meridional atmospheric pressure difference (curve 1) and the

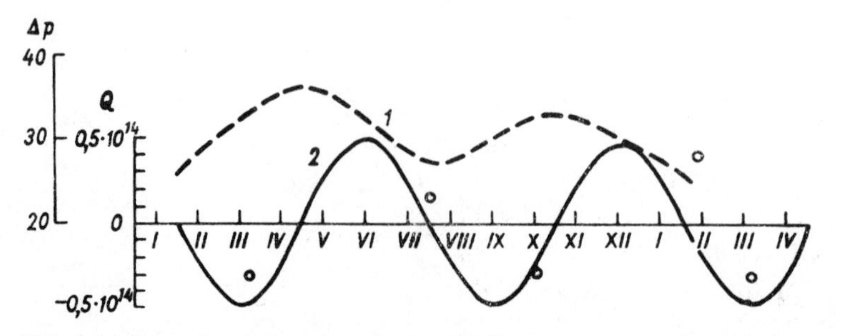

FIG. 6-9 Seasonal variations of atmospheric pressure difference (in mb) (1), of calculated flow rate of the Antarctic Circumpolar current (in cm^3/sec) (2), and of observed flow rates (circles) (from Ivanov [15]).

flow rate in the current on a section between the Antarctic continent and the Cape of Good Hope, as determined by the dynamic hydrological measurements from the ship *Discovery II* (circles). Both curves exhibit a well-defined semiannual harmonic, with the phase of the flow rate variations lagging that of the variations of the atmospheric pressure difference by approximately $\pi/2$ (i.e., by one and a half months). To describe the flow rate variations analytically, Ivanov used a simple linear model in which the components S_x and S_y of the total flow along the x and y axes were determined from

$$\frac{\partial S_x}{\partial t} - fS_y = -\frac{\tau_x}{\rho}; \quad \frac{\partial S_y}{\partial t} + fS_x = gH \int_0^t \frac{\partial^2 S_y}{\partial y^2}\, dt, \tag{6-4}$$

where τ_x is the wind stress at the sea surface and H is the sea depth. These equations were used to determine the flow rate $\int_0^L S_x\, dy$ of the current from the meridional pressure gradient. The result (curve 2 in Fig. 6-9) was in good agreement with the flow rates determined by the dynamic method.

The seasonal variations of the currents in the northern half of the Indian Ocean are undoubtedly generated by the dominant atmospheric circulation of the monsoons, namely by the northeast monsoon in winter and the stronger and longer southwest monsoon in (Northern Hemisphere) summer. Most of the available data on the seasonal variability of the currents in these waters were collected by the International Indian Ocean Expedition of 1959–1964. The current studies under this program were summarized by Duing [16]. They indicate that the monsoon effect can be traced from the north down to 10°S, to a depth of 400 m in the western part of this region, and to less than 100 m in its central and eastern parts. Figure 6-10 shows curves of the seasonal current-velocity variations at the surface and those of the wind speed in four zones of the Indian Ocean (*a* and *b* are zones off the African coast from 8° to 2°S, and 5° wide along longitude; *c* and *d* are zones along 92.5°E from 7°N to 9°S). These curves show that the seasonal variability of both the wind and the currents is much stronger in the west than in the east, and that the summer peak is longer than the winter peak. Duing's dynamic charts show that the greatest unsteadiness of geostrophic currents is observed during the transitional periods between monsoons.

Working with the data of the International Indian Ocean Expedition and deep-water hydrological measurements of earlier years, Neyman [17] plotted a series of dynamic charts of the Indian Ocean (with the reference surface at a depth of 1500 m) for several depths in the summer and winter seasons. The seasonal current variations show up only on charts of the northern half of the ocean. The following variations were noted: south of Ceylon, the monsoon current, which flows westward during the winter, turns to the east in the summer and merges with the Equatorial countercurrent. The winter cyclonic circulation system in the Arabian Sea is replaced in the summer by an anticyclonic gyre that is elongated along 10°N. The strong (4 to 6 knots) Somali current is southwestward in winter, but northeastward in summer (see also [18]). An interesting study by Lighthill [19] is devoted to the shaping of the Somali current under the action of the southwest monsoon (see Sec. 5-2). The seasonal

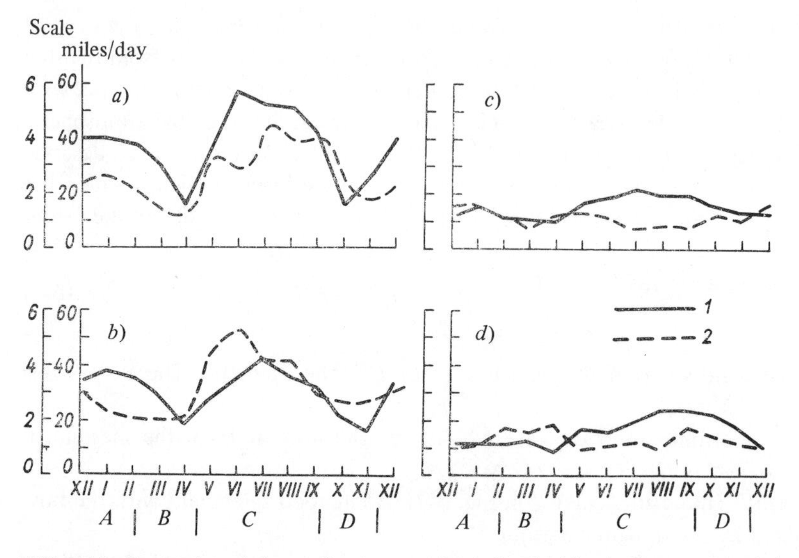

FIG. 6-10 Seasonal variations of current and wind velocities in four zones of the Indian Ocean (from Duing [16]). 1)Wind force (Beaufort scale); 2) current velocity in miles/day. *a*–8–4°N, 50–55°E; *b*–4°N–2°S, 47–52°E; *c*–7–3°N, 90–95°E; *d*–3°N–9°S, 90–95°E. *A*) Winter; *B*) spring transition; *C*) summer; *D*) fall transition.

variability in the Indian Ocean has been modelled numerically by Cox [20] (see Chap. 9). Gill and Niiler [21] undertook a broad-range study of the seasonal variability in the North Atlantic and northern Pacific.

REFERENCES

1. Panfilova, S. G. Seasonal variability of temperatures of ocean surface water, Okeanologiya, **12**, No. 3, 394–406 (1972).
2. Turner, J. S. and E. B. Kraus. A one-dimensional model of the seasonal thermocline, I, A laboratory experiment and its interpretation, Tellus, **19**, No. 1, 88–97 (1967).
3. Filyushkin, B. N. Thermal characteristics of the upper water in the northern Pacific Ocean, Okeanologicheskiye issledovaniya, No. 19, 22–69 (1968).
4. Kitaygorodskiy, S. A. and B. N. Filyushkin. The temperature discontinuity layer in the ocean, Trudy Inst. Okeanolog. Akad. Nauk SSSR, **66**, 3–28 (1963).
5. Phillips, O. *The Dynamics of the Upper Ocean*, Cambridge University Press (1966).
6. Kitaygorodskiy, S. A. and Yu. Z. Miropol'skiy. Contribution to the theory of the active layer of the open ocean, Izv. Akad. Nauk SSSR, Fizika atm. i okeana, **6**, No. 2, 177–179 (1970).
7. Miropol'skiy, Yu. Z. Nonstationary models of the convective wind mixing layer in the ocean, Izv. Akad. Nauk SSSR, Fizika atm. i okeana, **6**, No. 12, 1284–1294 (1970).
8. Fuglister, F. C. Annual variations in current speeds in Gulf-Stream system, J. Marine Res., **10**, No. 1, 119–127 (1951).
9. Iselin, C. Preliminary report on long-period variations in the transport of the Gulf-Stream system, Pap. Phys. Oceanogr. Meteor., **8**, No. 1, 40 (1940).
10. Kutalo, A. A. Seasonal variations of the circulation in the North Atlantic, Izv. Akad. Nauk SSSR, Fizika atm. i okeana, **7**, No. 3, 317–327 (1971).
11. Pavlova, Yu. V. Seasonal variations of the Kuroshio current, Okeanologiya, **4**, No. 4, 625-640 (1964).
12. Byshev, V. I. Annual and semiannual variations of certain properties of the marine

atmospheric surface layer, Izv. Akad. Nauk SSSR, Fizika atm. i okeana, **4**, No. 5, 540–547 (1968).
13. Ichiye, T. On the variation of oceanic circulation in the adjacent seas of Japan. In: Proc. of the UNESCO Sump. of Phys. Oceanography, Tokyo, 116–129 (1955).
14. Fedorov, K. N. Annual and semiannual variations of the general circulation of the oceans, Dokl. Akad. Nauk SSSR, **116**, No. 3, 393–396 (1957).
15. Ivanov, Yu. A. Seasonal variability of the Antarctic circumpolar current, Doklady Akad. Nauk SSSR, **127**, No. 1, 74–77 (1959).
16. Duing, W. *The Monsoon Regime of the Currents in the Indian Ocean*, E. W. Center Press, Honolulu (1970).
17. Neyman, V. G. New current charts for the Indian Ocean, Dokl. Akad. Nauk SSSR, **195**, No. 4, 948–952 (1970).
18. Leetman, A. The response of the Somali current to the southwest monsoon of 1970, Deep-Sea Res., **19**, No. 4, 319–325 (1970).
19. Lighthill, M. J. Dynamic response of the Indian Ocean to onset of the southwest monsoon, Phil. Trans. Roy. Soc. London, **A265**, No. 1159, 45–92 (1969).
20. Cox, M. A. A mathematical model of the Indian Ocean, Deep-Sea Res., **17**, No. 1, 47-75 (1970).
21. Gill, A. E. and P. P. Niiler. The theory of the seasonal variability in the ocean, Deep-Sea Res., **20**, No. 2, 141–177 (1973).

7 THE YEAR-TO-YEAR VARIABILITY

The year-to-year variability of the oceanological fields exhibits spatial scales comparable to the dimensions of the ocean basins as a whole. The principal causes of the year-to-year variability of such fields are rearrangements of the air-sea interaction patterns, including self-oscillatory processes in the air-sea-continent system. Analysis of the mechanisms driving the year-to-year variability of these fields is very important for the development of long-term weather and sea forecasting in the most important regions of the ocean, including ice conditions in seasonally icebound seas.

Multiyear observations. Very little is known about the year-to-year variability of the ocean, primarily because multiyear data series are available only for a few areas of the ocean. Among these areas is the northernmost (north of 50°N) zone of the Atlantic Ocean. Here, beginning in 1902, the Hydrographic Committee of the International Council for the Exploration of the Sea has systematically collected *en route* hydrometeorological observations from ships. Since 1926, systematic observations have been made in the north Atlantic from weather ships of the various countries bordering the Atlantic basin. The series of these observations are the most important source materials for study of the large-scale air-sea interaction and the long-term variability in the northern zone of the Atlantic Ocean. Their most significant deficiency is that they pertain only to the surface layer of the sea. Multiyear deep-water hydrological observations in the Atlantic basin have been made by expeditions of various countries, chiefly England and Norway, on the standard hydrological section across the Faeroe-Shetland and Faeroe-Iceland channels. Very important annual observations have been made on these standard sections during 1927–1939 and from 1946 to the present. Systematic deep-water hydrological observations are being made by the Soviet Polar Research Institute for Sea Fisheries and Oceanography on a standard hydrological section along the Kola meridian (from the Kol'skoye Poberezh'ye to 73°N) in the Barents Sea (these series extend over the period from 1929 to the present, with a forced interruption for the Second World War of 1941–1945). Beginning in 1929, Canadian and American oceanographers have made deep-water observations on three standard sections through the Labrador and West Greenland currents.

In the Pacific Ocean, an extensive program of annual (2 to 4 times a year) deep-water hydrological observations has been conducted by Japanese oceanographers since 1924 on seven standard sections through the Kuroshio current. In addition to seasonal observations of water temperature and salinity, the Japanese investigators also conduct systematic measurements on these sections with electromagnetic current meters (EMCM). In the Gulf of Alaska, Canadian investigators have run annual observations on a hydrological section between the weathership *Papa* and Vancouver Island since 1956. Since 1949, American oceanographers have run systematic observations on five hydrological sections through the California current. Apart from multiyear hydrometeorological observations from coastal and sea-island stations, which are generally nonrepresentative of open areas of the ocean, the examples cited above constitute all that we have in more or less continuous (over 10 to 15 years) series of hydrological observations in the ocean.

Variability of temperature and salinity. The most complete statistical workup of a multiyear series of observations (1876 through 1965) of temperature in the surface layer of the North Atlantic Ocean was recently completed by Potaychuk [1]. The spectral densities of water temperature fluctuations over a 90-year period that he calculated for 13 squares can be used with high confidence to identify the following periods in the year-to-year variability: one year, just over two years, 4 to 5, and 30 to 35 years. As an example, Fig. 7-1 shows plots of the spectral density for squares *J* and *H*. Potaychuk found that variations with these periods were intermittent over the time of the observations, for example, the 4 to 5-year periods were dominant in 1890 to 1925 and 1945 to 1955. The amplitudes of the annual temperature variations are 2 to 3°C, and those of the 4 to 5-year variations approximately 0.7°C. Potaychuk also obtained similar characteristics for the multiyear series from the Kola meridian. He also concluded that the multiyear variability of the temperature field in the North Atlantic is practically simultaneous despite the great complexity of the current system in this area. Potaychuk's data on the amplitudes of the multiyear water temperature variations are statistical averages, so that the variations between specific years may be considerably wider. Off Florida, for example, the annual average water temperature variations between 1941 to 1945 reached 6°C at station *Extra*, and in the northwestern Pacific they reached 4.5°C in the period 1937 to 1941.

Shishkov [2] plotted charts of the monthly average and annual average water

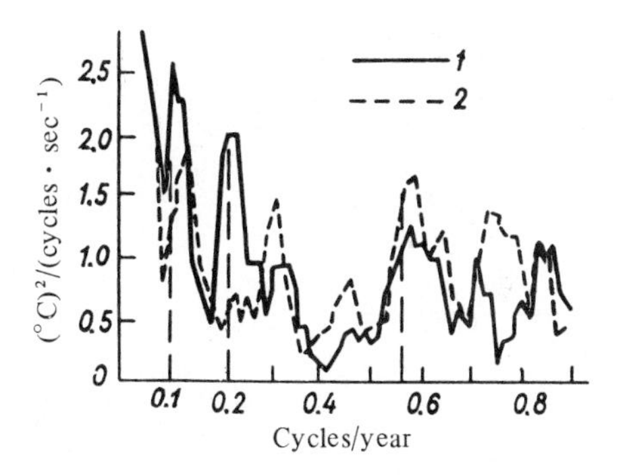

FIG. 7-1 Curves of spectral density of fluctuations of water temperature in the North Atlantic for squares *J* (1) and *H* (2), from Potaychuk [1].

temperature anomalies in the upper 100 meters of the North Atlantic over 14 squares for the period from 1946 through 1955. We see from Fig. 7-2, which shows the distribution of the annual average water-temperature isanomals, that regions with anomalies of the same sign occupy large areas of the ocean. These charts indicated that the zones of the major water-temperature anomalies are not associated with the same geographic regions and therefore, in Shishkov's opinion, are not related to the existing current system. An analysis of the atmospheric circulation over the ocean led to the conclusion that the principal cause of development of the oceanic temperature anomalies is meridional heat transport in the lower troposphere. In Shishkov's view, winter heat transfer between the sea and the atmosphere is a significant factor in generation of anomalies of the mean annual temperature of surface water.

Arkhipova [3] made an interesting study of the year-to-year variability of the heat balance in the North Atlantic Ocean for 1948–1957. She found below-average heat release over the entire North Atlantic in 1948–1949, with the largest anomalies in the areas of the Gulf Stream and the North Atlantic current and the smallest ones in the areas of the East Greenland, Labrador, and Canary currents. During the years in which the heat release was highest in the zone of the Gulf Stream and North Atlantic current (for example, in 1953), the heat-balance anomaly was of the opposite sign in the East Greenland and Canary currents. Heat release in the Gulf Stream region exhibits strong year-to-year variations. In 1949, for example, it was 60 to 100 kcal/cm^2 · yr, and in 1953 it rose to 160 kcal/cm^2 · yr. Arkhipova remarks that the

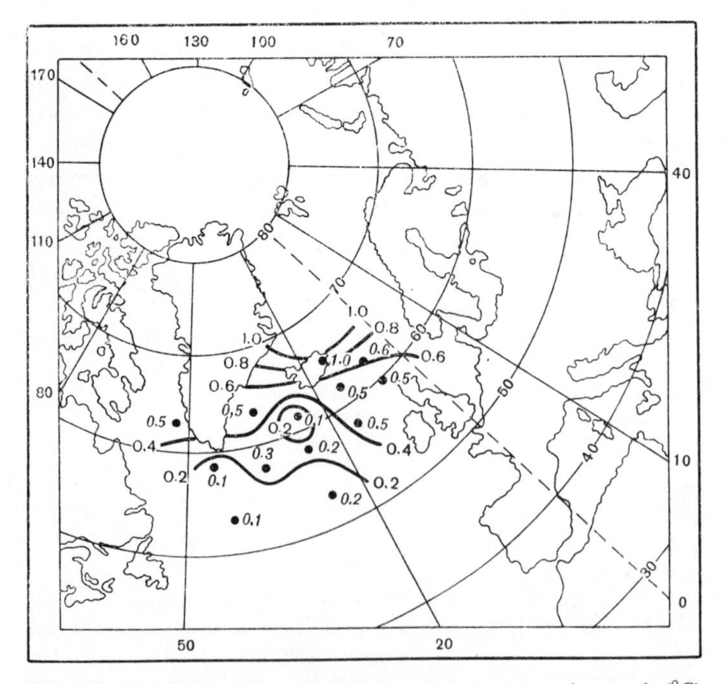

FIG. 7-2 Distribution of mean annual water-temperature isanomals (°C) in the North Atlantic Ocean (from Shishkov [2]).

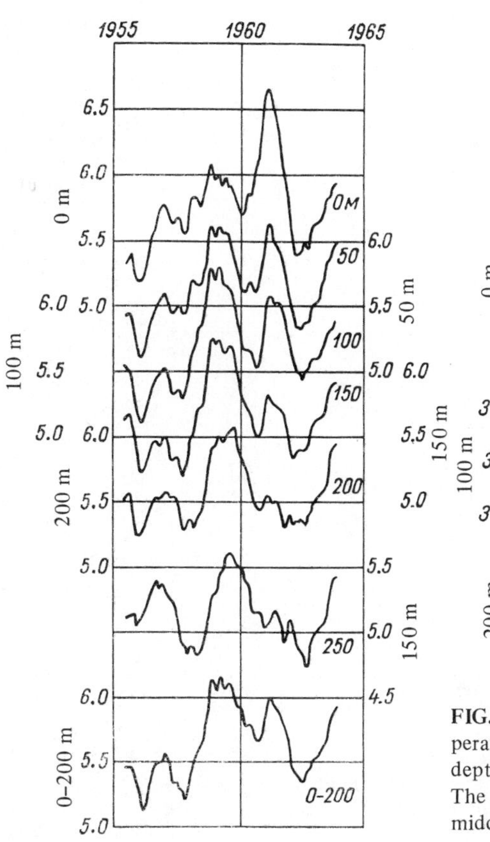

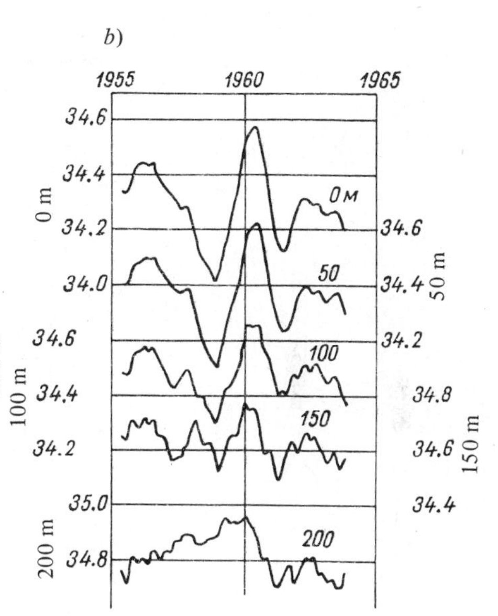

FIG. 7-3 Multiyear trend of mean average temperature (*a*) and salinity (*b*) off Nordkapp for depths from zero to 200 m (from Midttun [4]). The depths indicated on the curves refer to the middle of the ordinate scale for each curve.

largest long-term variability of the heat balance is observed in the Gulf Stream and North Atlantic currents, that is, in zones with the highest rates of heat advection.

Figure 7-3, which we cite from Midttun [4], shows the long-term variation of the mean average water temperature and salinity off the Nordkapp, at various depths from the surface to 250 m. The figure shows that even in high latitudes, the year-to-year variability of water temperature and salinity may be substantial in both the surface and subsurface waters. An even greater year-to-year variability of water temperature and salinity is observed when we compare the same seasons. Figure 7-4, cited from Panfilova [6], shows the year-to-year water-temperature variability for the month of August in a one-degree square in the Kuroshio zone over the period from 1933 through 1940, and Fig. 7-5 shows the total amplitude of the year-to-year salinity variation, also for August, at four stations in the northwestern Pacific. It is seen that the year-to-year variations of temperature and salinity in a given season of the year can reach 10°C and 1.5 to 3.0‰, respectively. In this case the largest amplitudes of the year-to-year variability are observed deep in the ocean, an indication of long-term fluctuations of the transport of heat and salts by the currents.

Panfilova [6] also compared the seasonal and year-to-year water temperature and

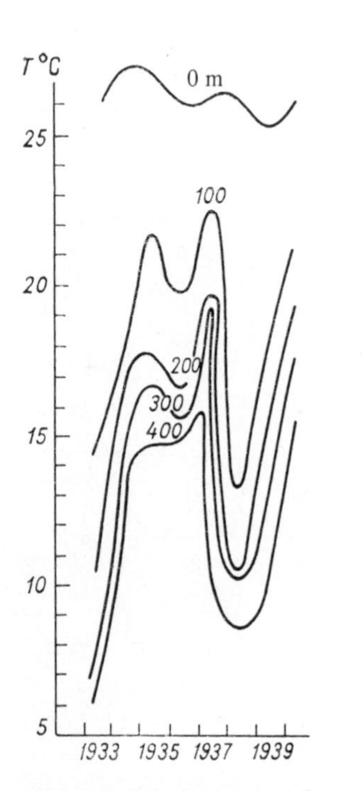

FIG. 7-4 Year-to-year variations of water temperature (August) at various depths in the square of 35–36°N, 150–151°E from 1933 through 1940 (from Panfilova [5, Chap. 3]).

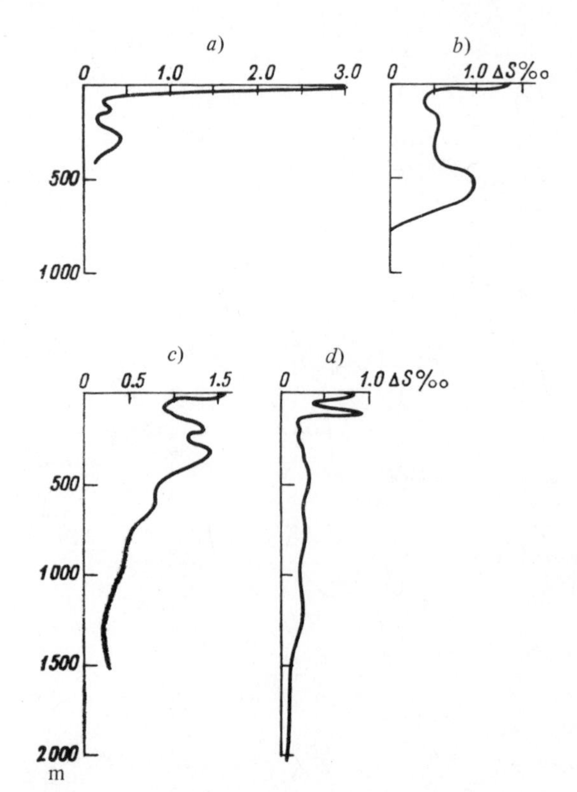

FIG. 7-5 Total amplitude of year-to-year salinity variations at various depths (August) at four stations in the northwestern Pacific (from Panfilova [5, Chap. 4]). *a*–50°N, 156°E; *b*–38°N, 162°E; *c*–35°N, 142°E; *d*–29°N, 134°E.

salinity variations at several stations in the Northern Pacific in terms of the ratio suggested by Kolesnikova and Monin [7]:

$$\mu = \frac{a}{A}\,\%,$$

where a is the amplitude of the year-to-year variations and A is the amplitude of the annual trend averaged over the period of the observations.

Figure 7-6 shows the variations of μ for the temperature over the year in three zones, a, b, and c, for which multiyear observation series (covering more than 30 to 40 years) were available. Off the Canadian coast (zones a and b), the largest value of μ is observed during the winter (January), ranging above 50% at most of the stations. The winter μ for Cape St. James is 95%, and reaches 75% at the ocean station *Papa*. Off the Japanese islands (zone c), μ varies from 20 to 55%, with a maximum during the summer (June–July); all four stations in this area are coastal and are strongly influenced by the presence of land. Figure 7-7 shows the annual trend of μ for the salinity at seven stations on the west coast of Canada. The year-to-year salinity variation significantly (by as much as 185%) exceeds the seasonal variations at the offshore stations, and at the onshore stations it occasionally even reaches 600 to 650%. These results demonstrate once again the importance of the year-to-year variations in the overall spectrum of the physical variability of the sea.

Variability of the North Atlantic circulation. The Faeroe-Shetland Channel is perhaps the only area of the ocean for which a multiyear series of deep-water observations of water-mass transport has been accumulated. Here, systematic deep-water hydrological measurements have been made since 1927. In recent years, they have been accompanied by instrumental current measurements from buoy stations. Tait has summarized the observations from 1927 through 1952 [8]. The volume

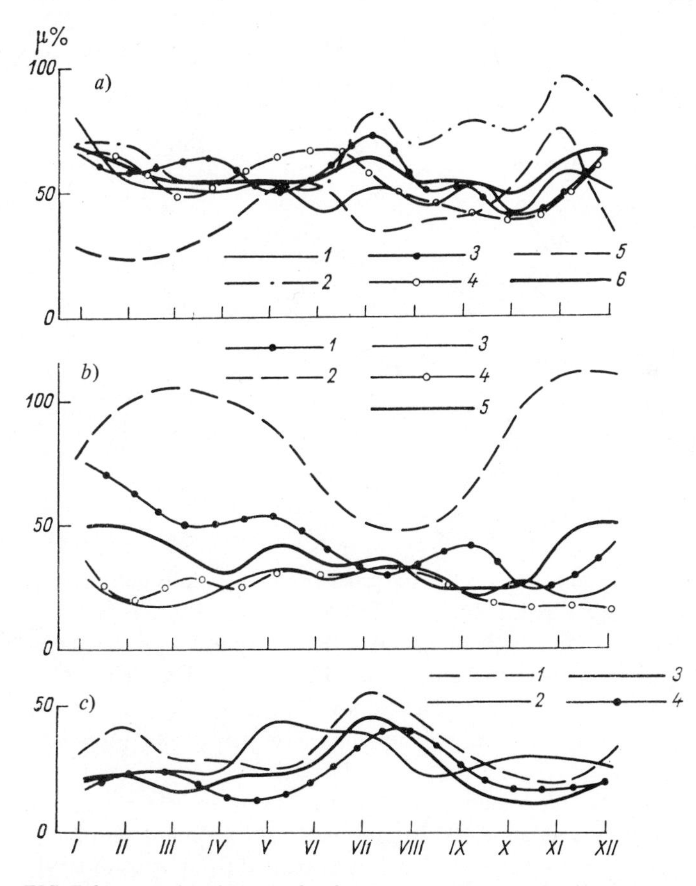

FIG. 7-6 Annual variations of μ for water temperature in three zones of the Northern Pacific (from Panfilova [6]). *a* and *b*–Canadian coast and Gulf of Alaska; *c*–Japan. *a*) 1–Langaren; 2–Cape St. James; 3–Kaighn; 4–Amphitrite; 5–Station *Papa*; 6–average; *b*) 1–Triplets; 2–Pine; 3–Entrance; 4–Departure; 5–Rose Rocks; *c*) 1–Aburatsubo; 2–Hosojima; 3–Oshoro; 4–Wajima.

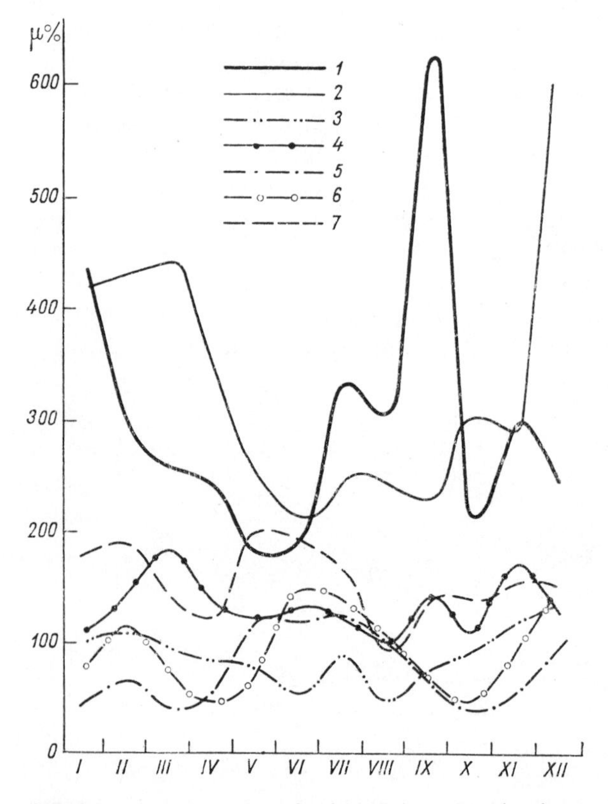

FIG. 7-7 Annual variation of μ for salinity at island and coastal stations on the Pacific coast of Canada (from Panfilova [6]). 1–Lagaren; 2–Cape St. James; 3–Kaighn; 4–Amphitrite; 5–Entrance; 6–Departure; 7–station *Papa.*

transport of North Atlantic waters through the passage varied during this period from 1.4 to 23.4 km^3/hr (the average was 8.2 km^3/hr). The lowest flow rate was observed in August of 1927. Two months later, a phenomenon extremely unusual for the region was recorded in the North Sea, that is, the appearance of an iceberg. The highest flow rate was reported for December 1951, and a slightly lower peak of 19.9 km^3/hr was observed in June 1947 (this was a season of an unusually strong and prolonged warming in Europe).

Cold deep Arctic basin water drains into the northwestern Atlantic Ocean via the Faeroe-Shetland Channel below depths of 400 to 500 m. This deep countercurrent is also subject to wide variations in time. Figure 7-8, which is cited from Grasshoff [9], indicates the year-to-year variations of temperature and salinity in the deep, polar water flowing over the sill of the Faeroe-Iceland Channel. The figure shows how complex the pattern of the space-time year-to-year variability of the current is in this channel.

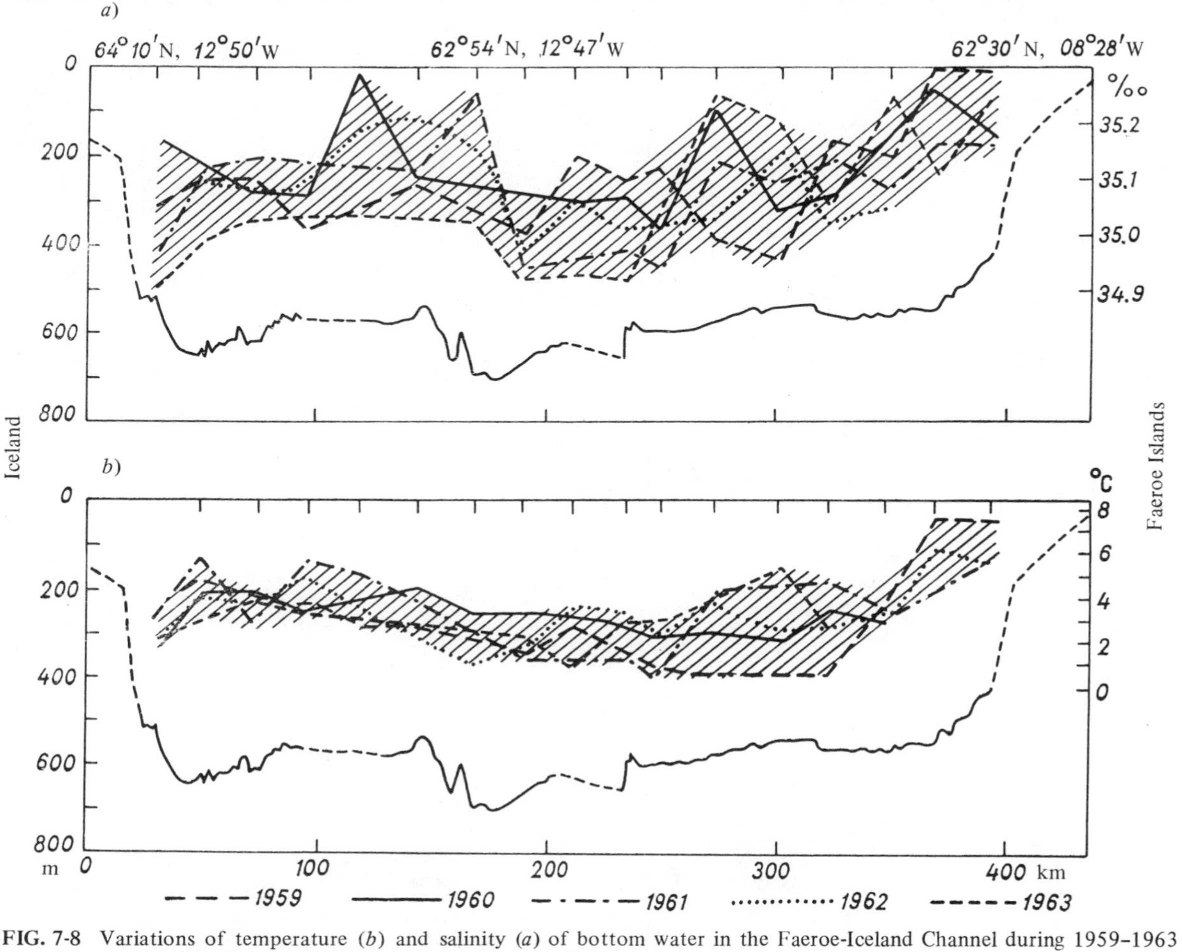

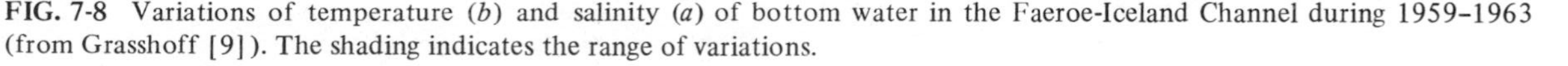

FIG. 7-8 Variations of temperature (*b*) and salinity (*a*) of bottom water in the Faeroe-Iceland Channel during 1959–1963 (from Grasshoff [9]). The shading indicates the range of variations.

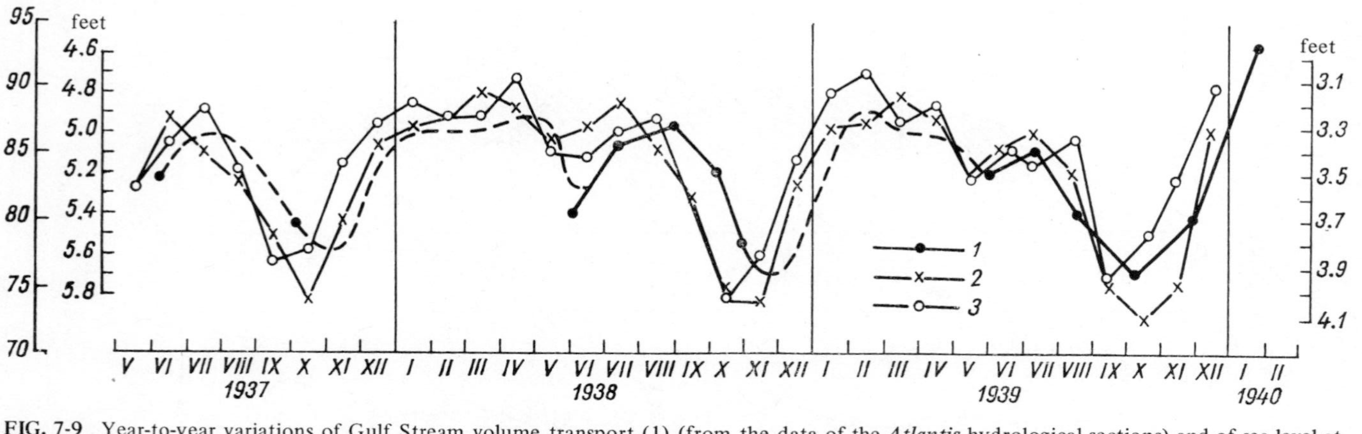
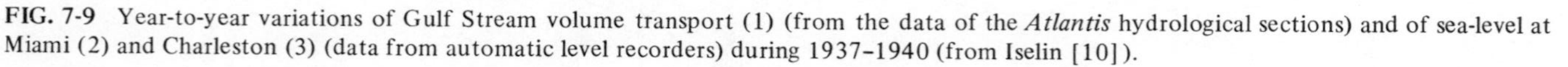

FIG. 7-9 Year-to-year variations of Gulf Stream volume transport (1) (from the data of the *Atlantis* hydrological sections) and of sea-level at Miami (2) and Charleston (3) (data from automatic level recorders) during 1937–1940 (from Iselin [10]).

The North Atlantic current is a continuation of the Gulf Stream, and one would think that their variations would be a single process. Unfortunately, we have no systematic multiyear observations of the variability of Gulf Stream volume transport. During the 1930's, such observations were begun by Iselin [10], but in the years that followed they were repeated only sporadically. Figure 7-9 shows the variations of Gulf Stream volume transport for the period from 1937 to 1940 as measured from the *Atlantis*, and the annual variation of sea level at Miami and Charleston. The range of seasonal variations of the volume transport was as high as $15 \cdot 10^6$ m^3/sec, and the range of the year-to-year variation was of the order of $5 \cdot 10^6$ m^3/sec during this short period. One of the first theories attempting to account for the pattern of the large-scale fluctuations of hydrological characteristics of circulation in the North Atlantic was that of Iselin [10]. He assumed that there exist certain normal values for the quantities of heat and salts in the North Atlantic circulation system (whose principal component is the Gulf Stream), and that this system is in geostrophic equilibrium. When the circulation intensifies, so does the effect of the Coriolis force. The latter is compensated by an increase in the radial density gradient from the center of the anticyclonic circulation to its periphery. The net result is that the diameter of the circulation system shrinks and the thermocline plunges to greater depths in the center zone of the circulation system. By contrast, when the circulation weakens, the diameter of the circulation system increases. For these reasons, the latitudes reached by the Gulf Stream when it flows at high velocity are lower than during periods in which its flow is not so swift. This theory found some support in the data of Stommel [11, 12], who observed that between 1920 and 1950 the surface temperature in the Norwegian sea increased by 2°C and that, simultaneously, the 10°C isotherm in the Sargasso Sea (the center of the above circulation system), an isotherm that normally runs in the thermocline layer, rose to 50 meters.

In a study of interactions in the air-sea-continent system, Shuleykin [12] concluded that this is a self-oscillatory system. The example of an elementary self-oscillatory system that he analyzed (jointly with Yershova) was the interaction of the North Atlantic circulation with the circulation of the Arctic basin. Figure 7-10 is a schematic representation of this system; here F stands for Florida, N for Newfoundland, P for the polar (Arctic) basin, G for the Gulf Stream, L for the cold water flow of the Labrador and East Greenland currents, and A for the warm North Atlantic current. The thick black arch represents the polar ice, the curved arrows mark the zone of mixing of the cold and warm waters, and the dots show the zone in which some fraction of the mixed waters plunges to the bottom without joining the North Atlantic current system. Any change in the regime of the cold waters L, which must affect in one way or another the thermal state

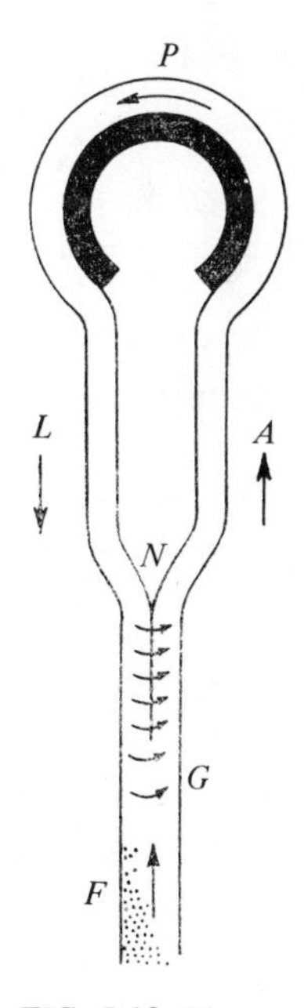

FIG. 7-10 Diagram of self-oscillatory Atlantic-Arctic basin system (according to Shuleykin [12]).

of the North Atlantic current A, will inevitably propagate along A, penetrating farther and farther from the mixing region N. Upon entering the Arctic basin, this thermal "pulse" will affect the equilibrium between the intruding warm waters and the ice drifting on the surface. In turn, a change in the amount of ice will result in a change in the regime of the cold flow L. Upon mixing in zone N, this change will have an effect on the state of the warm North Atlantic current, and so forth. Shuleykin's approximate description of this process is

$$\frac{dI}{dt} = -m(Q - Q_0);$$

$$\frac{dQ}{dt} = n(I - I_0), \tag{7-1}$$

where I is the quantity of ice in the Arctic basin, Q is the heat stored in the North Atlantic current, and m and n are empirically defined proportionality factors.

The period of this oscillatory system is given by

$$\tau = \frac{2\pi}{\sqrt{mn}}. \tag{7-2}$$

From Yershova's model experiments (see [12]), this period is approximately 3.5 years.

Duvanin [13] constructed a somewhat different scheme of the self-oscillations in the North Atlantic circulation system, assuming that all of the principal ocean currents of this system become stronger or weaker synchronously. A strengthening of the circulation produces penetration of positive water-temperature anomalies into the warm-current areas and of negative anomalies into cold-current areas. This should intensify the zonal circulation of the atmosphere which, in turn, would cause still stronger circulation of the ocean currents. Finally, Duvanin assumed that at maximum circulation the positive (or negative) temperature anomalies begin to penetrate into the zones of cold (or, respectively warm) currents, and so the circulation begins to weaken. This, in turn, changes the atmospheric processes so that the ocean currents are weakened even further. When the circulation drops to a minimum, the accumulation of heat in warm-current regions and of cold in cold-current regions will produce conditions conducive to a repetition of the cycle. The net result will be long-period oscillations of the intensity of the atmospheric and oceanic circulations. Duvanin tested this hypothesis on an electromechanical model simulating the conditions characteristic of the North Atlantic and produced a self-oscillatory process with a period in the range from 4 to 7 years.

Variability of the North Pacific circulation. From a series of observations of water temperature off the coasts of Japan and California for 1911 through 1941, Uda [14] found that the phase difference between the year-to-year variations in the western and eastern parts of this circulation system is 3 to 4 years (the time required for a temperature anomaly to travel the length of the southern circulation branch). These variations are produced by intrusion of warm anomalies on the Japanese side and cold

anomalies on the American side. Uda sees the basic cause of t
of the north Pacific circulation in large-scale fluctuations of
positions of the Siberian and Hawaiian atmospheric highs a
opinion, large-scale variations of the advection of interme
also stimulate the year-to-year variability of the north Pacif...

By analyzing the variations of the position of the northern boun... Kuroshio, as well as the frequency and strength of its meandering, and the times of appearance and disappearance of the cyclonic water gyre to the south of Japan, Ichiye [15] found periods of 4.5 and 9 to 10 years for the year-to-year variability of the Kuroshio. The 4.5-year period controls the variations of the position of the northern boundary of the Kuroshio, and the 9 to 10-year period of variations of the quasistationary cyclonic gyre. Ichiye attributes both of these processes to large-scale fluctuations of the atmospheric circulation over the northern Pacific. By analysis of periodograms of monthly average water temperatures in the active layer of the ocean in the Kurile (Oyashio)-Kuroshio current system, Fukuoka [16] detected periods of 2 to 3, 5 to 6, and 9 to 11 years; like Uda and Ichiye, he attributed the long-term variations in the thermal state of the north Pacific current system to long-term variations of the atmospheric circulation over the ocean. Batalin [17] found periods of 3, 4 to 5, 5 to 6, and 10 years in the variations of intensity of meandering of the Kuroshio, and ascribed these long-period variations to fluctuations of the Siberian, Aleutian, and Hawaiian atmospheric pressure centers.

Kort [18] attempted to define the year-to-year variability of the north Pacific circulation by analysis of deep-water hydrological observations on standard oceanographic sections through the Kuroshio and California currents. In Kort's opinion, the heat content Q calculated from

$$Q = c\rho \iint_S T(h, L)\, dh\, dL, \tag{7-3}$$

where c is the heat capacity of the water, h and L are the depth and width of the current, and S its cross-sectional area, is a fairly representative indicator of the variations of advection of heat by the currents. The advantage of this indicator is that it does not require knowledge of the characteristics of turbulent exchange in the sea or of the current velocity, direct measurements of which were begun only in recent years. Kort used this method to evaluate the 1954 through 1964 deep-water temperature data from the Shiono-misaki section. This section cuts the Kuroshio along 135°40′E from the Japanese coast to the southern boundary of the current. The results are plotted in Fig. 7-11. This figure indicates the seasonal (thin line) and annual average (heavy line) variations of heat content in the layers from 0 to 10 meters (*a*), from 0 to 200 meters (*b*), and from 0 to 1000 meters (*c*). The 1000-meter depth is the approximate lower boundary of the Kuroshio on this section, as well as the depth of the baroclinic layer in this area. The results indicate that the variations of the heat content in the upper (0 to 10 meter) and active (0 to 200 m) layers exhibit a pronounced seasonal pattern with a one-year period controlled by the insolation pattern. The mean yearly heat contents of these layers exhibit year-to-year variations

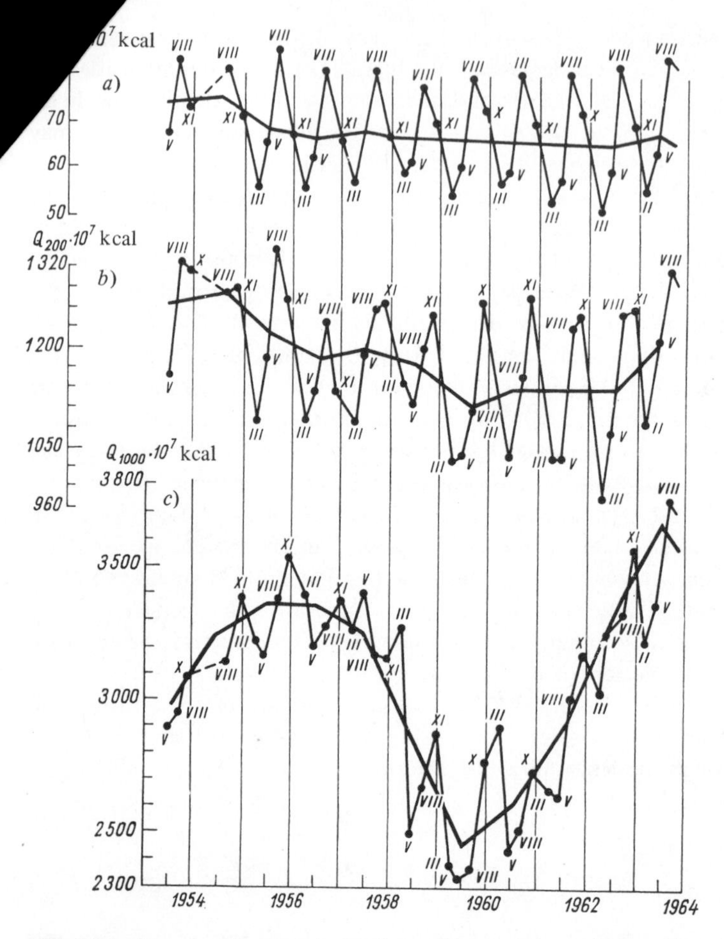

FIG. 7-11 Annual and seasonal variations of heat content of the surface (*a*), active (*b*), and baroclinic (*c*) layers of the ocean on the Shiono-misaki section (from Kort [18]).

with amplitudes smaller than those of the seasonal variations. But the principal component of the heat-content fluctuations in the baroclinic layer (to the depth of Kuroshio's lower boundary) is the year-to-year variation, whose amplitudes are greater than those of the seasonal variations. The period of the year-to-year heat-content fluctuations in the baroclinic layer is approximately (pending receipt of longer data series) estimated at 5 to 7 years, which is very close to the periods derived by other authors and agrees satisfactorily with the circulation period of the water masses in the north Pacific current system. Large-scale fluctuations of advection of heat by the Kuroshio result in 3 to 4° variations of the mean temperature of the baroclinic layer in the region of this current.

Kort also calculated the year-to-year variability of heat content in the California current by this method, and found a period of 5 to 6 years. Comparison of the

Kuroshio and California currents indicates that the extreme heat contents in the Kuroshio at the Japanese coast occur an average of 4 to 6 years after those in the California current. This time difference is almost two years longer than the time of flow of the water masses (and, consequently, of the temperature anomaly) along the southern branch of the north Pacific circulation system. These two years are apparently lost in forming the temperature anomaly in the baroclinic layer near the California coast. When we consider the relation, observed by Fukuoka [16], between the latitude of the axis or northern boundary of the Kuroshio (in the zone of the polar oceanological front) and the position of the peak of the westerly wind velocity, we can represent the self-oscillations of the oceanic and atmospheric circulations in the northern Pacific by the diagram shown in Fig. 7-12. The travel times of the temperature anomalies along the southern and northern circulation branches are indicated on the circulation lines. The time required for production of positive temperature anomaly in the eastern branch of the circulation is indicated in the dashed oval. The time difference between the northward displacements of the Kuroshio's northern boundary (or of the polar oceanological front) and of the zone of peak westerly wind velocities upon an increase in the velocity of the Kuroshio (and southward on its decrease) is indicated over the dashed lines in the northwestern corner of the diagram, and the mean time required for transport of cold air masses from the Arctic regions to the tropics is indicated in the northeastern corner of the figure.

The sea-air interactions in this area can be represented as follows. As the heating of the air over the north Pacific polar oceanological front increases (when the heat content in the Kuroshio is at maximum), the intensity of the westerly transport in the atmosphere increases. This occurs because northward movement of the oceanological front or, which is the same thing, the axis of the Kuroshio, produces a similar shift in the atmospheric polar front. Accordingly, the distance between it and the Arctic front is reduced, so that the meridional atmospheric pressure gradients in this region increase. The net result is intensification of the west-to-east air transport in the northern Pacific Ocean. This, in turn, increases the flow in the cold California current, so that cold water begins to pile up in the eastern part of the Pacific tropical zone,

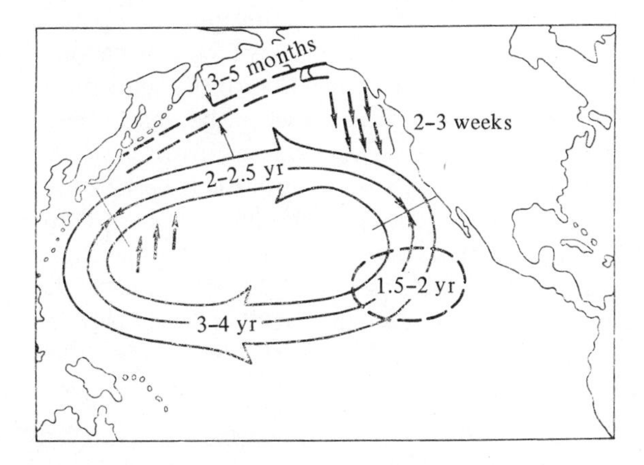

FIG. 7-12 Diagram of large-scale interaction of north Pacific circulation with westerly transport in the atmosphere (after Kort [18]).

generating a negative heat-content anomaly in the northern Pacific anticyclonic circulation. Cooling of the surface waters in the eastern part of the ocean (as a result of influx of cold water and air masses from the north) promotes the dominance in this region of virtually cloudless skies with little precipitation and hence enhanced warming of the water by the unimpeded influx of solar heat. This produces a new positive anomaly of heat content in the north Pacific circulation. A complete travel cycle of this temperature anomaly in the north Pacific circulation system takes an average of six years, as we indicated above.

Much time elapses between changes in heat content of the baroclinic layer and the corresponding changes in temperature of the lower layer of the atmosphere. On the other hand, the heat contents of the active and especially of the surface layer of the ocean fluctuate almost synchronously with changes in heat content of the lower atmosphere. The coherence between them is 0.87 to 0.90 with a phase difference $\varphi = -2.5$ to 7.1° (i.e., heat content changes in the lower layer of the atmosphere lead slightly the corresponding changes in the surface and active layers of the sea). Therefore, the variability of the temperature of the surface layer of the sea is controlled by fluctuations in the thermal state of the lower atmosphere, in full agreement with Shishkov's conclusions [2] that the meridional heat transport in the lower troposphere affects the temperature of the sea surface. This last finding sheds some light on the formation of the temperature anomalies over large areas of the sea surface that have recently been detected by several investigators. Indeed, if the thermal state of the active layer of the sea is controlled by an external heat cycle governed by the solar radiation budget, then the variations of the temperature field on the sea surface should also be large-scale. Other atmospheric processes that act on the surface of the ocean (atmospheric circulation, clouds, precipitation) are also of global scale. Given the above, we begin to understand why there are large areas of the sea with comparatively rapidly developing positive or negative temperature anomalies of the surface layers. In the deep layers, on the other hand, the redistribution of heat is controlled by currents. Since the velocity of the oceanic circulation is two orders lower than that of atmospheric circulation, the effect of advection of heat by currents on the heat exchange between sea and air manifests itself much more slowly, being significant only in long-period processes.

El Niño. For many years, oceanologists have been accumulating data on the so-called El Niño phenomenon. This is an anomalous advance of warm equatorial water (the southern branch of the Equatorial countercurrent) far to the south along the coast of South America (to 15°S in 1941), which occurs when the southeasterly trade wind weakens. Such deep incursions of warm waters sharply change the oceanological and weather conditions off Peru and Chile and result in large-scale kills of commercially valuable cold water fish, catastrophic rains, and storms of destructive force. Figure 7-13 shows the multiyear variation of water temperature near the Puerto Chicama station (Peru) and that of atmospheric pressure at Djakarta. The peaks of the positive water-temperature anomalies indicated on this figure are the times of onset of El Niño. They occurred in 1925–1926, 1930, 1932, 1939–1941, 1943, 1951, 1953, and 1957–1958. Thus, El Niño recurs with periods of 2, 4 to 5, and 8 years.

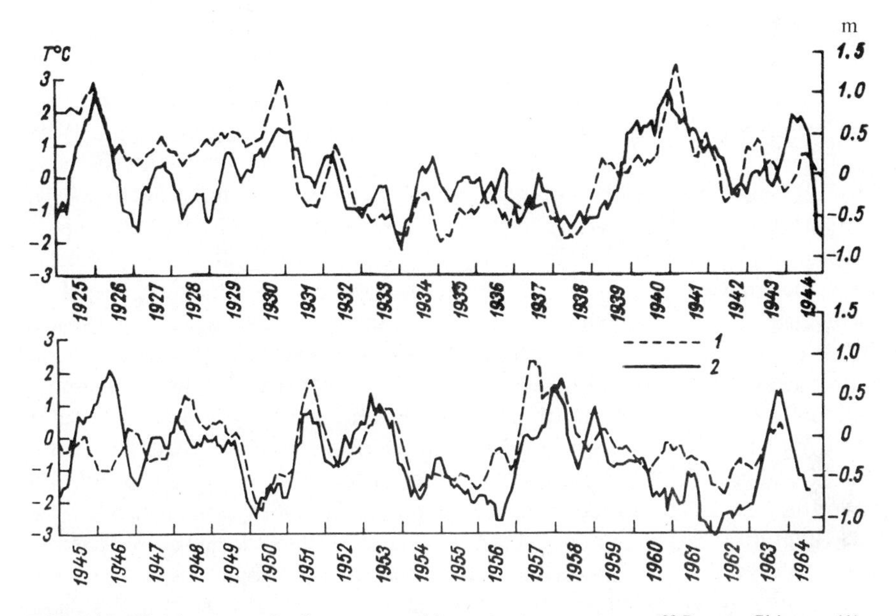

FIG. 7-13 Multiyear trend of mean monthly water temperatures off Puerto Chicama (1) and of atmospheric pressure at Djakarta (2) (from Berlage [19]).

A model of the interaction between the air and sea circulations and of their multiyear variability that sheds some light on the El Niño phenomenon was suggested by Bjerknes [20]. He starts with the fact that there exists a temperature contrast (as large as 8°C) between the western and eastern regions of the equatorial zone of the Pacific. Naturally, this causes a similar contrast in the lower atmosphere over the ocean. As a result, the thermal circulation of the atmosphere is such that the air masses descend in the east of the zone and rise in the west. This is a quasistationary part of the atmospheric circulation, and strongly influences north and south trade winds. It also results in intensified equatorial upwelling in the sea, an upwelling which is at peak in the eastern half of the equatorial zone. Thus, when the trade winds are intensified, the equatorial upwelling of deep waters is also intensified, and the temperature contrast between the western and eastern parts of the equatorial zone increases. Weakening of the trades is accompanied by a drop in the rate of equatorial upwelling, which leads to a decrease in the temperature contrast along the equatorial zone, the development of convergence of the warm surface waters from the north and south in the equatorial zone, and, as a consequence, even faster weakening of the trade winds. This weakening of the trade winds promotes pile-up of cold water in the eastern part of the equatorial zone (i.e., waters of the Peru current and those upwelling along the Peru and Chile coasts). The dominance of cold water in the east again increases the temperature contrast, and so the cycle begins to repeat itself.

This sea-air interaction in the equatorial zone is a self-oscillatory process with a period of about two years. However, as Bjerknes stresses, other macroprocesses in the atmosphere are superimposed on this simple mechanism, including the interaction with air circulation over the Indian Ocean. Figure 7-14 shows Bjerknes' diagram of this interaction on an equatorial section of the atmosphere. Streamlines indicate the vertical atmospheric circulation pattern over the equatorial zones of the Indian and Pacific Oceans, as derived from vertical wind profile data at the following stations: 1) Abidjan; 2) Nairobi; 3) Ghana; 4) Singapore; 5) Canton Island; and 6) Bogota. As the circulation over the Indian Ocean is displaced eastward (to 170°W), the zone of deep-water upwelling in the Pacific contracts sharply. As the Pacific Ocean circulation moves westward (to 130°E), upwelling is intensified in the equatorial zone of the Pacific Ocean. Bjerknes thinks that this mechanism is responsible for the two-year period in the interaction of the trade winds and the equatorial currents. The longer periods of El Niño (4 to 5 and 8 years) are probably generated by the circulation of temperature anomalies in the north and south Pacific anticyclonic ocean circulation systems. The large-scale variability of the atmospheric circulation at high latitudes of the Pacific and Atlantic Oceans and even the ice drift in the Arctic basin seem to be related to the year-to-year variations of the temperature in the equatorial Pacific.

Variability of the oceanological fronts. Fairly systematic long-term data are available only on the north Pacific polar oceanological front. The range of its latitudinal year-to-year motions, derived from observations of the 15°C isotherm between 1933 and 1964, is 5 to 6°, and the period of these oscillations averages 4 to 5 years. One interesting fact is that the latitudinal oscillations of the front of the subtropical convergence in the western tropical zone of the Pacific also have the range of 5 to 6°.

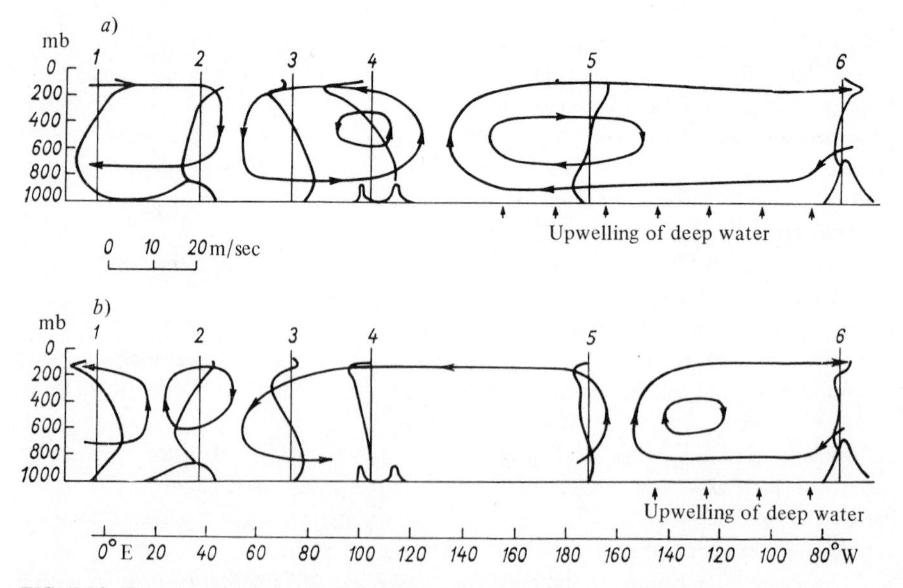

FIG. 7-14 Diagram of interaction of equatorial atmospheric circulations over the Indian and Pacific oceans (from Bjerknes [20]). *a*–November 1964; *b*–November 1965. The wind velocity scale is indicated.

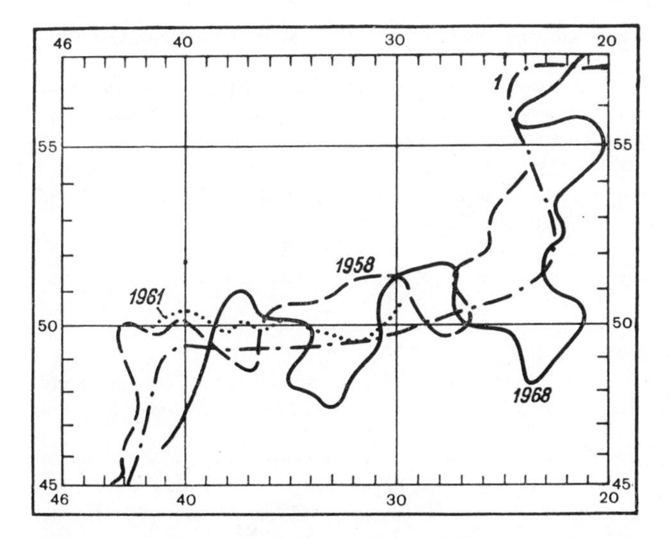

FIG. 7-15 Mean annual position of polar oceanological front in the north Atlantic in 1958, 1961, and 1968 (from Rossov and Kislyakov [21]). 1–Multiyear mean position of 10°C isotherm [23].

Using the data of the 1968 seasonal hydrological surveys made by the *Mikhail Lomonosov* and the *Atlantida*, plus the data of Dietrich and Muromtsev [22], Rossov and Kislyakov [21] analyzed the year-to-year variations of the position of the polar oceanological front in the north central and the north-northeast parts of the Atlantic. They showed that such variations may reach 200 miles, while the seasonal variations may be of the order of 100 miles. The line of the front follows the turns of the meanders of the Gulf Stream and North Atlantic current. Figure 7-15 shows the mean annual positions of the front in 1958, 1961, and 1968.

REFERENCES

1. Potaychuk, S. I. Some results of the statistical analysis of long-term variability of water temperature in the North Atlantic, Rapports et Proces-Verbaux des Reunions, **162**, 154–158 (1972).
2. Shishkov, Yu. A. Meridional transport of heat in the lower troposphere and temperature anomalies in the North Atlantic Ocean, Trudy Inst. Okeanol. Akad. Nauk SSSR, **57**, 156–199 (1962).
3. Arkhipova, Ye. G. Year-to-year variations of heat balance in the North Atlantic Ocean over the past decade, Trudy Gosud. Okeanogr. Inst., No. 54, 35–60 (1960).
4. Midttun, L. Variability of temperature and salinity at some localities outside the Coast of Norway. In: *Progress in Oceanography*, **5**, 41–54 (1968).
5. Panfilova, S. G. Water Temperature. In: Tikhiy okean (*The Pacific Ocean*), **II**, The Hydrology of the Pacific Ocean, Edited by A. D. Dobrovolskiy, Nauka Press, Moscow, 71–112 (1968).
6. Panfilova, S. G. The relation between the seasonal and year-to-year variabilities of hydrological quantities, Okeanologiya, **11**, No. 4, 588–598 (1971).
7. Kolesnikova, V. N. and A. S. Monin. The year-to-year variability of meteorological quantities, Izv. Akad. Nauk SSSR, Fizika atm. i okeana, **2**, No. 2, 113–120 (1966).

8. Tait, I. B. Hydrography of Faeroe-Shetland Channel, 1927–1952, Scot. Home Dept. Mar. Res., **2**, 309 (1957).
9. Von Grasshoff, K. Hydrographische Beobachtungen im Seegebiet des Island Faroe-Ruckens von 1959–1963 mit FFS Anton Dohrn (*Hydrographic Observations in the Waters of the Iceland-Faeroe Ridge from 1959 to 1963 on R/V* Anton Dohrn), Ber. Deutsch. Wiss. Romm. Meeresforsch., **XVII**, No. 1, 1–12 (1965).
10. Iselin, C. O. D. Preliminary report on long-period variations in the transport of the Gulf-Stream system, Pap. Phys. Oceanogr. and Meteor., **8**, No. 1, 40 (1940).
11. Stommel, H. *The Gulf Stream*, University of California Press (1965).
12. Shuleykin, V. V. Fizika moriya (*The Physics of the Sea*), USSR Academy of Sciences Press, Moscow (1968).
13. Dubinin, A. I. A model of the interaction between macroprocesses in the ocean and atmosphere, Okeanologiya, **8**, No. 4, 571–580 (1968).
14. Uda, M. Cyclic correlated occurrence of worldwise anomalous oceanographic phenomena and fisheries conditions. In: *J. Oceanography*, Soc. Japan, 20th Anniversary Volume, 368–376 (1962).
15. Ichiye, T. On the Variation of Oceanic Circulation in the Seas Adjacent to Japan. In: *UNESCO Symposium on Physical Oceanography*, Tokyo, 116–129 (1957).
16. Fukuoka, I. The variation of the Polar Front in the sea adjacent to Japan, Oceanogr. Mag., **6**, No. 4, 181–195 (1955).
17. Batalin, A. M. The state of the Kuroshio and fisheries problems, Trudy Soveshchaniya Ikhtiolog. Komissii Akad. Nauk SSSR, No. 10, 198–204 (1960).
18. Kort, V. G. Large-scale sea-air interaction (as illustrated by the behavior of the northern zone of the Pacific Ocean), Okeanologiya, **10**, No. 2, 222–240 (1970).
19. Berlage, H. P. The southern oscillation and world weather, Koninklijk Nederlands Meteorologisch. Instituut Mededelinger en Verhandlingen, No. 88, 1–142 (1966).
20. Bjerknes, J. Atmospheric teleconnection from the Equatorial Pacific, Monthly Weather Review, **97**, No. 3, 163–172 (1969).
21. Rossov, V. V. and A. G. Kislyakov. The Polar Front in the North Atlantic in 1968–69, Rapports et Proces-Verbaux des Reunions, **162**, 220–226 (1972).
22. Muromtsev, A. M. Osnovnyye cherty gidrologii Atlanticheskogo okeana (*The Principal Hydrological Features of the Atlantic Ocean*), Gidrometeoizdat Press, Leningrad (1963).

8 CLIMATIC VARIATIONS

The climate of the ocean shall be understood to mean the statistical mode of variations of the global oceanological fields with periods of less than a few decades, so that climatic characteristics are characteristics averaged over time intervals on the order of a few decades. They themselves may vary in time (with periods of more than a few decades), thus describing the variations of climate.

We encounter two difficulties in attempts to determine variations of climate in the past. The first is that the insolation, air temperature, atmospheric pressure, rainfall, the temperature and salinity of sea water, the velocities of ocean currents, etc., were not measured in the past (surviving manuscripts give only qualitative information on droughts, frosts, or ice on the polar seas). One is therefore forced to rely on unconventional and indirect information, such as thicknesses of the annual rings in trees and of annually deposited layers of certain sedimentary rocks (*varve analysis*), historical and geological data on the lengths of glacier valleys, ancient terraces on sea shores, fossils of thermophilic and psychrophilic organisms in cores from the ocean floor, and other geological, paleontological, and paleobotanical data.

The second difficulty is that local data may not be representative, and it may therefore be necessary to start with a global picture of the climate, for which the available data are certainly not adequate. Because of this, analysis of climatic variations requires highly specific research, so we shall confine ourselves here only to a few illustrations.

The ensemble of available data definitely indicates that there exist long-term variations of climate with time scales of less than a century, a typical example of which is the climate warming of the first half of the Twentieth century, as well as century-to-century variations. During the past 12,000 years since the end of the Würm glaciation, the latter included the warming to the "climatic optimum" in the Fortieth to Twentieth Centuries B.C., cooling to the "little climatic optimum" or the "Viking epoch" in the Eighth to Tenth Centuries A.D., cooling in the Thirteenth to Fourteenth Centuries, warming in the Fifteenth to Sixteenth Centuries, and cooling in the "Little Ice Age" of the Seventeenth to Nineteenth Centuries. During these periods, the background properties of the ocean-atmosphere system (most importantly, the distribution of solar heat over the upper boundary of the atmosphere and the configurations of the oceans and continents), as well as the composition of the atmosphere apparently remained unchanged, so that these variations of climate must be attributed to internal processes in the sea-air system, that is, to the interactions of the ocean and atmosphere. This means assuming that the background properties do not predetermine the climate, and that with any given set of these properties the ocean-atmosphere system may exist in a variety of states and undergo spontaneous transition from one to another.

Still longer term variations of climate are, however, probably due to variations of the background properties. Thus, the Pleistocene glaciation periods lasting 10^4 to 10^5 years are explained by Milankovich's astronomical theory as being due to changes in the distribution of solar heat on the upper boundary of the atmosphere as a result of perturbation of the earth's rotation by other planets. Climatic variations with periods of 10^7 to 10^8 years could have been caused by changes in the configuration of the oceans and continents, that is, by motion of the continents and the poles. We shall not discuss these problems further, referring the reader to [1].

Numerous instrumental measurements of characteristics of the atmosphere and ocean are available for purposes of analysis of the climatic warming of the first half of the Twentieth Century (although the direct characteristics of the sea-air interaction, such as heat, moisture, and momentum fluxes through the sea surface, which are important for analysis of the causes of the warming, have not yet been measured). But the difficulties of generalization over the entire planet are great, and different authors have proposed different interpretations of these data. Figure 8-1 presents plots of the

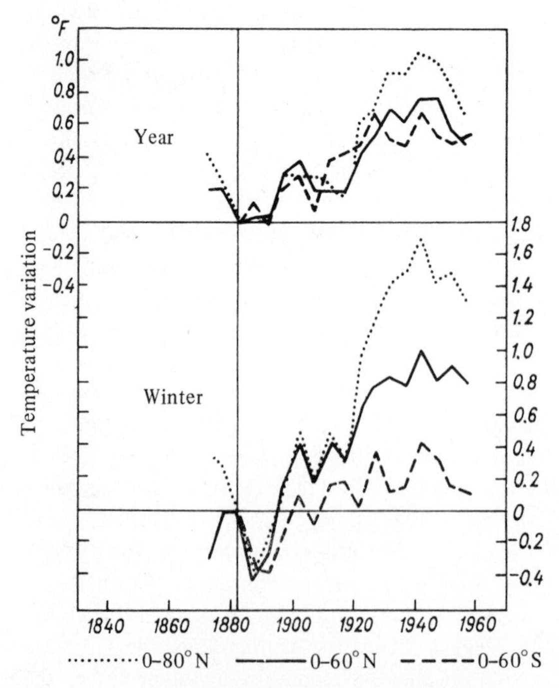

FIG. 8-1 Deviations of the five-year mean temperature in selected latitude belts over the past century from their values during the five years from 1880 to 1884 (from Mitchell [2]).

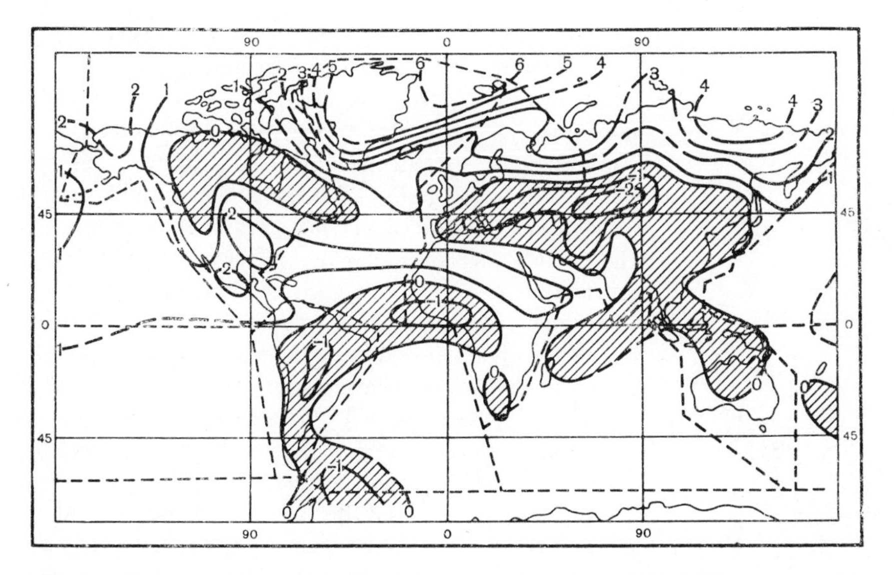

FIG. 8-2 Map of variations of mean winter temperatures from 1920–1930 as compared to 1900–1919 (from Mitchell [3]). Equal-temperature lines at 1°F intervals; zones of cooling are shaded.

increase in the mean air temperatures in various latitudinal belts over the period 1890 to 1950 [2]. They show that the greatest warming occurred in winter, especially in the Arctic. Figure 8-2 shows Mitchell's [3] global map of winter warming, from which we see that the warming trend generally occurred over the oceans, especially in the Arctic, while a slight cooling was generally observed on the continents. The difference in the effect of the oceans and continents is highly instructive.

During the period from 1910–1915 through 1950–1955, the amount of ice on the Barents Sea decreased by about 20% (but has begun to increase since the 1950's). The northern limit of iceberg occurrence in the Antarctic Ocean shifted to the south by an average of more than 15° of latitude (nearly 1700 km) during the period from 1889–1897 through 1954–1958. According to Butorin [4], the sea level has risen by an average of 6 cm during the last 50 years.

One of the main climatic variables is the state of the general atmospheric circulation. Dzerdzeyevskiy [5], who classified types of atmospheric circulation above the Northern Hemisphere over the past century, showed that during the warming, the recurrence of zonal circulation types was higher, while that of meridional types decreased (Fig. 8-3). Using the somewhat different classification of atmospheric circulation forms, developed by Girs [7], Vinogradov [8] found that the recurrence of the *W* form (westerly transport) decreased, while that of the *E* (easterly transport) and *C* (meridional transport) forms increased (Fig. 8-4). These results bring us closer to understanding of the climatic-warming mechanism.

An outstanding example of century-to-century climate variations is the cooling during the "Little Ice Age" of the Seventeenth to Nineteenth Centuries, which was

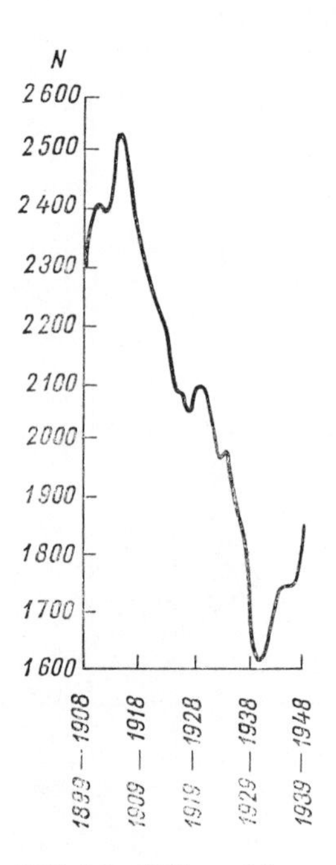

FIG. 8-3 Sliding 10-year sums of days with meridional type of circulation, from B. L. Dzerdzeyevskiy's classification (simultaneous Arctic air intrusions in two or more directions). Cited from [6, p. 265].

analyzed by Bjerknes [9]. The water-temperature anomalies of the North Atlantic [10] during this period are shown in Fig. 8-5. They were positive in the Sargasso Sea and negative near Iceland. According to Bjerknes, this distribution of water-temperature anomalies weakens the winter meridional atmospheric circulation in middle latitudes, which, in turn, causes an increase in these anomalies. The cooling that then develops in the northern Atlantic could result in a new ice age if, as Bjerknes assumes, the water-temperature anomalies were not smoothed by the simultaneously intensifying meridional circulation of Atlantic waters. This mechanism is, of course, qualitative and therefore hypothetical, but it is of interest because it represents a mechanism that is capable, in principle, of inducing prolonged and sharp climate variations.

The one property of the sea that is most useful in defining century-to-century climate variations is its level, whose past positions can be determined at many locations along the coast and on the bottom of the coastal zone from various geomorphological and geological indications, such as ancient terraces or the buried mouths of rivers, whose absolute or at least relative ages are known (or can be determined). If the shape of the bottom and coasts of the sea were not to vary with time, then it would be possible to attribute the drop in sea level (*regressions* of the coastline) to increased continental glaciation, while rises in the level (*transgressions* of the sea) would be ascribed to the melting of glaciers, on the assumption that the sum of the mass of water in the ocean and in land glaciers is constant. However, both the sea floor and the coasts may undergo vertical tectonic motions that differ at different places, and some of these motions may be directly due to processes in which water is transported from the sea into the continental glaciers and back. Thus when additional water enters the sea, its floor may sag under its weight, whereas (in the opposite case) blocks of the continental crust would tend to sink (into the mantle) under the weight of growing glaciers and rise when such glaciers melt (for example, Finland and Scandinavia have been experiencing a rise of this kind throughout the entire post-glacial period). This calls for careful analysis of local neotectonic motions of the coasts and global generalizations of data on the sea-level variations along various coasts.

The global generalizations offered by different authors differ in their details, but they all agree on one point: 15,000 years ago, the sea level was about 110 meters lower than it is now. Then it rose at a rate of about 2 cm per year until the "climatic optimum" of 5000 to 6000 years ago, after which the average rate of increase of the level dropped to 1 to 2 mm per year. Figure 8-6 shows curves of sea-level variation according to Fairbridge [11] and Curray [12], which clearly illustrate the enormous scale of the climatic variability of the sea.

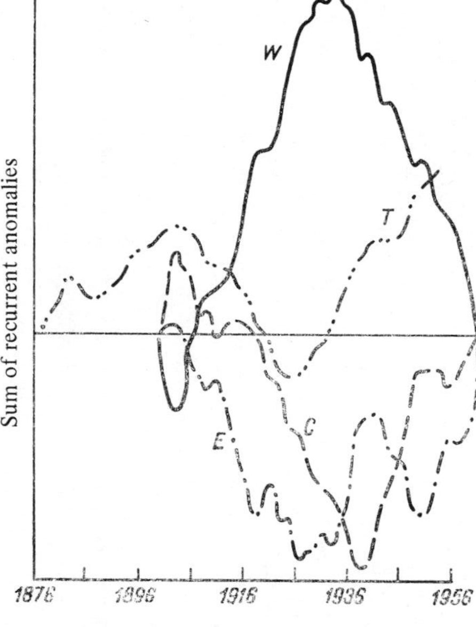

FIG. 8-4 Cumulative sums of anomalies in annual recurrence of atmospheric circulation forms W, E, and C and of water temperature T in the North Atlantic (from Vinogradov [8]).

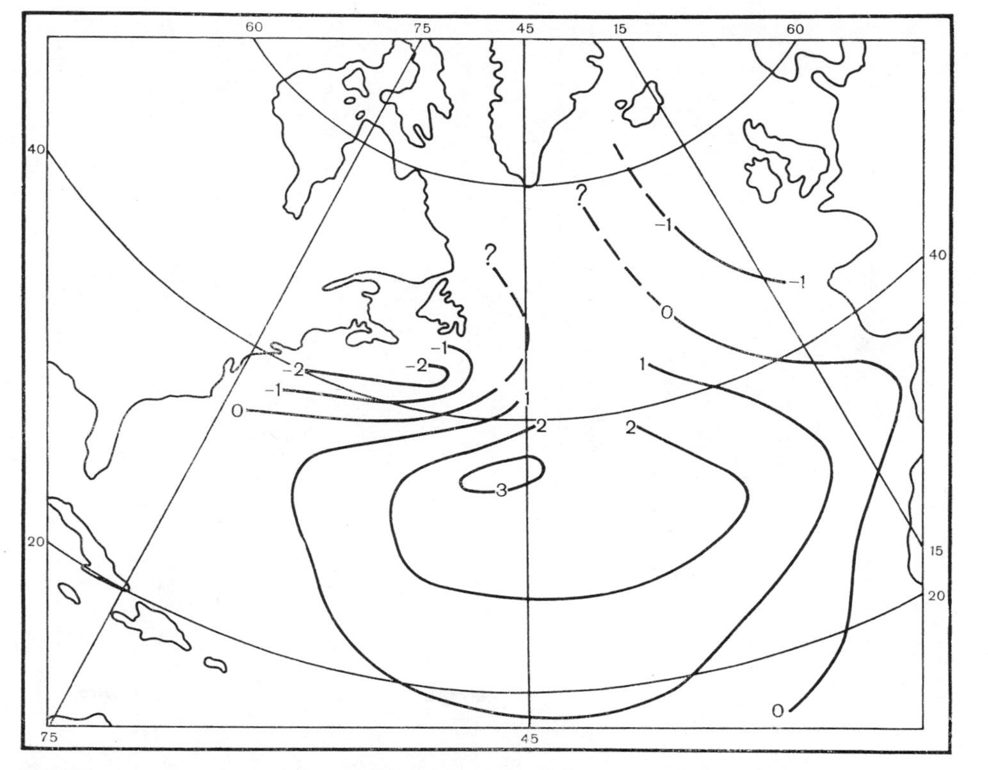

FIG. 8-5 Anomalies of surface-water temperatures in the North Atlantic from 1780 to 1820 relative to the norm and the averages for 1887–1899 and 1921–1938 (from Lamb and Johnson [10] (dashed curves of Bjerknes [9]).

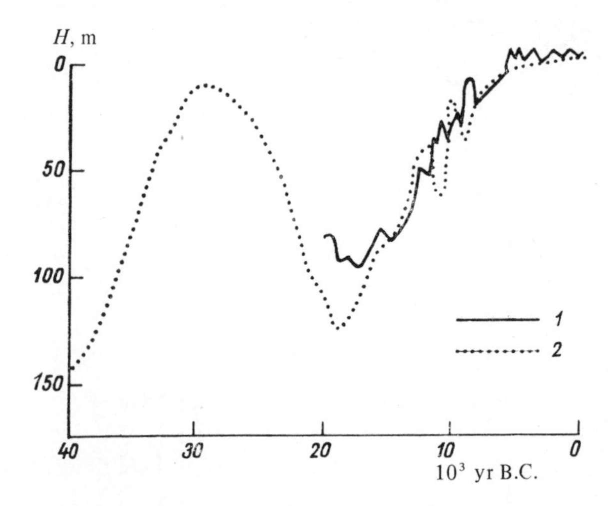

FIG. 8-6 Variations of sea level. 1–From Fairbridge [11]; 2–from Curray [12].

REFERENCES

1. Monin, A. S. Vrashcheniye Zemli i klimat (*The Rotation of the Earth and Climate*), Gidrometeoizdat Press, Leningrad (1972).
2. Mitchell, J. M. Recent secular changes of global temperature, Annals New York Acad. Sci., Art. 1, **95**, 235–250 (1961).
3. Mitchell, J. M. On the World-Wide Pattern of Secular Temperature Change. In: *Changes of Climate* (Arid Zone Research XX), UNESCO, 161–179 (1963).
4. Butorin, N. V. Vekovyye izmeneniya srednego urovnya Atlanticheskago okeana i ikh svyaz' s tsirkulyatsiyey atmosfeiy (*Secular Variations of Mean Atlantic Ocean Level and Their Relation to the Atmospheric Circulation*), USSR Academy of Sciences Press, Moscow-Leningrad (1960).
5. Dzerdzeevskii, B. L. Fluctuations of General Circulation of the Atmosphere and Climate in the Twentieth Century. In: *Changes of Climate* (Arid Zone Research XX), UNESCO, 285–291 (1963).
6. Rubashev, B. M. Problemy solnechnoy aktivnosti (*Problems of Solar Activity*), Nauka Press, Moscow-Leningrad (1964).
7. Girs, A. A. Mnogoletnyye kolebaniya atmosfernoy tsirkulyatssi i dolgosrochnyye gidrometeorologicheskiye prognozy (*Long-Term Variations of Atmospheric Circulation and Long-Term Hydrometeorological Forecasts*), Gidrometeoizdat Press, Leningrad (1971).
8. Vinogradov, N. D. Relation between atmospheric circulation forms and surface water temperature in the northern Atlantic Ocean, Problemy Arktiki i Antarktiki, No. 25, 44–53 (1967).
9. Bjerknes, J. Atmosphere Ocean Interaction During the "Little Ice Age" (Seventeenth to Nineteenth Centuries A.D.). In: *WMOJV 66 Symposium on Research and Development Aspects of Long-Range Forecasting*, WMO Tech. Note, No. 66, 77–88 (1965).
10. Lamb, H. H. and A. J. Johnson. Climate variations and observed changes in the general circulation, Geografiska Annaler, Nos. 2–3, 94–134 (1959); Nos. 3–4, 231–242 (1961).
11. Fairbridge, R. W. Unstatic Changes in Sea Level. In: *Physics and Chemistry of the Earth*, **4**, 99–185 (1961).
12. Curray, D. Late Quaternary History of the Continental Shelves of the USA. In: *The Quaternary of the United States*, **1** (Translated from the English), Mir Press, Moscow (1968).

9 NUMERICAL MODELING OF OCEANIC CIRCULATION

Formulation of the problem. We saw in the preceding chapters that the long-period variability of the ocean is shaped primarily by large-scale processes (with horizontal scales on the order of 1000 km), although, as will be shown below, the effect of small-scale motions that determine the "turbulent" momentum, heat, and salt fluxes is also significant. In this chapter, we shall present some ideas pertaining to methods of the analysis of this problem, which reduce to numerical modeling of the oceanic circulation. Such modeling procedures are patterned after the numerical modeling of the atmospheric circulation, which is much better developed at the present time.

Just as we did in the analysis of the small-scale and mesoscale processes in the sea, we shall begin with the basic laws of oceanic hydrothermodynamics, namely, Newton's laws of motion, the mass-conservation equation, the salt-diffusion equation, and the entropy transport equation (instead of the latter, we might have chosen the energy conservation equation or the equation of internal-energy transport). The principal feature of the large-scale motions is their turbulence. For this reason, the starting equations must be averaged, and instead of finding the *instantaneous* fields of the oceanological properties we consider the problem of determining *averaged* (in the statistical sense) fields of these properties. The following approximations are valid for the averaged motions:

1) the quasistatic approximation. It holds by virtue of the fact that for large-scale processes the ratio of the vertical to the horizontal scale is much smaller than unity. The filtering effect of this approximation has already been discussed in Sec. 4-1;

2) the Boussinesq approximation, whereby the density ρ is replaced by ρ_0 (the mean density) everywhere except in the term $g\rho$ of the equation of vertical motion (buoyancy effect) and whereby the mass-conservation equation is written as $\operatorname{div} \mathbf{u} = 0$ (the incompressibility conditions, which filters out acoustic waves, see Sec. 4-1). This approximation holds because $\delta\rho/\rho \sim 10^{-3}$ in the ocean;

3) the so-called traditional approximation (see Sec. 4-1) in the description of the Coriolis force. The justification for it is that $|w| \ll |u|$ for large-scale motions. It is possible, however, that retention of Coriolis force components incorporating the

factor $2\Omega \cos \varphi$ (φ is the geographic latitude) may be necessary in the case of the narrow equatorial belt;

4) the entropy transport equation is approximated by the equation of heat conduction for a moving medium. As we know, adiabatic temperature changes are neglected in this case;

5) the turbulent momentum, heat, and salt flows are assumed to be much larger than the corresponding molecular effects (which are neglected altogether);

6) it is assumed that the equation of state for the sea water can be represented as a relation between the *mean* values of the pressure p, density ρ, temperature T, and salinity s. It should be stressed that even in the case of instantaneous values of p, ρ, T, and s this relationship is known only approximately; the form of the equations proposed by Eckart [1] appears to be highly promising.

With allowance for approximations 1-6, the principal equations of the problem become

$$\frac{\partial u}{\partial t}+\frac{u}{a \cos \varphi} \frac{\partial u}{\partial \lambda}+\frac{v}{a} \frac{\partial u}{\partial \varphi}+w \frac{\partial u}{\partial z}-\frac{u v}{a} \tan \varphi-2 \Omega v \sin \varphi$$

$$=-\frac{1}{\rho_0 a \cos \varphi} \frac{\partial p}{\partial \lambda}+F_\lambda; \qquad (9\text{-}1)$$

$$\frac{\partial v}{\partial t}+\frac{u}{a \cos \varphi} \frac{\partial v}{\partial \lambda}+\frac{v}{a} \frac{\partial v}{\partial \varphi}+w \frac{\partial v}{\partial z}+\frac{u^2}{a} \tan \varphi$$

$$+2 \Omega u \sin \varphi=-\frac{1}{\rho_0 a} \frac{\partial p}{\partial \varphi}+F_\varphi; \qquad (9\text{-}2)$$

$$0=-\frac{\partial p}{\partial z}+g \rho; \qquad (9\text{-}3)$$

$$\frac{1}{a \cos \varphi} \frac{\partial u}{\partial \lambda}+\frac{1}{a \cos \varphi} \frac{\partial}{\partial \varphi}(v \cos \varphi)+\frac{\partial w}{\partial z}=0; \qquad (9\text{-}4)$$

$$\frac{\partial s}{\partial t}+\frac{u}{a \cos \varphi} \frac{\partial s}{\partial \lambda}+\frac{v}{a} \frac{\partial s}{\partial \varphi}+w \frac{\partial s}{\partial z}=\sigma; \qquad (9\text{-}5)$$

$$\frac{\partial T}{\partial t}+\frac{u}{a \cos \varphi} \frac{\partial T}{\partial \lambda}+\frac{v}{a} \frac{\partial T}{\partial \varphi}+w \frac{\partial T}{\partial z}=Q; \qquad (9\text{-}6)$$

$$\rho=\rho(T, s, p). \qquad (9\text{-}7)$$

Here we have used the designations

$$-F_\lambda=\frac{1}{a \cos \varphi} \frac{\partial \overline{u'^2}}{\partial \lambda}+\frac{1}{a \cos^2 \varphi} \frac{\partial}{\partial \varphi}(\overline{u'v'} \cos^2 \varphi)+\frac{\partial(\overline{u'w'})}{\partial z}; \qquad (9\text{-}8)$$

$$-F_{\varphi}=\frac{1}{a\cos\varphi}\frac{\partial}{\partial\lambda}(\overline{u'v'})+\frac{1}{a\cos\varphi}\frac{\partial}{\partial\varphi}\left(\overline{v'^{2}}\cos\varphi\right)+\frac{\partial(\overline{v'w'})}{\partial z}+\frac{\tan\varphi}{a}\overline{u'^{2}}; \tag{9-9}$$

$$\sigma=\frac{1}{a\cos\varphi}\frac{\partial}{\partial\lambda}(\overline{u's'})+\frac{1}{a\cos\varphi}\frac{\partial}{\partial\varphi}(\overline{v's'}\cos\varphi)+\frac{\partial(\overline{w's'})}{\partial z}; \tag{9-10}$$

$$Q=\frac{1}{a\cos\varphi}\frac{\partial}{\partial\lambda}(\overline{u'T'})+\frac{1}{a\cos\varphi}\frac{\partial}{\partial\varphi}(\overline{v'T'}\cos\varphi)+\frac{\partial(\overline{w'T'})}{\partial z}, \tag{9-11}$$

where λ is the longitude ($0<\lambda<2\pi$); φ the latitude ($-\pi/2\leqslant\varphi\leqslant\pi/2$); a is the radius of the earth, u, v, w, p, ρ, s (the salinity) and T are the mean values (this is not indicated as such in the equations in order to simplify the notation); the overbar indicates averaging, and the prime the fluctuations (of velocity, salinity, and temperature, respectively).

The boundary conditions at the sea floor ($z = H$) and at the coasts (which are assumed to be vertical walls) are obvious: nonslip (zero total velocity at the floor and zero horizontal velocity at the coasts) and vanishing of the turbulent heat and salt fluxes. The conditions that must be satisfied on the free surface of the ocean are the kinematic condition and three dynamic conditions (continuity of the pressure and of the stress components tangent to the surface), as well as the thermal balance (with allowance for the influx of shortwave solar radiation and long-wave scattered radiation, heat losses due to radiation from surface and to evaporation, contact heat exchange with the atmosphere), and the water balance (with allowance for rainfall, evaporation, and the formation and melting of ice).

The problem of closure of the set (9-1)-(9-7) is the principal difficulty in analysis of large-scale processes in the sea. This closure can now be accomplished only by appropriate parameterization of the small-scale processes responsible for the "turbulence" terms F_{λ}, F_{φ}, σ, and Q in (9-1)-(9-7). It is usually assumed that

$$F_{\lambda}=A_{L}\left(\frac{1}{a^{2}}\Delta_{s}u+\frac{\cos 2\varphi}{a^{2}\cos^{2}\varphi}u-\frac{2\sin\varphi}{a^{2}\cos^{2}\varphi}\frac{\partial v}{\partial\lambda}\right)+\frac{\partial}{\partial z}A_{H}\frac{\partial u}{\partial z}; \tag{9-8a}$$

$$F_{\varphi}=A_{L}\left(\frac{1}{a^{2}}\Delta_{s}v+\frac{\cos 2\varphi}{a^{2}\cos^{2}\varphi}v+\frac{2\sin\varphi}{a^{2}\cos^{2}\varphi}\frac{\partial v}{\partial\lambda}\right)+\frac{\partial}{\partial z}A_{H}\frac{\partial v}{\partial z}; \tag{9-9a}$$

$$\sigma=A_{sL}\frac{1}{a^{2}}\Delta_{s}s+\frac{\partial}{\partial z}A_{sH}\frac{\partial s}{\partial z}; \tag{9-10a}$$

$$Q=A_{TL}\frac{1}{a^{2}}\Delta_{s}T+\frac{\partial}{\partial z}A_{TH}\frac{\partial T}{\partial z}. \tag{9-11a}$$

Parameters $A_L, A_H, \ldots$ are known as the horizontal and vertical turbulent mixing coefficients, respectively. These coefficients are either specified *a priori* (on the basis of solutions of model problems and certain indirect arguments), or they are determined from auxiliary relationships (such as the equation for kinetic energy of the turbulent motion).

The form of relations (9-8a)–(9-11a) is based on the analogy to the processes of molecular friction, heat conduction, and diffusion and on certain symmetry arguments (see the paper [2] by Kamenkovich for the hypothesis of axial symmetry of the exchange tensors). Therefore, these relations may fail to describe several important processes. Thus, relations (9-8a) and (9-9a) make sense only when it is known *a priori* that the energy of the mean motion is converted to turbulent energy; in this case A_L and $A_H > 0$ (see the discussion of this problem in Monin and Yaglom [3, Sec. 6]). But there is also some evidence to the effect that conversions of turbulent energy back into energy of the mean motion may also occur (see [4, 5]).

Relations (9-10a) and (9-11a) are especially doubtful. Thus it is well known that a mixed layer that is approximately homogeneous with depth and about 100 m thick exists near the sea surface in all seasons. During the spring and summer, when the sea is warmed, mixing is accomplished for the most part by the wind; during this time, the thickness h of the homogeneous layer decreases (the effective Richardson number is positive). During the fall and winter, when the sea cools, mixing is more intense, since the effect of small-scale convection is added to that of the wind, and the thickness of the homogeneous layer increases. In parametric description of these processes one must employ an equation for the turbulent energy, as was done, among others, by Kraus and Turner [6] and Kitaygorodskiy et al. (see [7, Chap. 12]). We have already discussed this problem in Chap. 6 dealing with the seasonal variability.

Below the homogeneous layer there exists a fairly thin (5 to 10 m) layer in which the temperature and salinity change sharply with depth (especially in spring and summer). It appears that the effect of this layer, known as the seasonal thermocline or temperature-discontinuity layer, may be substantial in large-scale processes. However, the physical mechanism of formation of this layer is not well understood.

It is improbable that these processes could be described by relations of the type (9-10a) and (9-11a), even assuming variation of A_{TH} with depth (and perhaps even seasonal variations). Some authors, therefore (see, for example, [8]), attempt to parameterize vertical mixing with recourse to A_{TH}.

It would appear that processes in the upper homogeneous layer of the sea would best be analyzed with the aid of modified basic equations supplemented by certain relationships that give an approximate description of the small-scale mixing processes. Thus, depth averaging of Eq. (9-6) within the homogeneous layer eliminates A_{TH} but introduces other new parameters such as the thickness h of the upper homogeneous layer, the vertical turbulent heat flux q_h, and the vertical advection of heat (both quantities being specified on the lower boundary of the homogeneous layer). The vertical heat advection can be assumed equal to wT_s (w is the vertical velocity in the discontinuity layer, and T_s is the temperature at the ocean surface, which is virtually constant within the homogeneous layer) if the water is downwelling at the particular point, or equal to wT_0 (T_0 is the temperature at the lower boundary of the discontinuity layer) if the water is rising at this point. To determine h, q_h, and T_0 one must use some auxiliary relationships. These might include the integrated turbulent energy balance and certain *a priori* relationships that provide an approximate description of the small-scale processes under consideration (such as the Kraus-Turner relation [6] $q_h/c_p\rho = \Lambda(T_s - T_0)\,(\partial h/\partial t)$, where $\Lambda = 0$ if $\partial h/\partial t \leqslant 0$, and $\Lambda = 1$ if $\partial h/\partial t > 0$; c_p is the specific heat at $p =$ const).

In the case of the main baroclinic layer of the sea, one can probably consider coefficients A_{sH}, A_{TH}, A_{sL}, and A_{TL} of (9-10a) and (9-11a) to be constant, and neglect vertical turbulent exchange in (9-8a) and (9-9a). It is natural to take $T|_{z=0} = T_0$ as an upper boundary condition for the heat conduction equation (9-6), where T_0 is determined by combined analysis of the upper homogeneous layer, the discontinuity layer (which can be described parametrically), and the baroclinic layer.

The difficult problem in this approach is that of stating the horizontal boundary conditions for the heat conduction equation or, in other words, allowance for boundary currents (see [9]).

Finally, note that the most important oceanological parameter in numerical modeling of the sea-air interaction is the surface temperature T_s of the sea. It is this variable that determines the effect of the sea on the atmosphere.

So much for the general formulation of the problem. The origin of the theory of ocean currents dates back to the classic papers of Ekman [10, 11]. However, concentrated work on these problems started in the early 1950's. It seems to us that among the principal achievements of theory is Shtokman's demonstration of the importance of the spatial nonuniformity of the wind-generated shear stress field in shaping ocean currents, especially of the equatorial countercurrents (see [12, 13]). The other great accomplishment is the definition of the role of the so-called β-effect in shaping the strong boundary currents along the western coasts of the ocean, presented in the pioneering studies of Sverdrup [14], Stommel [15], and Munk [16]. These papers stimulated an extensive literature devoted to analysis of the structure of coastal boundary currents in the ocean at various relationships between viscous and inertial factors (see, for example, [17] and its bibliography), as well as to the problem of separation of boundary currents from the coast (see, for example, [18-20]), and to the now well-advanced theory of the meandering of strong narrow currents in the open ocean (which we discussed in Sec. 5-2). The third significant progress involved the theory of the main oceanic thermocline, which is essentially a thermal boundary layer in the sea (see the recent review by Welander [24]). The foundations of this theory were laid by Lineykin [21], Robinson and Stommel [22], Robinson and Welander [23], and others. Fourth and finally, we must include here the theory of equatorial undercurrents that explains the properties of the current field on the assumption of an *a priori* specified vertical thermohaline structure in the equatorial zone (see [25-27]).

Analysis of simplified theoretical models brings out the factors that determine the velocity, temperature, salinity, density, and pressure fields in the sea. This undoubtedly generates a basis for qualitative understanding of oceanic circulation and for derivation of models of large-scale processes in the actual ocean that would allow for *all the principal factors* that control these processes. Obviously, analysis of such a complex nonlinear problem would be impossible without modern numerical methods and large computers, and for this reason the model must of necessity be numerical. However, we must not overlook the fact that derivation of the numerical model (that is, the approximation of the basic equations in terms of difference equations, etc.) starts from the *preliminary qualitative* understanding of the processes involved.

So far, numerical modeling has been used in oceanology only for calculations of the stationary (mean annual) circulation of the ocean and its seasonal variability, given specified stationary or seasonally variable dynamic and thermodynamic atmospheric perturbations. However, the same general approach would also hold for computation

TABLE 9-1 Brief List of Numerical Models of the Barotropic and Baroclinic Circulation in the Ocean

Author	Year	Solution	Sea being modeled	Model	Model
Sarkisyan [28]	1954	Stationary	North Atlantic, depth constant	Barotropic	
Welander [29]	1959	Stationary	Global ocean bounded by the 50° N and 40° S parallels, depth constant	Barotropic	
Il'in et al. [3]	1969	Stationary	Global ocean, depth constant	Barotropic	
Takano [31]	1969	Stationary	Global ocean, depth constant	Barotropic	
Sag [32]	1969	Stationary	Global ocean, depth constant	Barotropic	
Laykhtman et al. [33]	1971	Stationary	Global ocean, realistic sea floor topography	Barotropic	
Kagan et al. [34]	1972	Nonstationary, seasonal variability	Global ocean, realistic sea floor topography	Barotropic	
Sarkisyan [35]	1966	Stationary	North Atlantic, realistic sea floor topography	Baroclinic	Diagnostic[1]
Bryan and Cox [36–38]	1967, 1968	Stationary	Right parallelepiped modeling Atlantic Ocean	Baroclinic	Prognostic
Bryan and Manabe (see [39])	1969	Stationary	Sea-air interaction model; ocean in form of right parallelepiped	Baroclinic	Prognostic
Vasil'yev and Fel'zenbaum [40–42]	1971	Stationary	Right parallelepiped modeling Atlantic Ocean	Baroclinic	Prognostic

Cox [43]	1970	Nonstationary, seasonal variability	Indian Ocean, realistic sea floor topography	Baroclinic	Prognostic
Gill and Bryan [47]	1971	Stationary	Antarctic circumpolar, ocean depth constant; depth of Drake Strait varied	Baroclinic	Prognostic
Holland [45]	1971	Stationary	Ocean in form of right parallel-epiped	Baroclinic	Prognostic
Sarkisyan et al. [46]; Kochergin et al. [47]	1972	Stationary	North Atlantic, realistic sea floor topography	Baroclinic	Prognostic
Wetherald and Manabe [48]	1972	Nonstationary, seasonal variability	Sea-air interaction model; ocean in form of right parallelepiped	Baroclinic	Prognostic
Holland [49]	1973	Stationary	Ocean rectangular in plan; sea floor topography characteristic of the western boundary current	Baroclinic	Prognostic

[1] If, in calculating velocities and sea level, the density is assumed to be known throughout the ocean, then the model is called diagnostic; when the density of sea water, as well as the level and current velocities is an unknown field, the model is called prognostic.

of the long-period variability of the sea, given specified unsteady atmospheric perturbations, and perhaps, later on, perturbations determined from a consistent unified model of the interacting air-sea system. It is this approach that should be used as a basis for forecasting of long-term weather changes.

The following brief review of numerical-modeling studies, which does not pretend to be complete, is given here in order to demonstrate the potential and limitations of such models. Given our previous discussion, we shall analyze only models that reflect as much as possible the actual conditions or, in other words, models that take account of the complex configuration of sea coasts (for example, the multiple interconnection of the oceans), the actual wind-stress, heat-flux and other fields, the real sea floor topography, and the baroclinicity of the ocean. At the least, the models should account for each of the above factors separately.

It is helpful to tabulate the studies to be discussed below (see Table 9-1).

Stationary solutions. We shall begin with barotropic (two-dimensional) ocean-current models capable of bringing out the important factors that shape the strong boundary currents at the western coasts of the oceans. Among these are the models of Sarkisyan [28] and Welander [29], in which solutions were obtained for realistic coast outlines and a realistic wind field over the sea. Later, the two-dimensional circulation in a global ocean with realistic shore contours (with allowance of the multiple-connectedness property of the ocean that results from the presence of such continents as the Antarctic) was computed by Il'in et al. [30]. A similar problem was also solved by Takano [31] and Sag [32]. The two-dimensional circulation in a global ocean was calculated by Laykhtman et al. [33] with allowance for the actual sea floor topography.

Allowance for the baroclinicity of sea water greatly complicates the problem, which then becomes three-dimensional (see Lineykin [50]), who closes the set of equations with the so-called density-diffusion equation). Sarkisyan [35] performed several interesting numerical experiments and concluded that the combined effect of sea-floor topography and baroclinicity of sea water is an important factor shaping the currents. Allowance for this factor produces major changes in the calculated sea level or overall volume transport distribution. Most studies on the effect of this factor, run by Sarkisyan and his followers, were done within the framework of diagnostic calculations of the sea level or the total stream transport function, starting from a density field in the ocean that was specified from observations. Recent work of Sarkisyan et al. [46] and Kochergin et al. [47] has extended the analysis of the combined effect of sea floor topography and baroclinicity to prognostic models. Their calculations indicate that this factor is important throughout the entire depth of the ocean. We shall return to this problem in our discussion of the energetics of the ocean.

An important series of studies dealing with currents in a baroclinic ocean was made by Fel'zenbaum et al. (see [40–42, 51–53]). Specific calculations for the Atlantic Ocean were performed by means of models of its temperature and salinity distributions. Therefore, in solving the problem, only the parameters of these models that depend on the horizontal coordinates are determined. This approach is similar in concept to the familiar method of integral relations in classicial boundary-layer theory.

At present, the most advanced numerical experiments are those of Bryan et al., employing a three-dimensional ocean-current model (see [36–41, 45, 48, 49]).

Bryan's model is based on Eqs. (9-1)-(9-7) in which F_λ, F_φ, σ and Q are calculated from relationships of type (9-8a)-(9-11a), assuming constant coefficients A_L, A_H, A_{sL}, A_{sH}, A_{TL}, and A_{TH}. However, the static stability condition, in the form $\partial\rho/\partial z > 0$, is checked in each time step. If this condition is not violated anywhere, there is no convection, and the calculations are continued. In the layers in which there is convection ($d\rho/dz < 0$), the density (or temperature and salinity) is assumed constant along the z-coordinate. In experiments in which temperature is specified, the latter is assumed equal to the temperature at the surface, whereas when the heat flux through the sea surface is given, the temperature is computed from relationships specifying consistency of the heat content of the liquid column.[1]

The boundary conditions are obtained from the following expressions:

on the sea surface at $z = 0$

$$w = 0, \quad \rho_0 A_H \left(\frac{\partial u}{\partial z}, \frac{\partial v}{\partial z}\right) = -(\tau_\lambda, \tau_\varphi), \tag{9-12}$$

$$T = T_0(\lambda, \varphi, t), \quad s = s_0(\lambda, \varphi, t), \tag{9-13}$$

or

$$-A_{TH}\frac{\partial T}{\partial z} = \Pi; \quad -A_{sH}\frac{\partial s}{\partial z} = -s(R - E); \tag{9-14}$$

at the sea bottom

$$\frac{\partial T}{\partial z} = \frac{\partial s}{\partial z} = 0, \quad w = 0,$$

$$\rho_0 A_H \left(\frac{\partial u}{\partial z}, \frac{\partial v}{\partial z}\right) = (\tau_\lambda^b, \tau_\varphi^b); \tag{9-15}$$

at the shorelines, which are assumed to be vertical walls,

$$u, v = 0; \quad \frac{\partial T}{\partial n} = \frac{\partial s}{\partial n} = 0, \tag{9-16}$$

where $(\tau_\lambda, \tau_\varphi)$ is the wind stress vector, $(\tau_b^b, \tau_\varphi^b)$ is the bottom-friction vector, Π is the heat flux, which depends on the sea-surface temperature and the properties of the atmosphere, R is the rainfall, E the evaporation rate, and n the normal to the coastline.

Conditions (9-12)-(9-16) were written for a constant ocean depth. Their modifications for the case of a variable depth are obvious. The quantities τ_λ^b and τ_φ^b are calculated from certain approximate expressions of the theory of the bottom

---

[1] For an interesting discussion of this problem, see Ivanov [54].

boundary layer or are simply set equal to zero (assumption of slip). This formulation of the boundary conditions at the sea bottom is more convenient for numerical solution than nonslip conditions. The error thereby induced appears to be small.

Note that the condition $w = 0$ at $z = 0$ [see (9-12)] is an approximation in the case of nonstationary problems. It filters out surface gravity waves, as we saw in Sec. 4-1, transforms barotropic Rossby waves into nondivergent Rossby waves (the distortion is small; see the end of this chapter), and has almost no effect on internal and gyroscopic waves.

So far, Bryan's model has been used mainly for computation of stationary states of the ocean, assuming constant mean annual effect of the atmosphere. This is an important stage in the study of the potential of any model, since the bulk of available data does pertain precisely to the distribution of the mean annual characteristics in the sea. Let us briefly discuss some of the most significant results.

Figure 9-1 shows isolines of the transport function that were calculated from the three-dimensional model. This figure appears to indicate that the model is capable of describing the separation of a boundary current from the western shore of an ocean. Note the higher (by a factor of about 1.5) volume transport in the current, as compared to the two-dimensional model.

The calculation of the vertical (Fig. 9-2) velocities is of some interest. Note the sharp intensification of vertical motion in the boundary current. Finally, Fig. 9-3 shows the vertical structure of the boundary current at the western shore of the ocean (theory and observations). It is worth noting that the theory agrees qualitatively with observations (we refer here to the effect whereby high velocities are concentrated in the upper sea).

Manabe and Bryan's numerical solution of the air-sea interaction problem is also very interesting [39]. Naturally, conditions (9-14) are assumed for the sea-air interface. The possible formation of ice on the sea surface is taken into account, and

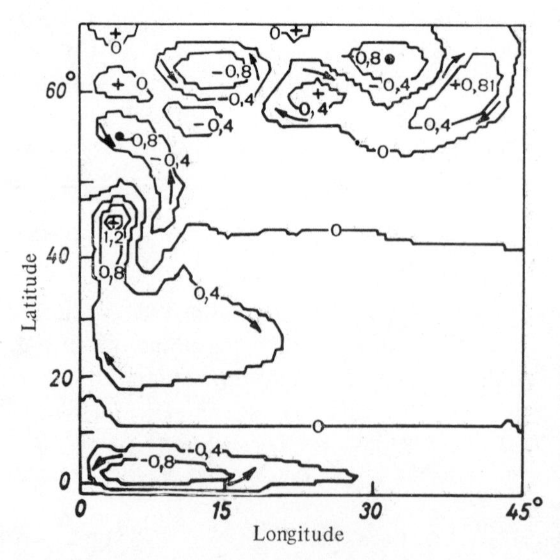

FIG. 9-1 Isolines of dimensionless transport function (after Bryan and Cox [37]). The temperature and wind for the sea surface were specified; the depth of the ocean was assumed constant. The parameters of the model were so chosen that inertial terms in the western boundary layer would be significant.

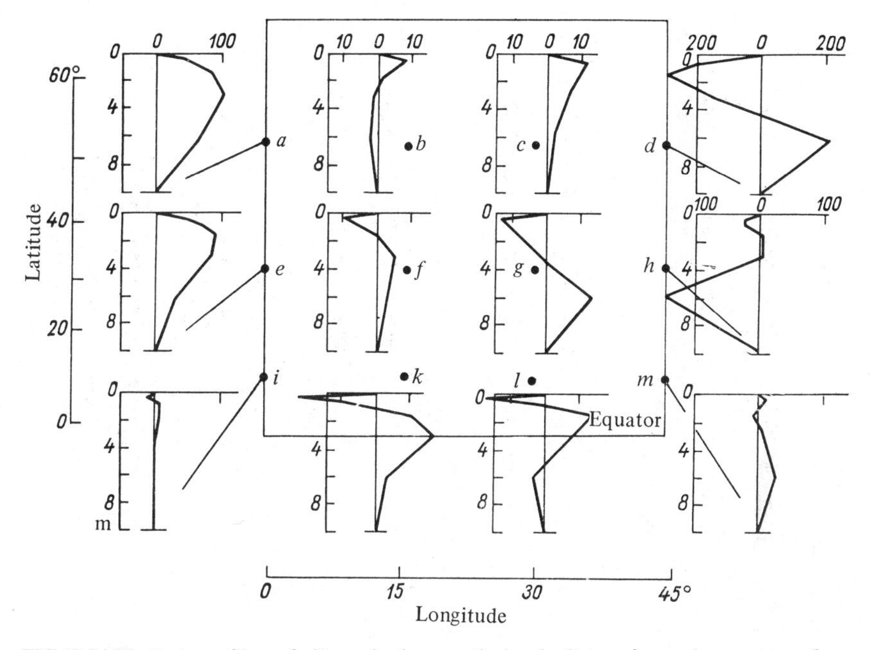

FIG. 9-2 Vertical profiles of dimensionless vertical velocity w for various zones of sea (after Bryan and Cox [37]). The scale for w is v^*d/a ($v^* = g\alpha\delta\vartheta d/2\Omega a$, where $d = 2\Omega A_H a^2/g\alpha\delta\vartheta$, $\delta\vartheta$ is the meridional temperature gradient on the sea surface, and α is the thermal expansion coefficient). The temperature and wind were specified for the sea surface, and the depth was assumed constant.

this somewhat modifies conditions (9-14). The balance equation which is written allows calculation of the ice-cover variations. Figure 9-4 compares the initial and final (stationary) states of the ocean, and Fig. 9-5 shows the meridional circulation in the basin. We call attention to the fact that this model does not show an Antarctic convergence zone and to the result that, off the Antarctic continent, water downwells to a depth of 1 to 2 km, but not to the bottom.

This model enabled Gill and Bryan [44] to analyze the effect of basin geometry on circulation in the Antarctic ring (Figs. 9-6 and 9-7). The existence of Drake Strait is responsible for the breakup (along the meridional section) of a single circulation system (Fig. 9-6*a*) into two circulation systems (Fig. 9-6*b*). We note the downwelling that Gill and Bryan computed for the area north of Drake Strait (Fig. 9-6*b*), which can serve as a model of the Antarctic convergence (see the discussion of Fig. 9-5). The streamlines in Figs. 9-6*c* and *b* are similar. Note, however, that in reality the downwelling in the zone of the Antarctic convergence occurs within a top layer about 1 km thick, while the results computed from the model indicate that the water descends to the very bottom of the sea.

Figure 9-7 demonstrates an interesting effect: in the case of a shallow Drake Strait (Fig. 9-7*c*), the rate of circulation is three times as high as in the case of a deep Drake Strait (Fig. 9-7*b*) because of formation of a strong meridional pressure gradient.

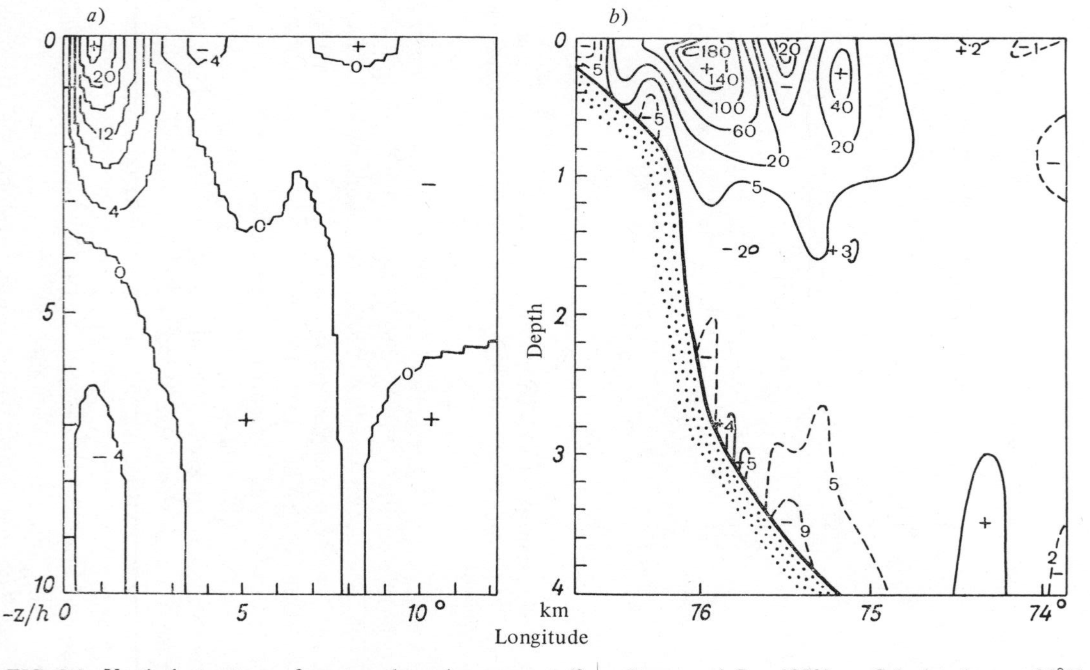

FIG. 9-3 Vertical structure of western boundary current (from Bryan and Cox [37]). *a*–Calculated, $\varphi = 28°N$, velocities indicated in dimensionless form (see Fig. 9-2 for v^* scale); *b*–observations of Swallow and Worthington in the Atlantic [55], velocities indicated in cm/sec. Temperature and wind on sea surface specified, depth assumed constant.

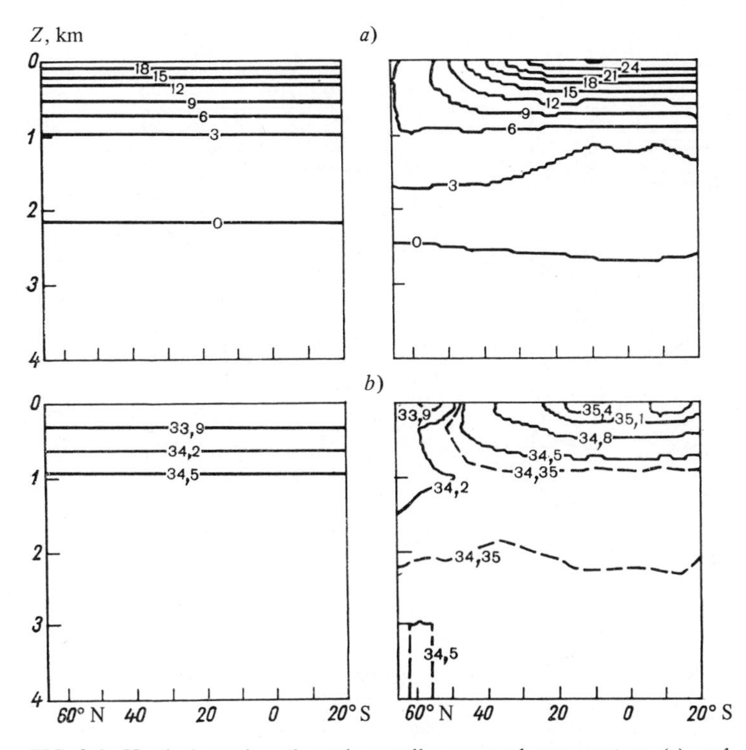

FIG. 9-4 Vertical section through zonally averaged temperature (*a*) and salinity (*b*) fields (after Bryan [39]). The initial state is shown on the left, and the calculated stationary state on the right. The depth of the ocean is assumed constant. The conditions of heat- and moisture-exchange between the ocean and the atmosphere, as well as the wind on the free surface of the ocean were specified.

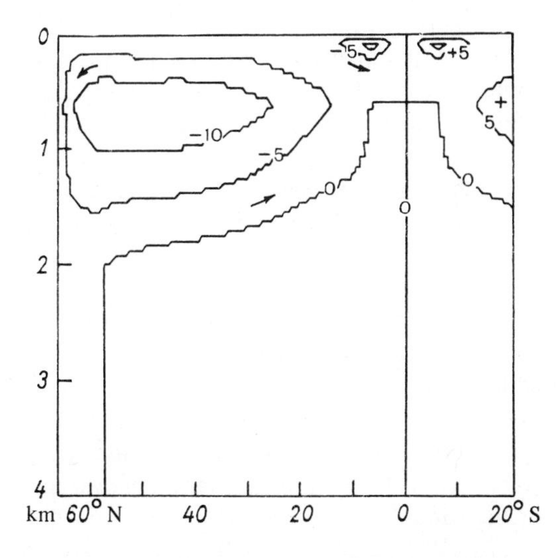

FIG. 9-5 Zonally averaged meridional circulation rate (from Bryan [39]). One circulation rate unit is 10^6 m^3/sec. Ocean depth assumed constant. Conditions of heat- and moisture exchange with the atmosphere and the wind assumed specified on the ocean surface.

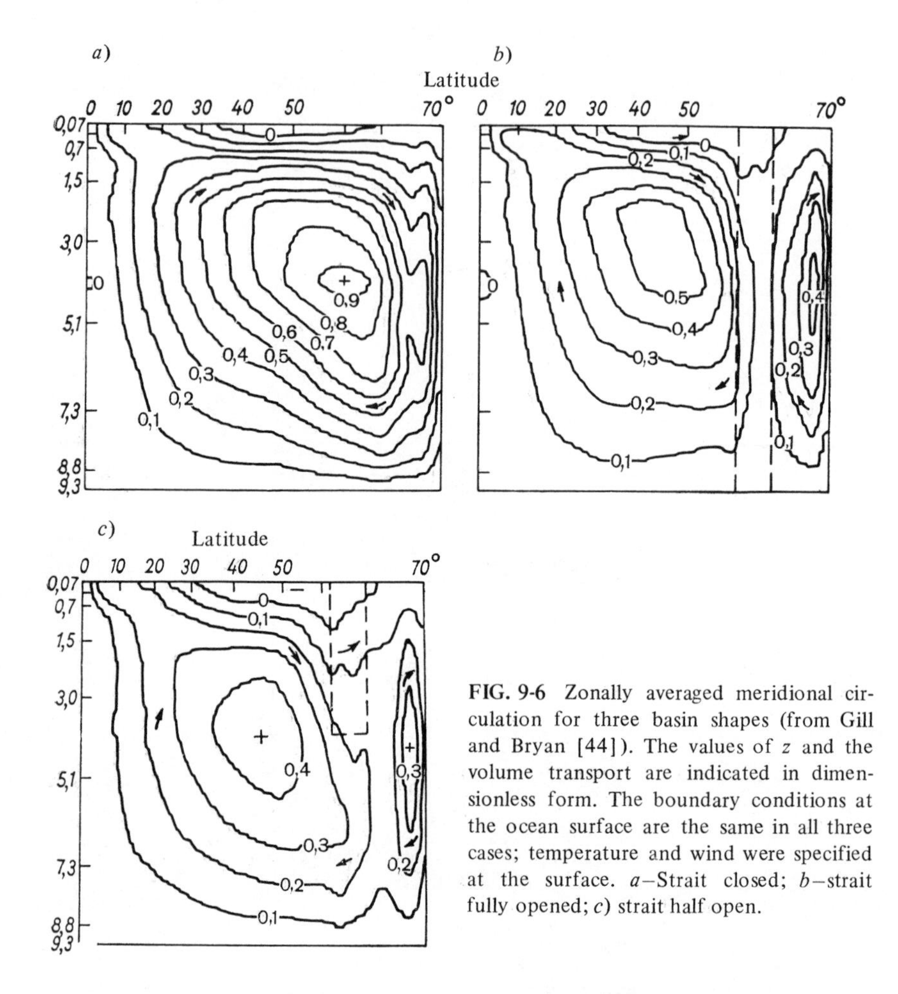

FIG. 9-6 Zonally averaged meridional circulation for three basin shapes (from Gill and Bryan [44]). The values of z and the volume transport are indicated in dimensionless form. The boundary conditions at the ocean surface are the same in all three cases; temperature and wind were specified at the surface. a—Strait closed; b—strait fully opened; c) strait half open.

Nonstationary solutions. Very few such solutions have so far been obtained, even though derivation of nonstationary solutions does not, in principle, require any modifications to the models described above (of course, the amount of computational work is much higher).

The first paper we would like to cite is that by Kagan et al. [34] (a natural extension of [33]), which analyzes the seasonal variability of the two-dimensional circulation in a global ocean with realistic coastlines and bottom topography as a function of the annual variation of atmospheric pressure at sea level. Figures 9-8 and 9-9, cited from this paper, present integrated-circulation streamlines for December and July. As one would expect, the seasonal variability is manifested most clearly in the Indian Ocean (this problem was discussed in Chap. 6).

A very important paper is that by Cox [43] on the seasonal variations of the circulation in the Indian Ocean. The temperature T, the salinity s, and the wind stress (τ_λ, τ_φ) on the surface of that ocean were specified. The strongest variation over the year was assumed to exist in the case of the wind, and the smallest in the case of the temperature. The salinity was assumed independent of time. These variables were written as

$$T = T_s + T_r \cos \alpha;$$
$$(\tau_\lambda, \tau_\varphi) = (\tau_{\lambda s}, \tau_{\varphi s}) + (\tau_{\lambda 1}, \tau_{\varphi 1}) \cos \alpha + (\tau_{\lambda 2}, \tau_{\varphi 2}) \sin \alpha,$$

where α is the phase angle and T_s, T_r, τ_s, τ_1, and τ_2 were selected so as to fit

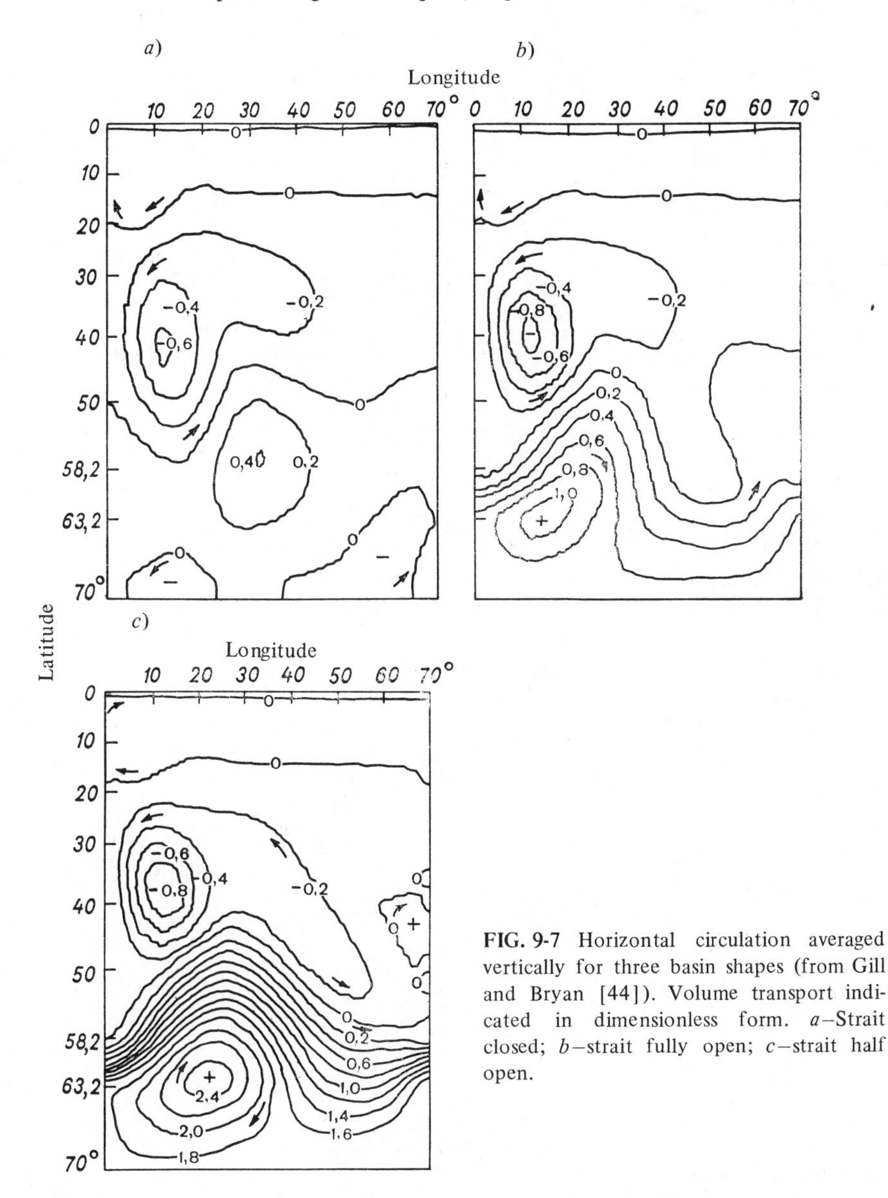

FIG. 9-7 Horizontal circulation averaged vertically for three basin shapes (from Gill and Bryan [44]). Volume transport indicated in dimensionless form. *a*—Strait closed; *b*—strait fully open; *c*—strait half open.

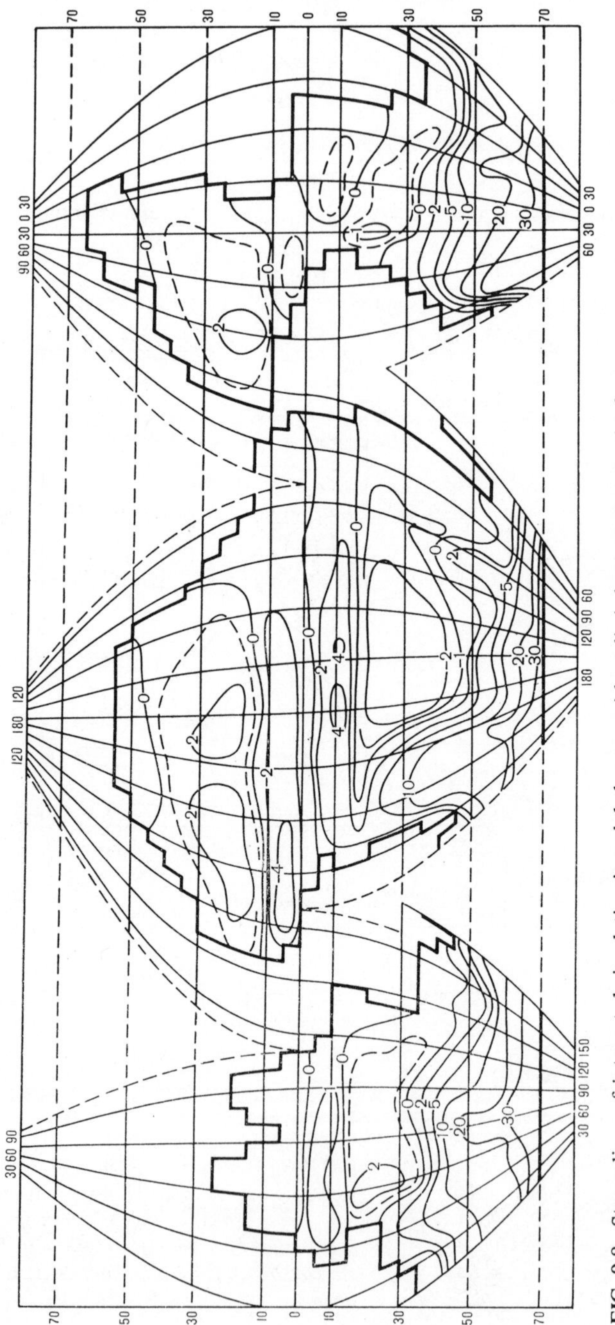

FIG. 9-8 Streamlines of integrated circulation in a global ocean with realistic bottom topography for December (from Kagan et al. [34]). To determine the water transport (in cm^3/sec) between adjacent isolines, multiply the difference of values indicated on the isolines by $6.37 \cdot 10^{12}$.

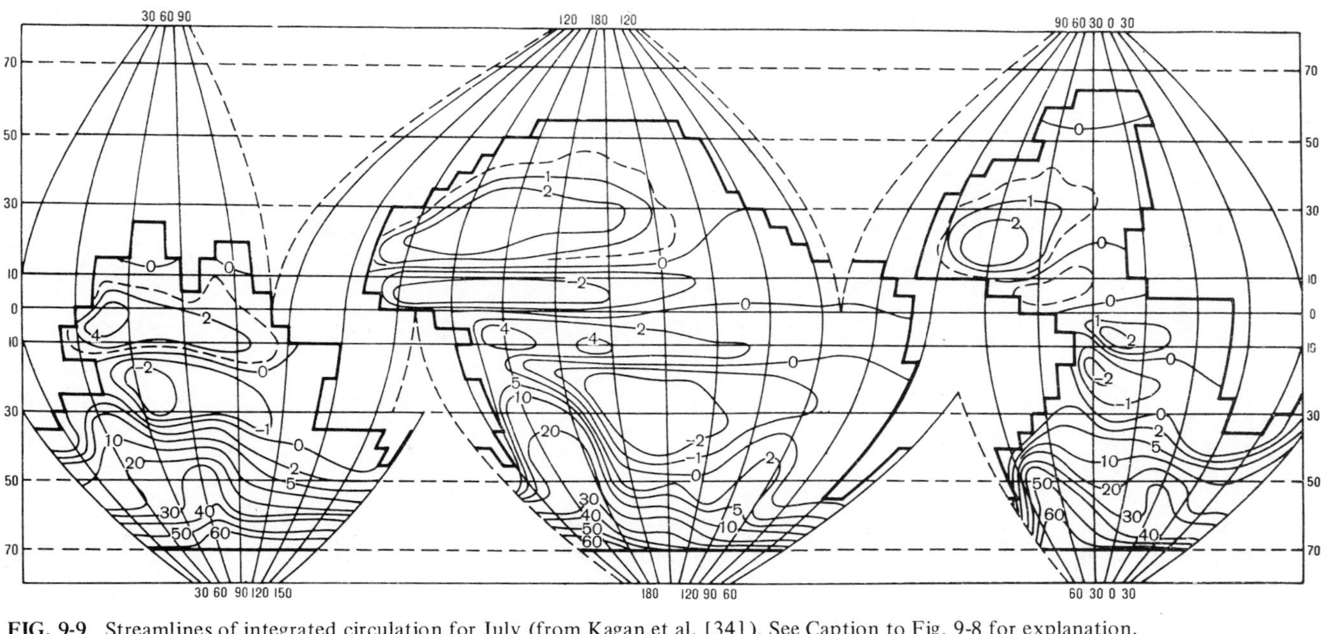

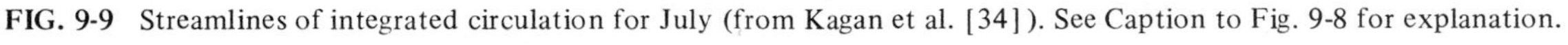
FIG. 9-9 Streamlines of integrated circulation for July (from Kagan et al. [34]). See Caption to Fig. 9-8 for explanation.

observations (Fig. 9-10). The temperature and salinity were specified on the liquid boundary of the region (at all levels), and the total stream function was set equal to zero.

The Cox model gives a satisfactory description of the seasonal variations in the Indian Ocean (Figs. 9-11 and 9-12). Let us now discuss its most conspicuous feature, namely, the Somali current. During the summer, the southwest monsoon produces the strong Somali northward flows along the African coast at velocities of about 3.5 m/sec (see Fig. 9-11). The results indicate an interesting relationship, namely, that upwelling along the African coast due to the southwesterly monsoon and the coastal "cooling"

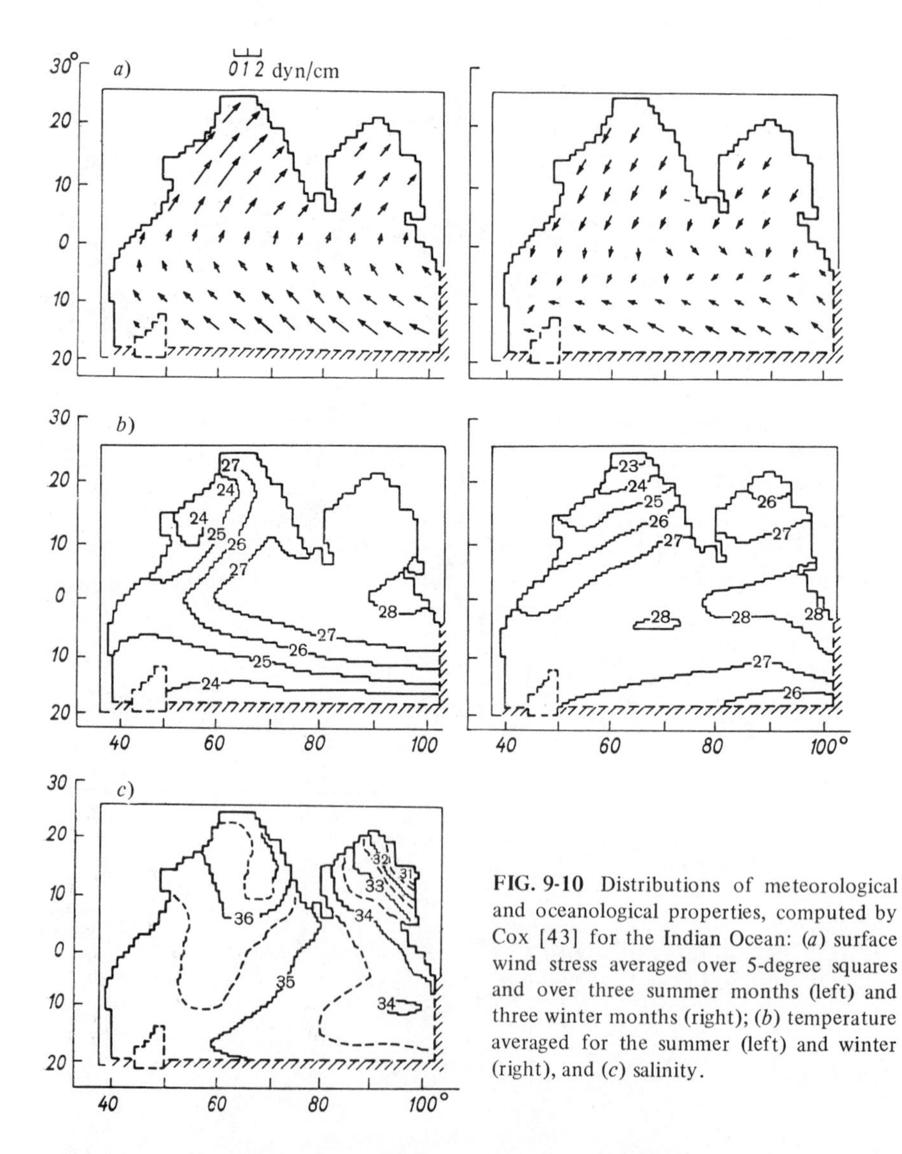

FIG. 9-10 Distributions of meteorological and oceanological properties, computed by Cox [43] for the Indian Ocean: (*a*) surface wind stress averaged over 5-degree squares and over three summer months (left) and three winter months (right); (*b*) temperature averaged for the summer (left) and winter (right), and (*c*) salinity.

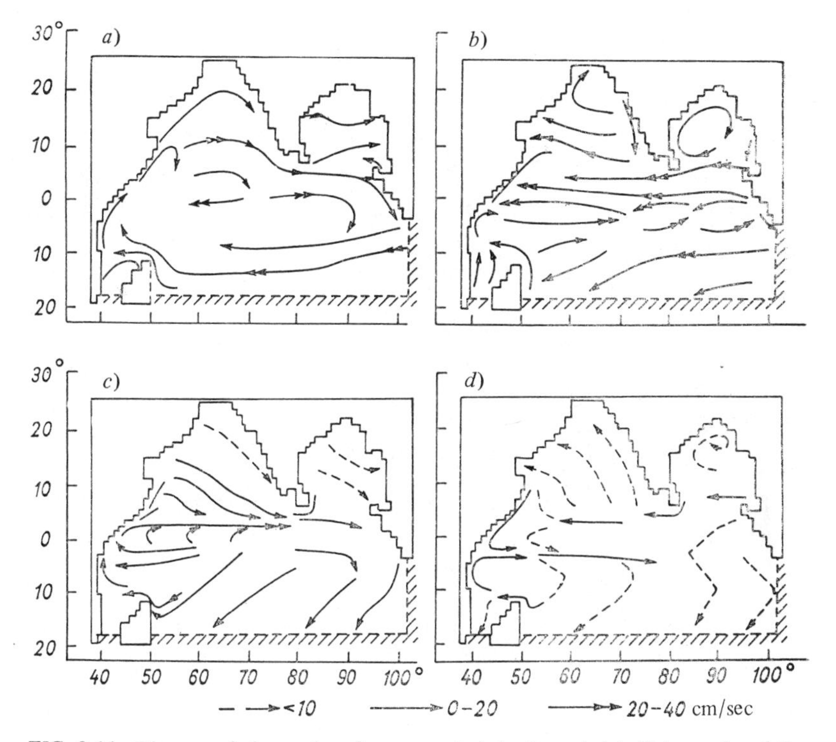

FIG. 9-11 Diagram of observed surface currents (*a* in August, *b* in February) and the corresponding currents calculated from the model (averaged over the top 50-meter layer of water) (*c* in August, *d* in February) with allowance for a realistic bottom topography (from Cox [43]).

due to this upwelling play an important role in shaping the Somali current. The totality of this effect intensifies the pressure gradient perpendicular to the coast, and it is this gradient that produces the strong longshore current, by virtue of the geostrophicity condition. Cox's estimates indicate that this baroclinic effect is controlling north of 3°N (the principal zone of formation of the Somali current); south of 3°N, the barotropic mechanism ("squeezing" of the oceanic circulation against the western coast due to the β effect) is controlling (Figs. 9-11, 9-13, and 9-14).

The southwesterly monsoon weakens and reverses direction in the fall. The Somali current also becomes weaker (Figs. 9-11, 9-14) and the strong currents shift to the south. This is in large part due to the fact that with the change in the direction of the monsoon in winter, the coastal upwelling gives way to the weaker coastal downwelling, so that the pressure gradient perpendicular to the coast is reduced. The wind-controlled seasonal variations of the Somali current are clearly seen in Fig. 9-15. The phase difference between the current velocity variations and the changes in the wind is very short (about 2 weeks). Cox likens the seasonal variations of the temperature field, which control those of the Somali current to a considerable extent, to a kind of

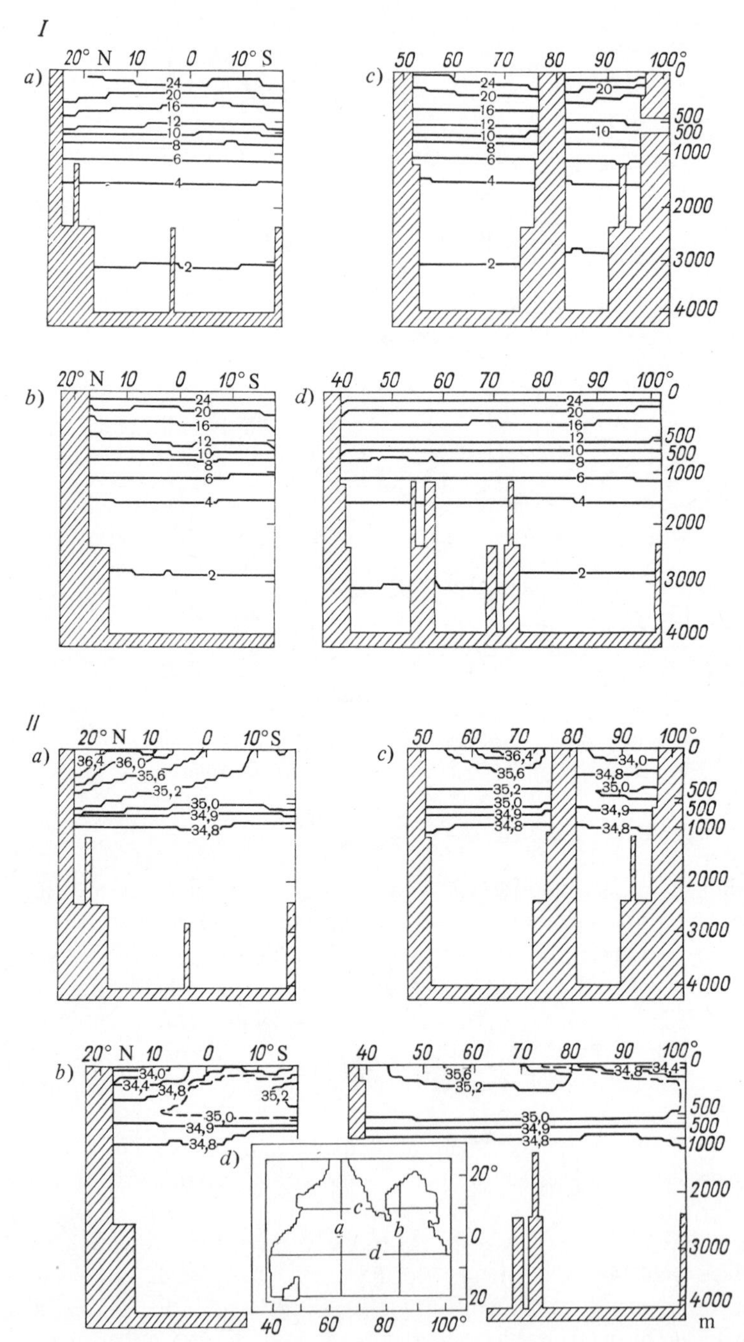

FIG. 9-12 Fields of temperature (*I*) and salinity (*II*), calculated by Cox [43] for winter. The coordinates of the sections are indicated at the bottom. Calculation made with allowance for a realistic bottom topography.

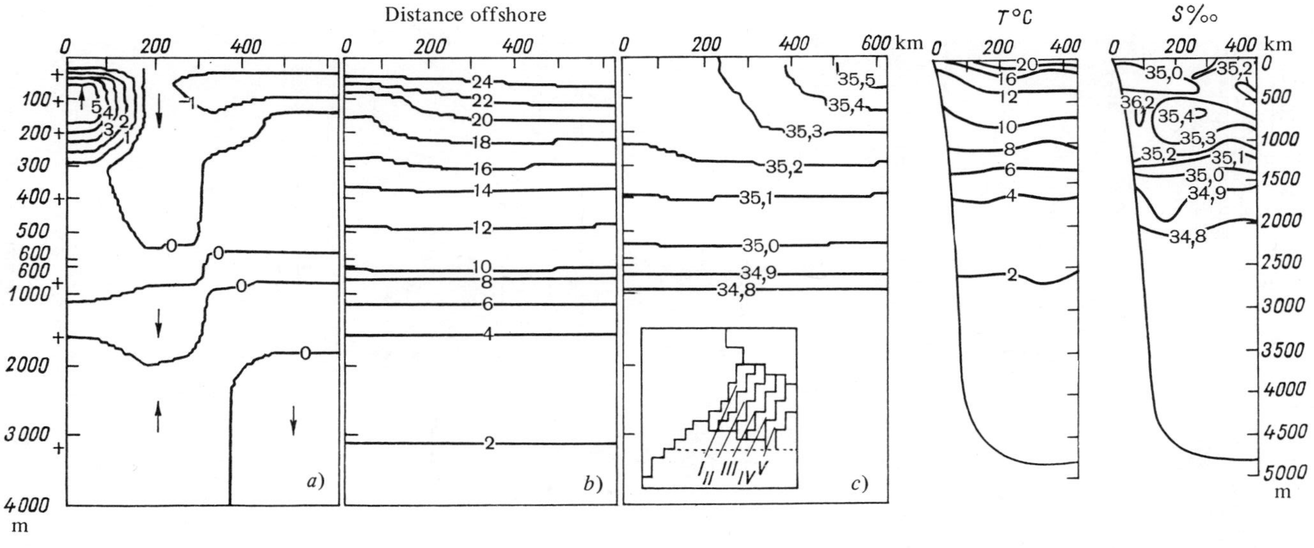

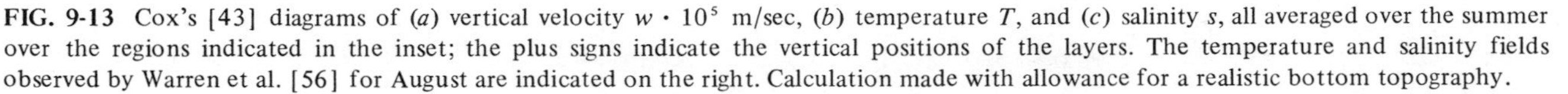

FIG. 9-13 Cox's [43] diagrams of (*a*) vertical velocity $w \cdot 10^5$ m/sec, (*b*) temperature T, and (*c*) salinity s, all averaged over the summer over the regions indicated in the inset; the plus signs indicate the vertical positions of the layers. The temperature and salinity fields observed by Warren et al. [56] for August are indicated on the right. Calculation made with allowance for a realistic bottom topography.

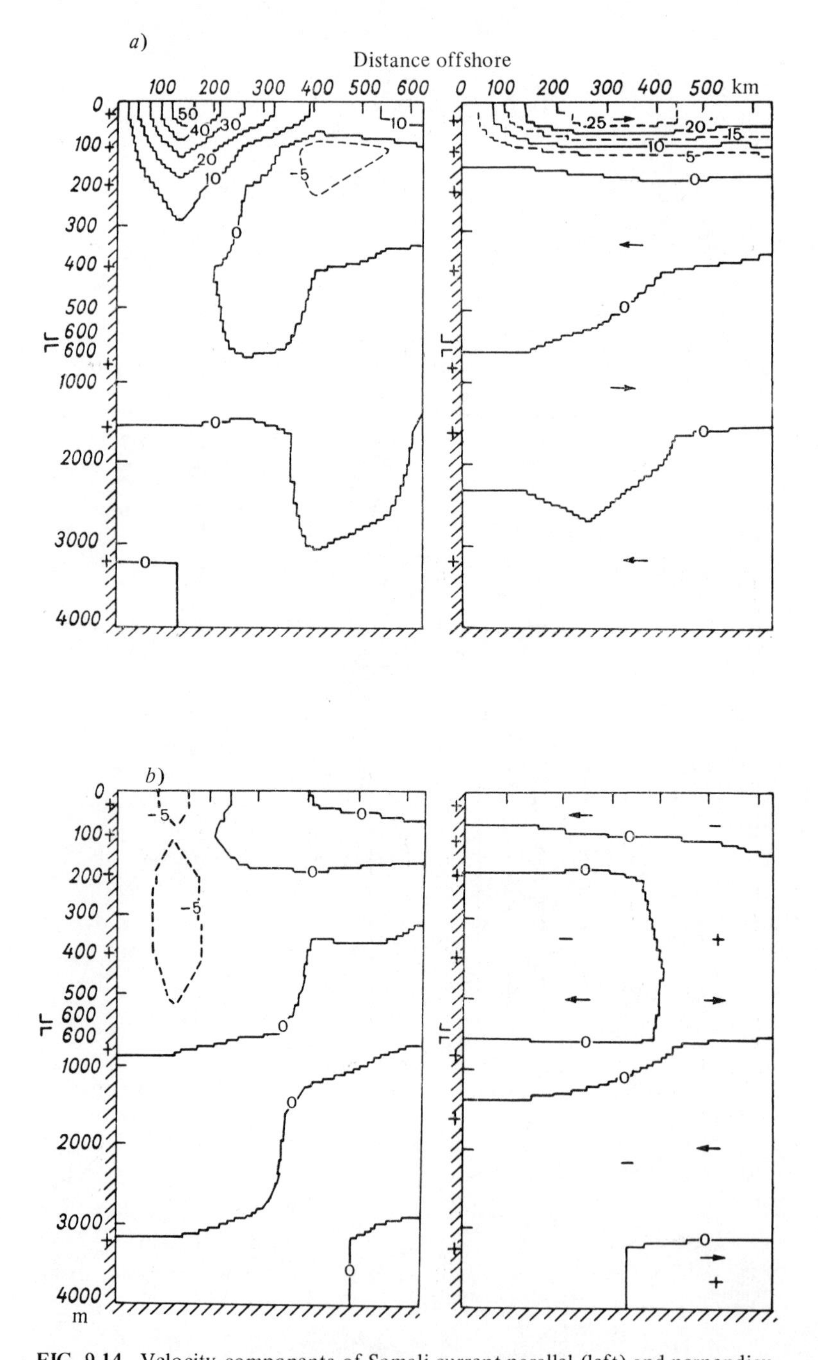

FIG. 9-14 Velocity components of Somali current parallel (left) and perpendicular (right) to the coast (from Cox [43]). *a*–Summer average; *b*–winter average. Values computed from model with realistic bottom topography; velocities in cm/sec.

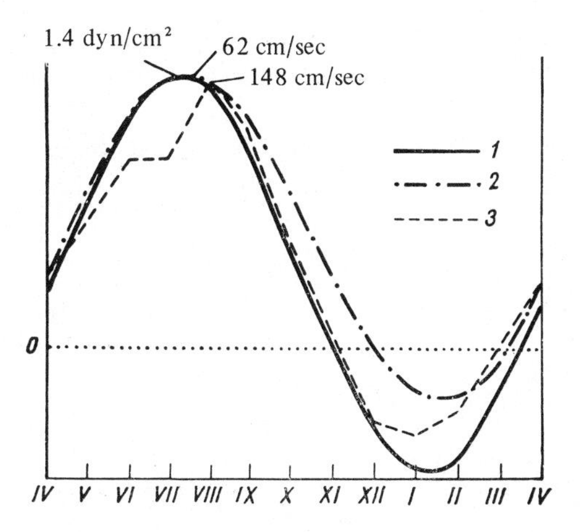

FIG. 9-15 Time curves of wind stress (1) assumed in the model of the current velocity averaged over the first difference layer (2), and observed current velocity (3) (from Cox [43]). All quantities pertain to components along the Somali coast, averaged over region *I* of the inset in Fig. 9-13.

oscillator: the oscillations are generated by upwelling, and damping is applied by horizontal advection and mixing.

Let us now briefly consider an experiment performed by Wetherald and Manabe [48] using Bryan and Manabe's ocean atmosphere model. Taking the stationary ocean-atmosphere state computed by Bryan [39] as the initial state, Wetherald and Manabe extended the integration over 1.5 years, assuming seasonal variations of insolation. Figure 9-16 shows the seasonal variations of sea temperature at various latitudes and depths (25, 50, 100, and 300 m). The results were as follows: Latitude 60°: the range of the temperature variations decreases with increasing depth so that at 300 m the variations vanish for all practical purposes. Mixing and convection are significant in this case. An interesting fact is that the average seasonal temperature at the surface is much higher than the annual average temperature obtained by solving the stationary problem under the assumption of the mean annual insolation. This nonlinear effect is due entirely to small-scale convection and mixing. In fact, during the period of cooling of the sea surface (fall and winter), the temperature of the latter will be higher than in the case of no convection, the reason being heat influx due to convective mixing of the upper layer. Similarly, the surface layers should have temperatures higher than the annual average during the spring and summer.

Latitude 31°: in this case the amplitude of the temperature fluctuations decreases more sharply with depth than in high latitudes, because convection extends to shallower depths. Obviously, the above-described effect of the average seasonal warming of the surface waters is not as conspicuous in this case.

Latitude 3°: in this case, advective processes (wind-governed upwelling and downwelling) are controlling, and mixing and convection are of secondary importance. This is probably the reason why the amplitudes of temperature variations vary little with depth. It is interesting that the seasonal average temperature at the surface will be lower than the mean annual temperature obtained by solving the stationary problem.

These properties of the temperature field are clearly in the pattern of the meridional circulation (Fig. 9-17). The mean seasonal meridional circulation weakens in middle and high latitudes (because of the mean seasonal warming in high latitudes and the resulting decrease in the meridional temperature gradient), but becomes slightly stronger in the tropics (due to the mean seasonal cooling in the equatorial zone).

Energetics of the ocean. Multiplying Eq. (9-1) by u and (9-2) by v, adding the results, and substituting (9-3), we have

$$\frac{\partial}{\partial t}\left(\rho_0 \frac{u^2+v^2}{2}\right) = -\operatorname{div}\left[\left(\rho_0 \frac{u^2+v^2}{2} + p\right)\mathbf{v}\right] + g\rho w + uF_\lambda + vF_\varphi .$$

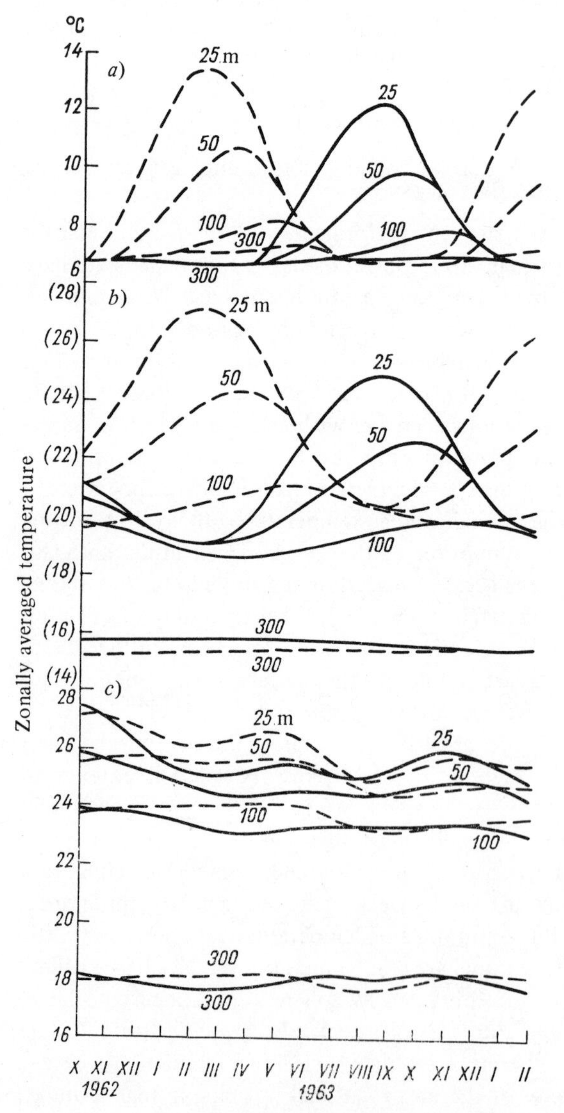

FIG. 9-16 Time curves of zonally averaged temperature for various depths in both hemispheres at latitudes 60° (*a*), 31° (*b*), and 3° (*c*) (from Wetherald and Manabe [48]). The mean values for the Northern Hemisphere are represented by the solid lines, and those for the Southern Hemisphere by the dashed lines. The numbers indicate the depths.

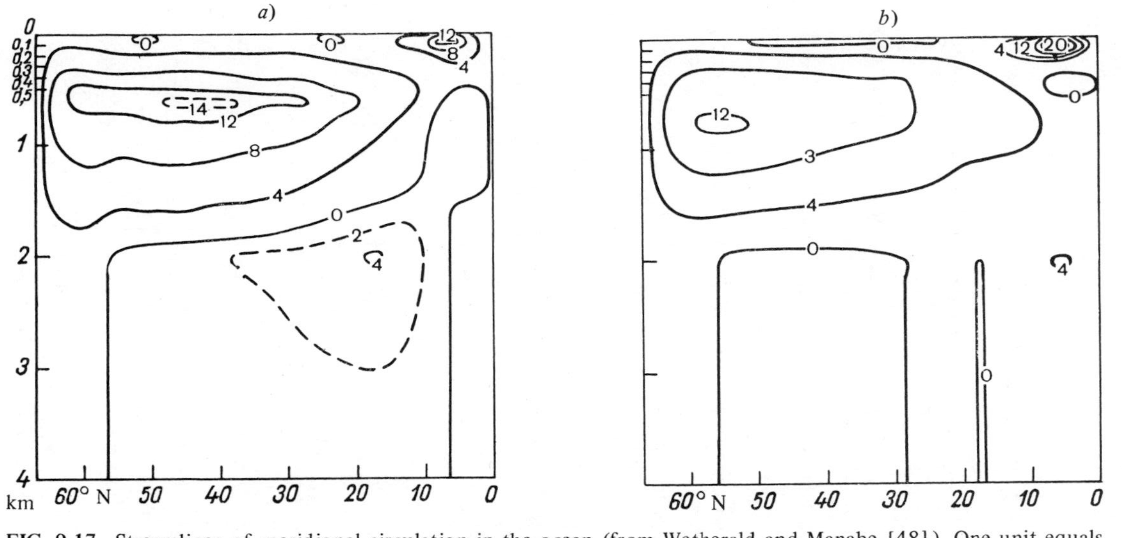

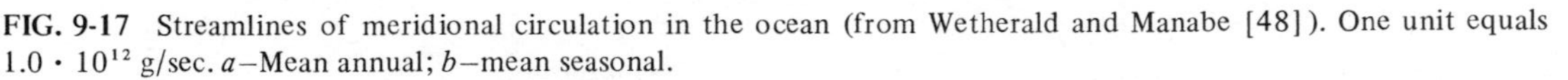

FIG. 9-17 Streamlines of meridional circulation in the ocean (from Wetherald and Manabe [48]). One unit equals $1.0 \cdot 10^{12}$ g/sec. *a*—Mean annual; *b*—mean seasonal.

Assuming the coasts of the basin to be vertical walls and integrating the above equation over the entire water volume V, we obtain by virtue of (9-12), (9-15), and (9-16)

$$\frac{\partial E}{\partial t} = W + B + D, \tag{9-17}$$

where the meaning of E, W, B, and D is as follows, if we denote the area of the ocean surface by Σ: $E = (1/\Sigma) \iiint_V (\rho_0/2)\,(u^2+v^2)\, dV$ is the mean kinetic energy of the ocean, $W = (1/\Sigma) \iint_\Sigma (u\tau_\lambda + v\tau_\varphi)\, d\Sigma$ is the mean work of wind stress forces, $B = (1/\Sigma) \iiint_V g\rho w\, dV$ is the mean work of buoyant forces, and $D = (1/\Sigma) \iiint_V (uF_\lambda + vF_\varphi)\, dV - W$ is the mean kinetic energy dissipation. To give some idea of the magnitudes involved, let us cite the values for the stationary solution [39]: $W = 0.92$ erg/(cm^2 · sec), $B = -0.31$ erg/(cm^2 · sec), and $D = -0.61$ erg/(cm^2 · sec).

Following Holland [57], we now determine the barotropic component $(\overline{u}, \overline{v})$ of the horizontal current velocities as the average of the true current velocities over the depth of the ocean: $(\overline{u}, \overline{v}) = (1/H) \int_0^H (u, v)\, dz$ and the baroclinic component $(\hat{u}, \hat{v})$ as the difference between the true velocity $(\overline{u}, \overline{v})$ and the barotropic component $(\overline{u}, \overline{v})$: $(\hat{u}, \hat{v}) = (u, v) - (u, v)$. Further, let $\overline{E} = (1/\Sigma) \iiint_V (\rho_0/2) \times (\overline{u}^2 + \overline{v}^2)\, dV$ be the mean kinetic energy of the barotropic components of the motion, and let $\hat{E} = (1/\Sigma) \iiint_V (\rho_0/2)\,(\hat{u}^2 + \hat{v}^2)\, dV$ be the mean kinetic energy of its baroclinic component. An equation for the variation of $\overline{E}$ is easily derived by first averaging Eqs. (9-1) and (9-2) over z from 0 to H and then, after multiplying the averaged equations by $\overline{u}$ and $\overline{v}$, respectively, adding the results:

$$\frac{\partial \overline{E}}{\partial t} = N_e + W_e + B_e + D_e, \tag{9-18}$$

where the term N_e is governed by the effect of nonlinear terms (an exact expression for it is easily derived), $W_e = (1/\Sigma) \iint_\Sigma (\overline{u}\tau_\lambda + \overline{v}\tau_\varphi)\, d\Sigma$ is the work of the wind stress forces in barotropic motion,

$$B_e = -(1/\Sigma) \iiint_V \left\{ (1/H) \left[\bar{u} \int_0^H \frac{\partial p}{a \cos \varphi \partial \lambda} dz + \bar{v} \int_0^H \frac{\partial p}{a \partial \varphi} dz \right] \right\} dV$$

and

$$D_e = (1/\Sigma) \iiint_V \left\{ (1/H) \left[\bar{u} \int_0^H F_\lambda dz + \bar{v} \int_0^H F_\varphi dz \right] \right\} dV - W_e.$$

Subtracting Eq. (9-18) from (9-17) and applying the readily verified identity $E = \bar{E} + \hat{E}$, we find

$$\frac{\partial \hat{E}}{\partial t} = N_i + W_i + B_i + D_i, \tag{9-19}$$

where $N_i = -N_e$, $W_i = W - W_e$, $B_i = B - B_e$ and $D_i = D - D_e$ (the physical meaning of these terms is obvious).

Let us analyze B_e. Since N_e is usually of secondary importance, it is B_e that describes the interaction of the barotropic and baroclinic components of the motion (naturally, on the average over the entire ocean). After simple transformations, the relation for B_e can be written as

$$B_e = -\frac{1}{\Sigma} \iint_\Sigma \left(\frac{1}{H} \int_0^H p' \, dz - p_b' \right) \times \left(\bar{u} \frac{\partial H}{a \cos \varphi \partial \lambda} + \bar{v} \frac{\partial H}{a \, \partial \varphi} \right) d\Sigma,$$

where p' is the deviation of the pressure from the hydrostatic value $g\rho_0 z$, and p_b' is the value of p' at the bottom of the sea.

It follows from the above formula that if ρ = const, then $(1/H) \int_0^H p' \, dz = p_b'$ and $B_e = 0$ even in the case of an incised sea floor, but if $\rho \neq$ const and the depth of the ocean is constant, B_e also vanishes. Thus, an interaction of the barotropic and baroclinic components can exist only in a baroclinic ocean with an incised sea floor (combined effect of sea-floor topography and baroclinicity); if $B_e = 0$, the energy $\bar{E}$ of the barotropic component of the motion is practically completely controlled by the work W_e of the wind stress forces and by the dissipation D_e.

The above quantities were calculated by Holland [57] in several numerical experiments. Figure 9-18 shows the seasonal variation of these quantities for the Indian Ocean, as reported by Cox [43]. It is readily seen that the energy flux N_e governed by nonlinear effects can be completely neglected. However, the two fluxes

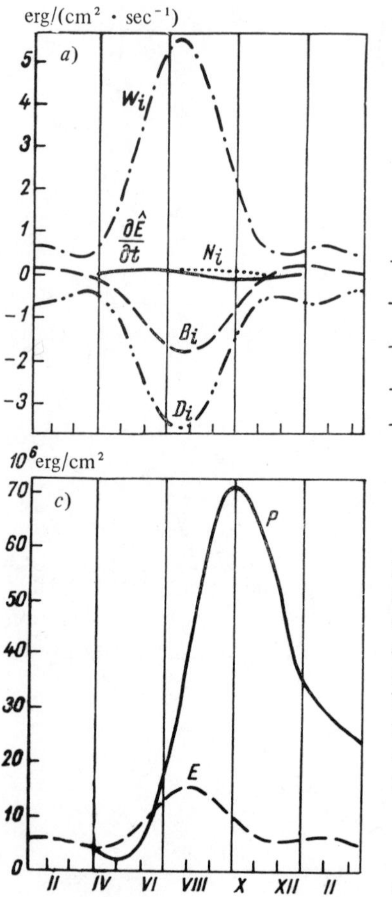

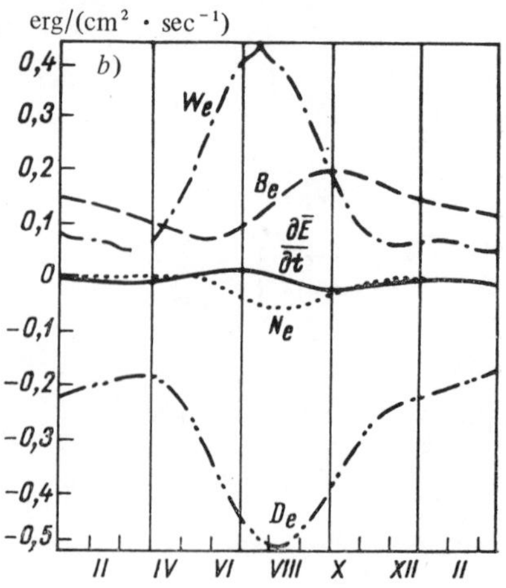

FIG. 9-18 Energetics of the Indian Ocean, after Holland [57]. *a*–Energetics of baroclinic component of flow, see Eq. (9-19) and the explanation of $\hat{E}$, N_i, W_i, B_i, and D_i given in the text; *b*–energetics of the barotropic component, see Eq. (9-18) and the text explanation for $\bar{E}$, N_e, W_e, B_e and D_e; *c*–variations of mean kinetic energy E (see text) and mean potential energy

$$P=(1/\textstyle\sum)\iiint_V (g\rho z)\, dV$$

over the year.

B_e and W_e are of comparable magnitude and apparently equally important in analysis of the barotropic motion.

Features of numerical models. Time scales. Difference schemes for study of the global circulation have been discussed in many publications (see, for example, Marchuk's monograph [58, 59], Bryan's paper [60], and the papers cited above). All we need to say here is that in derivation of the difference schemes it is usually required that they satisfy the appropriate conservation laws (in difference form) in the adiabatic case.

Estimates of the characteristic time scales involved in this problem are of considerable interest. To illustrate, let us consider briefly Rossby waves in a sea of constant depth (see Sec. 4-1). Since Rossby waves have low frequencies, the quasistatic approximation holds in their case, and the characteristic curves describing the vertical structure of the oscillations will be $\varepsilon_n(\omega) = \text{const}$, $\varepsilon_n > 0$ (see Sec. 4-1). It is therefore convenient in studying each mode to introduce the "equivalent depth" $h_n = 1/g\varepsilon_n$. It can be shown that any forced oscillation can be represented in the form of a linear superposition of "normal" modes, where each mode represents oscillations of a homogeneous layer of liquid of constant depth h_n that can be described by the Laplace tidal equations. In fact (see [61]), one needs only to take into account the first two "normal" modes, that is, the barotropic (surface) and first baroclinic (internal) modes. It is easily found that h_0 is for all practical purposes equal to the average sea depth (4000 m), while h_1 is of the order of 1 meter (this follows from the estimates indicated in Sec. 4-1 and from analysis of a two-layer model of the liquid (see [62]).

We are principally interested in waves with horizontal wave numbers $k \sim 2\pi/1000$ km (the numbers m and n on the sphere are of the order of 10). For such wavelengths, we can find an asymptotic formula for the characteristic curves of the type 2 Laplace tidal equations (for a detailed WKB analysis, see [63]). This relation takes the form

$$\omega_n = -\frac{\beta k_x}{k_x^2 + k_y^2 + f^2 \varepsilon_n}, \tag{9-20}$$

where β is the latitudinal variation of the Coriolis parameter (we note that exactly the same relation can be derived in the β-plane approximation).

Since $\varepsilon_n(\omega) = 1/gh_n$ = const, a dispersion relation for the Rossby waves follows immediately from (9-20):

$$\omega_n = -\frac{\beta k_x}{k_x^2 + k_y^2 + f^2/gh_n}. \tag{9-21}$$

The Rossby-wave dispersion curves in Fig. 4-1-5 are really plots of this relationship.

From (9-21) we obtain an expression for the group velocity

$$(c_{gx}, c_{gy}) = -\left[k_y^2 - k_x^2 + \frac{f^2}{gh_n},\ 2k_x k_y\right] \frac{\beta}{k_x^2 + k_y^2 + f^2/gh_n}. \tag{9-22}$$

Let us begin with a barotropic Rossby wave. It is natural to assume that the wave numbers k_x and k_y are governed by the characteristic horizontal scales of the background atmospheric disturbances ($\sim 2\pi/1000$ km). But then the term f^2/gh_0 in the denominators of (9-21) and (9-22) can be neglected in the barotropic mode,[2] and the dispersion relation (9-21) represents, for a given ω, a circle on the k_x, k_y plane with its center at $(-\beta/2\omega, 0)$ and a radius of $\beta/2\omega$. The group velocity vector points along the normal to this circle at the point (k_x, k_y), and its absolute value is β/k^2. Since a chord is always smaller than a diameter, we have

$$k\omega < \beta. \tag{9-23}$$

---

[2] The presence of the term f^2/gh_0 in the denominator of (9-21) is due to the free-surface effect; thus, the solid-wall approximation (see Sec. 4-1) has almost no distorting effect on barotropic Rossby waves.

The importance of relation (9-23) is stressed by Lighthill in his review [64]. Thus, if $\omega \sim 1/\text{week}$, $k < 1/100$ km. Relation (9-23) describes the class of external perturbations capable of exciting significant barotropic motions in the ocean.

The question as to the characteristic features of the external perturbations is by itself an extremely important one. Following Pedlosky [65] and Lighthill [61], we shall consider almost zonal ($k_x^2 \ll k_y^2$) external perturbations. But then, even for the excited barotropic waves, $k_x^2 \ll k_y^2$ and

$$c_{gx} \simeq -\frac{\beta}{k_y^2}, \quad c_{gy} \simeq 0. \tag{9-24}$$

Hence if $k_y = 2\pi/1200$ km, $c_{gx} \simeq 1$ m/sec.

It is important to note that, according to (9-24), all wave packets "carry" energy toward the western coast of the ocean (intensification of currents at the western coasts). The problem of the formation of nonstationary coastal boundary layers has been treated in several papers [61, 66, 67]. It has been shown that the "thickness" x_t of a nonstationary boundary layer at the western coast of the ocean is given by

$$x_t = \frac{1.4}{\beta t}.$$

We can readily calculate that about a week after the disturbance has reached the western coast ($t = 0$), the "thickness" x_t will have decreased to the actually observed 100 km. This estimate is also valid for the baroclinic mode.

Finally, let us consider the baroclinic mode. Since $h_1 \simeq 1$ m and k_x and k_y are controlled by the external perturbations, we have $f^2/gh_1 \gg k^2$ and relations (5-2-2) and (5-2-3) become

$$\omega \simeq -\frac{k_x \beta g h_1}{f^2};$$

$$c_{gx} \simeq -\frac{\beta g h_1}{f^2};$$

$$c_{gy} \simeq 0. \tag{9-25}$$

We see that baroclinic Rossby waves propagate practically without dispersion. In middle latitudes, $c_{gx} \simeq 1$ cm/sec, at $\varphi = 20°$, $c_{gx} \simeq 10$ cm/sec, and at $\varphi = 6°$, $c_{gx} \simeq 1$ m/sec.

The region around the equator must be described in greater detail for purposes of analysis (since the Coriolis parameter f can no longer be assumed constant), and is of great interest in connection with studies of the Somali current. Such an analysis was made by Lighthill [61], who showed that the characteristic time of formation of the Somali current under the action of the southwesterly monsoon is one month. This time consists of the three weeks that it takes for perturbations produced by the entire monsoon belt (zonal dimensions of about 2000 km) to reach the western shore, plus the week that is needed for concentration of the perturbations (once they have arrived) into a boundary current about 100 km thick (note that the characteristic group velocities of barotropic and baroclinic Rossby waves are the same in the equatorial region).

Such, in general terms, are the features of propagation of Rossby waves in the ocean (see also [68]). It is natural to assume that the orders of magnitude of the velocities of propagation of the perturbations in the ocean are given correctly by the above linear theory (although the basic equations of the numerical model are nonlinear and contain dissipative factors).

Let us now return to the numerical models. These problems are generally solved by convergence methods, and Bryan's model probably converges to a steady state in several hundred years. It is interesting that this convergence time is controlled in Bryan's model by the behavior of the density field and readily follows from analysis of the simple heat conduction equation $\partial T/\partial t - A_{TH}(\partial^2 T/\partial z^2) = 0$, $t^* = H^2/A_{TH}$. This very long time causes serious difficulties in numerical experiments (the number of time steps required to achieve convergence of the problem is about 100,000; this requires one week of machine time on the Univac 1108). It appears that t^* should depend strongly on the form in which the dissipative factors are taken into account in the model. However, it is evident that the convergence time cannot be *shorter* than the time required by the *baroclinic* Rossby wave to cover a distance equal to the characteristic dimension of the basin. According to the estimates just obtained, this value is in the *tens of years* in middle latitudes (see [64]).

It is no surprise, therefore, that convergence was not reached in all of the numerical experiments of Bryan and his co-workers. For example, solution of the problem was broken up into three stages by Cox [43]. Table 9-2, which is taken from that paper, gives a good idea of the parameters of the computation (let us add that the number of levels along the vertical was 6 to 10).

Because of the presence of strong narrow currents, their seasonal variations, and their separation from the coast, as well as the importance of coastal upwellings and downwellings, it is evidently necessary to use difference schemes whose step in space is variable. For "special" areas, Gill [69] indicates that a 20-km horizontal step is necessary; since the spatial positions of these areas change with time, Lagrangian grids

TABLE 9-2 The Three Stages in the Cox Numerical Experiment [43]

	I	II	III
Horizontal step	4×4^0	2×2^0	1×1^0
A_L (momentum)	$2 \cdot 10^9$ cm²/sec	$2 \cdot 10^8$ cm²/sec	$5 \cdot 10^7$ cm²/sec
A_L (density)	10^8 cm²/sec	10^8 cm²/sec	$5 \cdot 10^7$ cm²/sec
A_{TH}	1 cm²/sec	1 cm²/sec	1 cm²/sec
Integration range	0–130 yr	130–185 yr	185–192 yr
Time step	0.6 day	0.3 day	0.1 day
Machine time (Univac 1108) .	0.2 hr/yr	1.7 hr/yr	22 hrs/yr

that "meander" together with the boundary currents appear to be a highly promising technique. Other solutions of this problem are, of course, also possible—e.g., Il'in's grid [70] for the simplest case.

Another extremely important fact must also be kept in mind. In Chap. 5, we mentioned that synoptic eddies in the ocean exhibit very high energies. If it is found (which is quite possible; see [4, 5]) that energy can be transferred from synoptic eddies to large-scale motions, we will need a fundamentally different [as compared to (9-8a) and (9-9a)] parameterization of "turbulent" effects in the ocean. In any event, numerical modeling of the interaction of synoptic eddies (scale 100 km) and large-scale currents (scale 1000 km) for the *global ocean* is hardly feasible at the present time (due to the limited capacities of the computers).[3]

REFERENCES

1. Eckart, C. Properties of Water, Part 2, The equations of state of water and sea water at low temperatures and pressures, Amer. J. Sci., **256**, No. 4, 225–240 (1958).
2. Kamenkovich, V. M. Eddy diffusivity and viscosity in large-scale motions of the ocean and atmosphere, Izv. Akad. Nauk SSSR, Fizika atm. i okeana, **3**, No. 12, 1326–1333 (1967).
3. Monin, A. S. and A. M. Yaglom. *Statistical Fluid Mechanics*, Volume 1, Nauka Press, Moscow (English translation, MIT Press, 1971) (1965).
4. Webster, F. Measurements of eddy fluxes of momentum in the surface layer of the Gulf Stream, Tellus, **17**, No. 2, 239–245 (1965).
5. Ozmidov, R. V., V. S. Belyayev, and A. D. Yampol'skiy. Certain features of the transport and transformation of turbulent energy in the ocean, Izv. Akad. Nauk SSSR, Fizika atm. i okeana, **6**, No. 3, 285–292 (1970).
6. Kraus, E. B. and F. S. Turner. A one-dimensional model of the seasonal thermocline, II, The general theory and its consequences, Tellus, **19**, No. 1, 98–106 (1967).
7. Kitaygorodskiy, S. A. Fizika vzaimodeystviya atmosfery i okeana (*The Physics of Air-Sea Interaction*), Gidrometeoizdat Press, Leningrad (1970).
8. Kamenkovich, V. M. One model for determination of sea-surface temperature, Okeanologiya, **9**, No. 1, 38–43 (1969).
9. Lelikova, Ye. F. and B. V. Khar'kov. Calculation of sea-surface temperature, Okeanologiya, **12**, No. 6, 975–981 (1972).
10. Ekman, V. W. On the influence of the earth's rotation on ocean currents, Arkiv Mat. Astron. Physik, **2**, No. 11, 1–53 (1905).
11. Ekman, V. W. Uber Horizontalzirkulation bei winderzeugten Meerestromungen (*Horizontal Circulation with Wind-Driven Ocean Currents*), Arkiv Mat. Astron. Physik, **17**, No. 26, 1–74 (1923).
12. Shtokman, V. B. Ekvatoryal'nyye protivotecheniya v okeanakh (*Equatorial Countercurrents in the Oceans*), Gidrometeoizdat Press, Leningrad (1948).
13. Shtokman, V. B. Izbrannye trudy po fizike morya (*Selected Works in Marine Physics*), Gidrometeoizdat Press, Leningrad (1970).
14. Sverdrup, H. U. Wind-driven currents in a baroclinic ocean; with application to the Equatorial Currents of the Eastern Pacific, Proc. US Nat. Acad. Sci., **33**, No. 11, 318–326 (1947).
15. Stommel, H. The westward intensification of wind-driven ocean currents, Trans. Am. Geophys. Union, **29**, No. 2, 202-206 (1948).
16. Munk, W. H. On the wind-driven ocean circulation, J. Meteorol., **7**, No. 2, 79–93 (1950).

---

[3] Note that $\bar{N} \simeq 2 \cdot 10^{-2}$ sec^{-1} and $H \simeq 10$ km in the atmosphere, so that $L_R \simeq 1000$ km (see Sec. 5-2). This is why interaction between synoptic eddies and a zonal current in the atmosphere could be numerically modeled.

17. Kamenkovich, V. M. Contributions to theory of inertial-viscous boundary layer in a two-dimensional ocean-current model, Izv. Akad. Nauk SSSR, Fizika atm. i okeana, **2**, No. 12, 1274–1295 (1966).
18. Parsons, A. T. A two-layer model of Gulf-Stream separation, J. Fluid Mech., **39**, Part 3, 511–528 (1969).
19. Kamenkovich, V. M. and G. M. Reznik. Contribution to theory of stationary wind-driven currents in a two-layer liquid, Izv. Akad. Nauk SSSR, Fizika atm. i okeana, **8**, No. 4, 419–434 (1972).
20. Kamenkovich, V. M. and G. M. Reznik. Separation of a boundary current from a shore due to the effect of sea-floor topography (in a linear barotropic model), Doklady Akad. Nauk SSSR, **202**, No. 5, 1061–1064 (1972).
21. Lineykin, P. S. Determination of the thickness of the baroclinic layer of the ocean, Doklady Akad. Nauk SSSR, **101**, No. 3, 461–464 (1955).
22. Robinson, A. R. and H. Stommel. The oceanic thermocline and the associated thermohaline circulation, Tellus, **11**, No. 3, 295–308 (1959).
23. Robinson, A. R. and P. Welander. Thermal circulation on a rotating sphere; with application to the oceanic thermocline, J. Mar. Res., **21**, No. 1, 25–38 (1963).
24. Welander, P. The thermocline problem, Phil. Trans. Roy. Soc. London, **A270**, No. 1206, 415–421 (1971).
25. *Discovery, Observation, and Development of the Theory of the Lomonosov Current*, Collection edited by A. G. Kolesnikov, Izd. MGI AN UkrSSr, Sevastopol' (1968).
26. Fel'zenbaum, A. I. Dynamics of Ocean Currents (A Review). In: Itogi Nauki. Gidromekhanika (*Progress in Science, Fluid Mechanics*), USSR Academy of Sciences Press, Moscow (1968); VINITI (1970).
27. Fill, A. E. Models of equatorial currents, Proc. Symp. Durham (1972); US Nat. Acad. Sci., 181–203 (1975).
28. Sarkisyan, A. S. Calculation of stationary wind-driven currents in the sea, Izv. Akad. Nauk SSSR, Ser. geofiz., No. 6, 554–561 (1954).
29. Welander, P. On the Vertically Integrated Mass Transport in the Oceans. In: *The Atmosphere and the Sea in Motion*, Ed. B. Bolin, Rockefeller Inst. Press, Oxford, Univ. Press, N.Y. (1959).
30. Il'in, A. M., V. M. Kamenkovich, T. G. Zhugrina, and M. M. Silkina. Calculation of total transport in the ocean (Stationary problem), Izv. Akad. Nauk SSSR, Fizika atm. i okeana, **5**, No. 11, 1160–1172 (1969).
31. Takano, K. General circulation in the global ocean, J. Oceanogr. Soc. Japan, **25**, No. 1, 48-51 (1969).
32. Sag, T. W. Numerical Model for Wind Driven Circulation of the World's Ocean. In: *Pnys. of Fluids*, Suppl. II, 177–183 (1969).
33. Laykhtman, D. L., B. A. Kagan, L. A. Oganesyan, and R. V. Pyaskovskiy. Global circulation in a barotropic ocean of variable depth, Doklady Akad. Nauk SSSR, **198**, No. 2, 333–337 (1971).
34. Kagan, B. A., D. L. Laykhtman, L. A. Oganesyan, and R. V. Pyaskovskiy. A numerical experment on the seasonal variability of the global circulation in a barotropic ocean, Izv. Akad. Nauk SSSR, Fizika atm. i okeana, **8**, No. 10, 1052–1072 (1972).
35. Sarkisyan, A. S. Osnovy teorii i raschet okeanicheskikh techeniy (*Fundamentals of the Theory and Calculation of Ocean Currents*), Gidrometeoizdat Press, Leningrad (1966).
36. Bryan, K. and M. D. Cox. A numerical investigation of the oceanic general circulation, Tellus, **19**, No. 1, 54–80 (1967).
37. Bryan, K. and M. D. Cox. A nonlinear model of an ocean driven by wind and differential heating, Part I, Description of the three-dimensional velocity and density fields, J. Atm. Sci., **25**, No. 6, 945–967 (1968).
38. Bryan, K. and M. D. Cox. A nonlinear model of an ocean driven by wind and differential heating, Part II, An analysis of the heat, vorticity and energy balance, J. Atm. Sci., **25**, No. 6, 968–978 (1968).

39. Bryan, K. Climate and the ocean circulation, Part III, The ocean model, Monthly Weather Review, **97**, No. 11, 806–827 (1969).
40. Vasil'yev, A. S. and A. I. Fel'zenbaum. Steady circulation in a baroclinic ocean, numerical solutions including calculation of density (Slip at bottom), Morskiye gidrofiz. issled., No. 4, 37–62 (1969).
41. Vasil'yev, A. S. and A. I. Fel'zenbaum. Steady circulation in a baroclinic ocean, numerical solutions including calculation of density (Zero slip at bottom), Morskiye gidrofiz. issled., No. 4, 63–69 (1969).
42. Vasil'yev, A. S. and A. I. Fel'zenbaum. Steady circulation in a baroclinic ocean, Morskiye gidrofiz. issled., No. 1, 9–23 (1971).
43. Cox, M. D. A mathematical model of the Indian Ocean, Deep-Sea Res., **17**, No. 1, 47–75 (1970).
44. Gill, A. E. and K. Bryan. Effects of geometry on the circulation of a three-dimensional southern-hemisphere ocean model, Deep-Sea Res., **18**, No. 4, 685–721 (1971).
45. Holland, W. R. Ocean tracer distributions, Part I, A preliminary numerical experiment, Tellus, **23**, No. 4–5, 371–392 (1971).
46. Sarkisyan, A. S., V. P. Kochergin, and V. I. Klimok. Theoretical model and calculations of the density field in an ocean with arbitrary bottom topography, Izv. Akad. Nauk SSSR, Fizika atm. i okeana, **8**, No. 7, 740–751 (1972).
47. Kochergin, V. P., A. S. Sarkisyan, and V. I. Klimok. Numerical experiments to calculate density in the northern Atlantic Ocean, Meteorologiya i Gidrologiya, No. 8, 54–61 (1972).
48. Wetherald, R. T. and S. Manabe. Response of the joint ocean-atmosphere model to the seasonal variation of the solar radiation, Monthly Weather Review, **100**, No. 1, 42–59 (1972).
49. Holland, W. R. Baroclinic and topographic influences on the transport in western boundary currents, Geophys. Fluid Dynam., **4**, No. 3, 187–210 (1973).
50. Lineykin, P. S. Osnovnyye voprosy dinamicheskoy teorii baroklinnogo sloya morya (*Fundamental Problems of the Dynamic Theory of the Baroclinic Layer of the Ocean*), Gidrometeoizdat Press, Leningrad (1957).
51. Mikhaylova, E. N., A. I. Fel'zenbaum, and N. B. Shapiro. Computation of steady-state sea and ocean currents, Doklady Akad. Nauk SSSR, **168**, No. 4, 688–791 (1966).
52. Mikhaylova, E. N., A. I. Fel'zenbaum, and N. B. Shapiro. Computation of unsteady ocean currents and tides, Doklady Akad. Nauk SSSR, **175**, No. 5, 1041–1044 (1967).
53. Fel'zenbaum, A. I. and N. B. Shapiro. Use of the integrated stream function in ocean-current theory, Morskiye gidrofiz. Issled., No. 1, 39–76 (1969).
54. Ivanov, Yu. A. Relations between heat-exchange processes at the ocean surface and the rainfall-evaporation difference, Izv. Akad. Nauk SSSR, Fizika atm. i okeana, **3**, No. 7, 757–763 (1967).
55. Swallow, J. C. and L. V. Worthington. An observation of a deep countercurrent in the western North Atlantic, Deep-Sea Res., **8**, No. 1, 1–19 (1961).
56. Warren, B., H. Stommel, and J. C. Swallow. Water masses and patterns of flow in the Somali Basin during the southwest monsoon of 1964, Deep-Sea Res., **13**, No. 5, 825–860 (1966).
57. Holland, W. R. Energetics of baroclinic oceans, Proc. Durham Symp. (1972); US Nat. Acad. Sci., 168–177 (1975).
58. Marchuk, G. I. Chislennyye metody v. prognoze pogody (*Numerical Methods in Weather Forecasting*), Gidrometeoizdat Press, Leningrad (1967).
59. Marchuk, G. I. Chislennoye resheniye zadach dinamiki atmosfery i okeana na osnove metod rascheplenіya (*Numerical Solution of Problems in the Dynamics of the Atmosphere and Ocean on the Basis of a Splitting Method*), Nauka Press, Siberian Division, Novosibirsk (1972).
60. Bryan, A numercial method for the study of the circulation of the World Ocean, J. Comput. Physics, **4**, No. 3, 347–376 (1969).
61. Lighthill, M. J. Dynamic response of the Indian Ocean to onset of the southwest monsoon, Phil. Trans. Roy. Soc. London, **A265**, No. 1159, 45–92 (1969).
62. Kamenkovich, V. M. Normal oscillations of a multilayer rotating liquid, Izv. Akad. Nauk SSSR, Fizika atm. i okeana, **3**, No. 3, 284–290 (1967).

69. Gill, A. E. Ocean models, Phil. Trans. Roy. Soc. London, **A270**, No. 1206, 391–413 (1971).
70. Il'in, A. M. A difference scheme for a differential equation with a small parameter in the highest derivative, Matematicheskiye Zametki, **6**, No. 2, 237–248 (1969).
63. Phillips, N. Models for Weather Prediction. In: *Annual Rev. Fluid Mech.*, Palo Alto, Calif., **2**, 251–292 (1970).
64. Lighthill, M. J. Time-varying currents, Phil. Trans. Roy. Soc. London, **A270**, No. 1206, 371–390 (1971).
65. Pedlosky, J. A note on the western intensification of the oceanic circulation, J. Mar. Res., **23**, No. 3, 207–209 (1965).
66. Il'in, A. M. Asymptotic behavior of the solution of one boundary-value problem, Matematicheskiye Zametki, **8**, No. 3, 273–284 (1970).
67. Il'in, A. M. Behavior of the solution of one boundary-value problem as $T \to \infty$, Matematicheskiy Sbornik, 87(129), No. 4, 529–553 (1972).
68. Phillips, N. Large-scale eddy motion in the Western Atlantic, J. Geophys. Res., **71**, No. 16, 3883–3891 (1966).

INDEX